Astronomers' Universe

For further volumes:
http://www.springer.com/series/6960

Mario Bertolotti

Celestial Messengers:
Cosmic Rays

The Story of a Scientific Adventure

 Springer

Mario Bertolotti
University of Roma "La Sapienza"
Dipartimento di Scienze di Base e
Applicate per l'Ingegneria (SBAI)
Rome
Italy

ISSN 1614-659X
ISBN 978-3-642-28370-3   ISBN 978-3-642-28371-0 (eBook)
DOI 10.1007/978-3-642-28371-0
Springer Heidelberg New York Dordrecht London

Library of Congress Control Number: 2012943470

Printed on acid-free paper

Springer is part of Springer Science+Business Media (www.springer.com)

Introduction

The history of the research on cosmic rays is that of a scientific adventure. For just one century, researchers climbed mountains, flied in balloons and planes, and travelled all around the globe, trying to understand the nature of this radiation which comes from outer space. At present the research has extended out of our planet, in the open space where spacecrafts are sent to explore our solar system, and in the future, likely beyond.

But what are *cosmic rays*? They are a continuous rain of charged particles which move at nearly the speed of light and invest our planet at every moment from all directions. These particles are nuclei of common atoms—for the most hydrogen nuclei—stripped by their electrons, together with gamma rays, neutrinos, electrons, a few antiparticles, and may be something else. The term ray is therefore strictly speaking an oxymoron, the cosmic rays being particles and not radiation, but it has survived for historical reasons.

What makes them different from any other kind of radiation is their huge individual energy, which has allowed them to play a fundamental role in the development of modern physics. Before their discovery, the most energetic known particles were the ones emitted in the spontaneous decay of radioactive nuclei.

Some time later, in an effort to duplicate the effects of cosmic rays in controlled conditions in the laboratory, physicists began to develop accelerators that were able to get more and more high-energy particles, also if the highest energies of the cosmic ray particles will probably never be reached.

The story started when, at the end of the nineteenth century, people observed that an electrically charged body in air loses its charge slowly. Trying to understand this apparently uninteresting fact, which after all without an in-depth interpretation, could be attributed to the losses of the body supports, led to the discovery that pairs of ions of unknown origin always exist in the air, and later led to the conclusive proof that a radiation of extraterrestrial origin was responsible, together with natural radioactivity, for their existence.

However 20 years had to pass before the existence of an extraterrestrial radiation—the cosmic radiation—was universally accepted and a new and exciting field of research started which brought to the discovery of new constituents of matter and to what today constitutes high-energy physics.

Physicists discovered in cosmic rays particles of subatomic dimensions with energies thousands, million, and billion times larger than those of particles emitted by radioactive materials existing on the Earth. For the first time they witnessed processes in which energy materializes into particles which then suddenly disappear, originating other still unknown particles. The study of the new mysterious radiation evolved together with the fabrication of suitable instruments that are able to study its properties. We may say that our knowledge in this field has progressed checking the road, often starting from wrong hypotheses which were then corrected as the instrumentation and measurement techniques were implemented.

People initially believed the rays to be the bare product of the Earth's natural radioactivity, and only when Victor Hess performed measurements with sealed instruments at several altitudes in the atmosphere, it was proved that there was a component coming from outside the Earth. People believed this component to be made of ultrapenetrative gamma rays, but today we know that most of it is made of particles. When the positive electron was discovered, one tried to interpret it as a proton. When the muon was found, it was initially named meson and confused with the particle that Yukawa had introduced to explain the forces among nucleons in an atomic nucleus, and only later, in the pursuit of cosmic rays studies, it was discovered that the Yukawa's particle was something else, that is, the π-meson which originates the muon when decaying.

Quantum mechanics—which was developed in the years of these discoveries—found the explanation for the behaviour of cosmic rays at high energies and a proof bench and a sink of data whose understanding contributed to its formulation in the final form of nowadays.

Since the 1950s, in an attempt to develop a general theory of elementary processes, large accelerators that are able to produce a large number of high-energy particles suitable for experiments have been built; still, even today particles with super-energies may be found only in cosmic rays, a few ones but extremely useful since they help to discover secrets of the Universe that no accelerator may ever unveil.

The Hess discovery has in fact uncovered new scenarios in astrophysics and cosmology. The mysterious radiation carries important messages concerning the physical conditions of regions far away in the space and, in an effort to explain the origin of it, physicists have developed a great number of new ideas on the nature of the events that take place in stars and in the masses of diluted gas which fill the interstellar space. The last challenge has been the study of neutrino, and cosmic rays have given us, as we shall see, a good experimental data to work with.

However what is the impact on our everyday life? The particles which hit us continuously from space, interacting with living matter, contribute to genetic mutations which may stand at the origin of life itself and of its evolution; they influence the telecommunication systems, both on Earth and among satellites and spacecrafts, while the less energetic part, consisting of electrons emitted by our Sun during storms in its atmosphere, originates the marvellous phenomenon of boreal auroras.

The study of cosmic rays has contributed to the understanding of geophysical, solar, and planetary phenomena: they are the sole sample of matter coming directly to us from space. All other information we have concerning our Universe is indirect information reaching us through the study of the light emitted by stars. There are also surprising effects. It is well known that one of the methods used to date archaeological objects, old down to 100,000 years, employs the carbon isotope of mass 14. This isotope ^{14}C is radioactive and decays into nitrogen, with a lifetime of 5,568 years. Within such a short time, it is obvious that this isotope is continuously regenerated by some natural processes: the bombardment, by the neutrons generated in the high atmosphere by cosmic rays, of atmospheric nitrogen atoms which, by transmutation, produce it. The radioactive carbon so generated rapidly oxidizes into radioactive carbon dioxide which mixes with the normal carbon dioxide already existing in the atmosphere. The decaying and production processes are at equilibrium, so that the proportion of the radioactive isotope does not change in time. The age is derived from the abundance of carbon 14 in the sample under examination, assuming that the carbon is incorporated into the living sample (for example a tree) until there is an exchange with the atmosphere. When the sample dies, the exchange ceases and the decaying carbon 14 is no more replaced, so its content starts to decrease.[1]

In the following pages I will try to retrace the basic steps of the discovery of cosmic radiation up to the ideas of its mysterious origin, which has not yet been completely clarified and which therefore may still deserve surprising findings.

A complete exposition which takes into account all the contributions and all the subtle arguments that were made during the 100 years of research from the Hess discovery would be too long and complex; I will give a synthetic description comprehensive enough, however, to allow to unveil the fascination of these researches.

Bibliographic references pertaining to all described facts are provided. Short biographic notes of some of the more important people involved in the researches have also been included, in an attempt to outline the personalities of the involved scientists.

I apologize for any omission. Comments will be welcome. Finally I want to give a special thank to my wife Romana who allowed me to spend all my free time on this work, to my friends Bruno and Viera Crosignani who went through the text suggesting improvements, to Claus Ascheron and Deepak Ganesh for help in publication and editing the manuscript.

Rome, Italy   Mario Bertolotti

---

[1] This process was proposed by W.F. Libby [Phys. Rev. 69, 671 (1946)] who suggested that cosmic neutrons react with atmospheric nitrogen of mass 14 producing carbon, according to the reaction $_7N^{14} + n \rightarrow p + {}_6C^{14}$. The produced carbon then decays with β-emission $_6C^{14} \rightarrow \beta + {}_7N^{14}$ turning back to nitrogen and the process may start again.

Acknowledgements

Y. Suzuki, to allow me the use of Fig. 11.4, Itaru Shimizu to allow me the use of Fig. 11.5, A.M. Hillas to allow me the use of Fig. 13.2, F. Halzen to allow me to use Fig. 11.6, T. Gaisser to allow me the use of Fig. 12.10.

Contents

Chapter 1
Cosmic Rays: Prologue

1.1 The Beginning: A Charged Body Loses Spontaneously Its Charge

An apparently insignificant fact initiated our story in the eighteenth century. The French physicist Charles-Augustin de Coulomb (1736–1806)—to whom we owe the expression of the electrical force exerted between two point charges—observed that a charged metallic conducting body, placed in air, even if insulated from the surrounding, loses slowly its charge [1]. That was probably the first documented observation of electrical conduction in clear air, later confirmed by many other researchers. May be the most convincing evidence of this phenomenon was given at the end of the nineteenth century, by Sir Charles Vernon Boys (1855–1944), an experimenter of great ability and ingenuity, who built many precision instruments.

In an experiment on the insulating properties of glass and other substances he studied how a pair of very narrow gold electrically charged leaves, hanging from an insulating support, lose their charge when arranged as shown in Fig. 1.1. Finding that, even if insulated, they slowly lose their charge, he concluded [2]: "*It is more probable that the loss is due mainly to convection through the air*".

At the time many agreed with him that the charge losses were due to convection of charge by dust particles, but the evidences were not convincing. We may point out that the effect is very critical; the air is substantially non-conducting, and the experimental techniques, at the time very rudimental, were being developed at the same time as the observations themselves were being made. Therefore we must not be surprised that considerable disagreement existed among the first pioneering measurements. Refining the observation techniques, experiments made later with a *gold-leaf electroscope* (see Fig. 1.2), or with a similar instrument, confirmed it does not hold an electrical charge indefinitely.

An *electroscope* is an instrument which allows to observe the presence of a charge on a body and, with a suitable procedure, to measure its sign, and even with some incertitude, its value. It consists of a glass (or metallic with a glass wall) case

M. Bertolotti, *Celestial Messengers: Cosmic Rays*, Astronomers' Universe, DOI 10.1007/978-3-642-28371-0_1, © Springer-Verlag Berlin Heidelberg 2013

Fig. 1.1 Boys' arrangement.
A is a flat brass hook at which
is suspended the piece of
glass or quartz B, bent to the
form shown, so that it may
hang a piece of bent brass, C,
to which the leaves, D, are
attached. When the leaves are
charged they divert, but with
time close (from Boys[2])

Fig. 1.2 A gold-leaf
electroscope

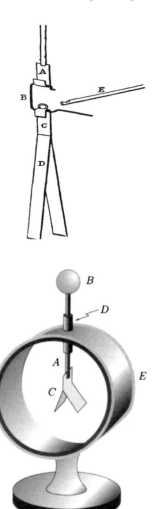

inside which are inserted a pair of thin metallic leaves hanging from a metallic rod
that exit from the case, being insulated from this (see Fig. 1.2).

When a charge is given to the electroscope, the two leaves fly apart. This happens
because both leaves have received the same charge (positive or negative) and like
charges repel each other. The leaves remain apart for some time, but eventually,
little by little, they drop to their initial position, as the electric charge gradually
disappears.

In an ordinary electroscope, the charge escapes mainly through the insulating
sleeve that separates the rod from the metallic case. Through special design these
losses can be reduced practically to zero; but even so, the electroscope will not
retain its charge indefinitely. If it is sealed, the final result is independent from the
filling gas.

By the end of the nineteenth century, the air's conductivity was the object of extended studies by many researchers, the most notable being C.T.R. Wilson (1869–1959), H.F. Geitel (1855–1923), and J. Elster (1854–1920). At that time, physicists knew enough about the structure of matter to be able to explain the spontaneous discharge of the electroscopes. They knew that matter is made of atoms, each chemical element representing a different kind of atom, and that atoms of one or more elements group together to form the molecules of the different chemical substances.

Even if they had no clear idea of how an atom look, they knew that ordinarily atoms contain equal amount of positive and negative charge, which makes them to be neutral, and that charges have a granular structure, that is they occur as simple multiples of an elementary charge. Indeed, in 1897, J.J. Thomson had discovered the electron,[1] a particle with a very small mass, which later was identified as the fundamental unit of negative charge. According to this picture, the discharge of the electroscope was explained by the fact that the air or the gas surrounding the leaves is always slightly *ionized*, that is some gas molecules occasionally lose their electrical neutrality. A molecule may lose an electron and thus be left with a positive charge excess. The electron may remain free or attach to a neutral molecule, making it negative. Therefore in the gas a small percentage of free electrons and charged molecules exist, which are called *ions*.[2] If, for example, the leaves are charged positive, they attract negative and repel positive ions. The positive charge which makes the leaves to repel each other is so gradually neutralized by the negative ions, and the leaves drop. Similarly, the negatively charged leaves attract positive ions and discharge themselves.

---

[1]The paternity of its "discovery" is usually attributed to the English physicist Thomson (1856–1940) who measured its physical characteristics (mass and charge). But already a few years before, the Irish physicist Stoney (1826–1911) in order to explain in a simple way the laws of electrolysis—expounded by Michael Faraday (1791–1867) in 1833—speculated on the existence of an *atom of electricity*, which later, in 1894, he called *electron* [3]. And still before, the celebrated English physiologist and physicist Helmoltz (1821–1894) in a conference at the London Chemical Society in 1881 said [4]:

"*if we accept the hypothesis that the elementary substances are composed of atoms, we cannot avoid concluding that electricity also, positive as well as negative, is divided into definite elementary portions which behave like atoms of electricity*".

However the electron in these considerations had an abstract nature, and no material substrate was attributed to it. Only later as a consequence of a series of experimental results obtained by various researchers and culminated with the J.J. Thomson's measurements [5], the final result was arrived at, then confirmed and completed by further researches, that the elementary negative charge is associated to a massive particle. So to fix the birth day of the electron in 1897, the year of the crucial Thomson's experiment, is somehow a mystification. In fact, the concept of the electron as a material particle, a universal constituent of all atoms, has grown gradually in a time span of at least 20 years around this date. The etymology is immediate: the name is a direct transliteration of $\eta\lambda\varepsilon\kappa\tau\rho o\nu$, the Greek name for amber, the material that assumes, when rubbed, what by convention was named a negative charge.

[2]Ιόν is the neutral past participle of the Greek ἰὲναί, go. The term was introduced by Faraday in 1834 to describe the movement of charged particles under an electric field in electrolytic solutions.

The ions existing in the air may be actually "seen" with an instrument invented by C.T.R. Wilson—the *cloud chamber*—which for many years was of enormous help in the experimental studies of nuclear physics and cosmic rays.

1.2   Wilson and the Cloud Chamber

Charles Thomson Rees Wilson (Fig. 1.3) was born on the 14th of February, 1869, in the parish of Glencorse, near Edinburgh. His father, John Wilson, was a farmer, and his ancestors had been farmers in the South of Scotland for generations.

At the age of four he lost his father, and his mother moved with the family to Manchester, where he was at first educated at a private school, and later at Owen's College—now the University of Manchester. Here, intending to become a physician, he took up mainly biology. In 1888, with a scholarship went on to Cambridge (Sidney Sussex College), where he took his degree in 1892. It was here that he became interested in the physical sciences, especially physics and chemistry. At Cambridge, Wilson started to work in the J.J. Thomson laboratory.

In 1900 he was made Fellow of Sidney Sussex College, and University Lecturer and Demonstrator. In 1913, he was appointed Observer in Meteorological Physics at the Solar Physics Observatory. In 1918, he was appointed Reader in Electrical Meteorology, and in 1925, Jacksonian Professor of Natural Philosophy. He was elected a Fellow of the Royal Society in 1900. He died on 15 November 1959.

Wilson remembers: "*In September 1894 I spent a few weeks in the Observatory which then existed on the summit of Ben Nevis, the highest of the Scottish hills [1,344 m]. The wonderful optical phenomena shown when the sun shone on the clouds surrounding the hill-top, and especially the coloured rings surrounding the sun (coronas) or surrounding the shadow cast by the hill-top or observed on mist or cloud (glories), greatly excited my interest and made me wish to imitate them in the laboratory*" [6]. People knew that oversaturated water vapour may condense on dust particles originating clouds. Therefore Wilson in order to reproduce the phenomena that had struck him so much employed a vessel filled with moist air. By increasing suddenly the volume of the vessel, supersaturated vapour was produced which condensed in minute water drops on the dust particles contained in the air. The drops, with the dust grain around which they formed, fall thereafter under the effect of gravity on the bottom of the vessel. In this way, by successively repeating the expansions, the air contained in the vessel becomes dust-free and all the dust particles depose on the bottom.

After some months of work at Cavendish Laboratory, Wilson realized that the few drops which appeared and appeared again every time the dust-free moist air volume was expanded, were probably the result of vapour condensation on some nuclei— may be the ions that caused the "residual" conductivity of the atmosphere—which are continuously produced. This hypothesis found a confirmation at the beginning of 1896, after that Wilson exposed a first primitive version of his cloud chamber, to the newly discovered X-rays.

Fig. 1.3 C.T.R. Wilson

At that time, J.J. Thomson was investigating the conductivity of air exposed to X-rays, and Wilson [7, 8] could use an X-ray tube built at Cavendish Laboratory by Everett, the Thomson's assistant. The huge increase of condensation which showed itself as a "*drop rain*", agreed perfectly with the observations made by Thomson [9] and Thomson and McClelland [10], immediately after the Roentgen[3] discovery, that the air was made conductive by the passage of X-rays. When in the summer of that year Thomson and Rutherford[4] firmly established [12] that the conductivity was due to the gas ionization, there was no more doubt that ions in gases could be detected and photographically registered, and then easily studied, thank to the Wilson method. Between 1896 and 1899, Wilson published four papers on the X-rays effects on the nucleation in the laboratory of small "clouds". In his paper "*On the condensation Nuclei Produced in Gases by the Action of Roentgen Rays, Uranium Rays, Ultra-violet Light, and other Agents*" [13], he emphasized the difference between ions and nuclei which do not carry electric charge and studied the way in which positively and negatively charged ions act as condensation nuclei

---

[3]W.C. Roentgen (1845–1923) discovered X-rays in 1895 [11].

[4]Ernest Rutherford has been one of the major experimental physicists. He was born in 1871 at Nelson, New Zealand, and after school he was awarded a scholarship to go to Cambridge (1895) where he worked under Thomson on gas ionization. In 1898 he was appointed professor at McGill University in Montreal, Canada, where he remained until 1907. In that period he demonstrated the existence of alpha and beta rays in the radioactive decay, and together with Frederick Soddy (1877–1956), a chemistry from Oxford, built the theory of radioactive transmutations. In 1907, he moved to Manchester University, Great Britain, where with Hans Geiger (1882–1949) invented the particle counter and studied the alpha particle scattering building his atomic model. He was awarded the chemistry Nobel prize in 1908. He was knighted in 1914 and in 1919 had the chair at Cambridge and the direction of the Cavendish laboratory. He died in London on 1937.

for water vapour, showing that an electric field was able to sweep out the drops which therefore should be electrically charged.

He also found that negative ions have a greater efficiency than the positive ones in the cloud formation [14].

The first Wilson [15] experiments on the condensation phenomena had proved that in the mist, dust-free air, are always present nuclei that in order that water should condense upon them, require exactly the same degree of supersaturation as the nuclei produced in enormously greater number by Roentgen rays, and he therefore concluded that the two nuclei were identical and probably were ions [16]. After publication of this result, Julius Elster and Hans Geitel [17, 18] showed that a charged conductor, exposed in the open air or in a room, lost its charge by leakage through the air and remarked that the phenomenon is readily explained on the supposition that positively and negatively charged ions are present in the atmosphere.

The German scientists Hans Geitel[5] and Julius Elster[6] were friends since the school time and became both teachers at Wolfenbuettel Gymnasium. When Elster married, Geitel went to live with the couple, and the two friends built a laboratory in the house were they lived. There they started their research (often supported by themselves) on photoelectric effect, spectroscopy, electrical conduction in gases, and especially atmospheric electricity, gaining international fame [19].

Their results on the loss of charge by a charged conducting body agreed also with some earlier experiments by Linss [20].

However, how and where the ions responsible of the atmospheric electrical conduction were produced was still an open question.

1.3 The Studies on Air Ionization

The problem was to know whether free ions were likely to occur in the atmosphere. After renouncing to use water vapour condensation, that he did not completely master yet, Wilson decided to use a purely electric method to detect ionization [21]. "*Attacked from this side the problem resolved itself into the question, Does an insulated-charged conductor suspended within a closed vessel containing dust-free air lose its charge otherwise than through its supports, when its potential is well below that required to cause luminous discharges?*" [22]

---

[5]Hans Geitel was born on July 6, 1855, in Braunschweig and studied at Heidelberg and Berlin Universities (1875–1879). In the fall of 1879 he went to teach at Wolfenbuettel Gymnasium where he died on 15 August 1923.

[6]Julius Elster was born in Blankenburg on 24 December 1854. He studied sciences, especially physics at Heidelberg and Berlin Universities (1875–1879) gaining his PhD in 1879 at Heidelberg. He was appointed a teacher at the Gymnasium of Wolfenbuettel in 1881 and remained in that school, notwithstanding he received better offers until he retired in 1919 due to an illness. He passed away in Wolfenbuettel on 8 April 8 1920.

The experiments started in July, 1900, and immediately led to positive results that were read before the Cambridge Philosophical Society on 26 November [23, 24]. Almost simultaneously a paper by Geitel on the same subject appeared on Physikalische Zeitschrift [25], in which identical conclusions were arrived at in spite of the great differences in the method employed.

The results obtained by Wilson were that if a charged conductor is suspended in a vessel containing dust-free air, there is a continuous leakage of electricity from the conductor through the air which does not decrease even after many weeks. This leakage takes place in the dark at the same rate as in the diffuse daylight. Therefore it does not depend on light. The rate of leak is the same for positive as for negative charges and does not depend on the initial potential of the body. It is approximately proportional to the pressure and the loss of charge per second is such as would result from the production of about 20 ions of either sign in each cc per second, in air at atmospheric pressure. Wilson at the end concluded that ions are produced continuously in air, in agreement with Geitel's experiments.

Geitel pointed out that the Italian chemist, physiologist, and physicist Carlo Matteucci [26] (1811–1868) had already arrived at the conclusion that the electricity loss rate is independent of the potential and decreases with lowering the pressure, this last property being observed later also by Warburg [27].

Wilson performed also experiments *"carried out to test whether the continuous production of ions in dust-free air could be explained as being due to radiation from sources outside atmosphere, possibly radiation like Roentgen rays or like cathode rays, but of enormously greater penetration power. The experiments consisted in first observing the rate of leakage through the air in a closed vessel as before, the apparatus being then taken into an underground tunnel and the observations repeated there"* [28]. The Caledonian Railways tunnel near Peebles was used. The expected decrease of leakage was, however, not observed and therefore Wilson concluded, in agreement with Geitel, that the ionization is a property of air itself.

At the end, on 5th December 1901, Wilson [29] presented some results on the rate of production of ions in air and in other gases, and used the term, that is now commonly used, *"ionization"*, to describe the rate of production of ions.

The final result of all these studies was that, notwithstanding any instrumental precaution was taken, an electrical conductivity remained in the air.

1.4   The Wilson Cloud Chamber

The most important result obtained by Wilson was that ions are able to serve as condensation centres for the supersaturated water vapour. This phenomenon is utilized in the *Wilson chamber* to make visible the trajectory of a single ionizing particle. The instrument consists of a chamber filled with air saturated with water vapour in which a sudden expansion is produced. The pressure decrease produced by the expansion results in a lowering of temperature and the water vapour becomes supersaturated. Then the water vapour condenses in the form of small cloud drops on

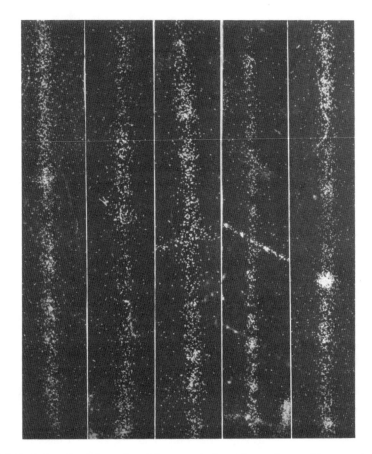

Fig. 1.4 Typical tracks of charged particles as can be found in cosmic rays. The expansion of the chamber was delayed so to give time to the drops to move apart and therefore to be more visible. In the photo the tracks have been magnified (Reprinted with permission from R.B. Brode, Rev. Mod. Phys. **11**, 222 (1939). Copyright 1939 by the American Physical Society)

dust particles, on cold walls, and electrified centres. The degree of super-saturation is very important for the drops formation because if it is too large the drops form everywhere. If a ionizing particle traverse the chamber at the moment of expansion, the ions produced along its path serve, when the apparatus is suitably regulated, as a support to the drops. A fine cloud line shows the passage of the particle and so its trajectory can be photographed. Electrons give little wisps and thread-like trajectories and alpha particles or fast atomic nuclei produce a much thicker, usually rectilinear, track.

Among other things, the method allows to study the collisions between particles, to know the emission direction of electrons and atomic nuclei and to have precious information on angular directions of trajectories. Figure 1.4 shows some tracks of charged particles of cosmic rays.

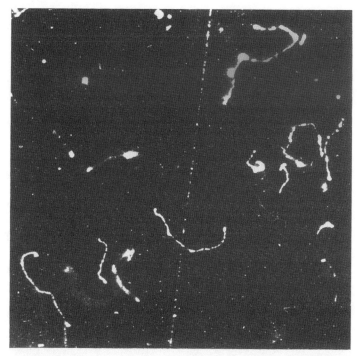

Fig. 33. C. T. R. WILSON, Proc. Roy. Soc., London (A) **104**, **192** (1923).

Fig. 1.5 An example of tracks of electrons produced by X-rays ionization. At the centre also a rectilinear track of a very fast electron is visible (from C.T.R. Wilson, Proc. Roy. Soc. London A **104**, 192 (1923))

By counting the number of drops along the trajectory, the number of produced ions can be derived and, by knowing the energy necessary to spend to form an ion, the energy that the particle has lost in a given length of its trajectory is calculated.

We show here two examples of trajectories. Figure 1.5 shows the sparse drops produced by the ionization due to X-ray, and in Fig. 1.6 the rectilinear marked tracks produced by alpha rays are visible.

Fast particles with different electric charge and different velocities produce tracks of different density in the gas. For a given velocity the density or number of ions per unit length in the track increases by increasing the charge. This we may easily understand because the electric forces exerted on the electrons of a molecule by the particle which passes nearby are proportional to the electric charge of the particle itself. Therefore particles with a larger charge cause a disturbance more violent than those with a lower charge and therefore ionize a larger number of molecules.

For a given charge, by increasing speed, the density of ions decreases. This too can be easily understood. The interaction between particle and molecule is effective only when the two bodies are sufficiently near to each other, and therefore the time

Fig. 1.6 Tracks produced by alpha particles from a radioactive sample

of interaction is longer if the particle moves slowly than if it is fast and, obviously, the more this time is longer the larger will be the effect of forces disturbing the molecule.

When the kinetic energy of the particle increases, its velocity increases but it cannot be greater than the light speed, in agreement with the Einstein (1879–1955) relativity theory. At very high velocities near the light velocity, the mass of the particle begins to increase and therefore when energy is given to the particle, a smaller and smaller fraction of it increases the speed and a larger and larger fraction increases the mass. The density of the track left by the particle first decreases by increasing its energy and then tends to a practically constant value, characteristic of a particle which moves at the light speed. The track with the smallest possible density is the one produced by an electron or proton moving at a velocity about the same as that of light.

As an example, β rays[7] with an energy of the order of 1 MeV[8] move at a speed that is about 94 % the light speed and in the air produce 50 ion pairs per centimetre of path. Alpha particles (which have a double charge[9]) emitted by Polonium have an energy of 5.3 MeV and a speed which is only 5.4 % the light speed, producing about 24,000 ion pairs per centimetre of path.

In the case of photons things go differently.

At the time of the first Wilson's experiments, people did not speak of photons and generically believed that the electromagnetic wave disturbed atoms and extracted

---

[7]β-Rays are fast electrons emitted by radioactive substances.

[8]The energy of atomic and nuclear particles is very small with respect to the energies of our macroscopic world. For this reason a particular unit is used called electronvolt (in short eV) that is the energy an electron gains when it is accelerated under a potential difference of 1 V. One electronvolt is equivalent to 1.6×10^{-19} J. Often multiples are used, KeV $= 1,000$ eV, MeV $= 10^6$ eV, GeV $= 10^9$ eV, TeV $= 10^{12}$ eV, PeV $= 10^{15}$ eV.

[9]Alpha particles are nuclei of helium emitted by some radioactive substances. They have a double charge and a mass that is about 7,000 times the electron mass. Therefore, the alpha particle moves with a speed that is much smaller than that of one electron of the same kinetic energy.

some electron. Although the concept of a photon was introduced by Einstein[10] already in 1905, discussing the photoelectric effect as an example, people were very reluctant to believe it was a real concept. Millikan's [34–37] measurements, ten years later, gave eventually an experimental confirmation of Einstein's formula for the photoelectric effect, and in 1922 Compton (1892–1962) [38] discovered the effect, which today bears his name, that he correctly interpreted as the proof of a collision between a photon and an electron. At this point, in the 1920 s, the interaction of photons with the matter was explained on the ground of two processes only:

In the first process, called *photoelectric effect*, a photon hits on an atom and transfers all of its energy to one of its electrons which is expelled, taking the whole energy of the incident photon less the small quantity spent to come out from the atom.

The second effect is *Compton effect*, called also *Compton scattering*. In this case a photon collides with an electron, in the same way as a ball knocks-on elastically another one of different dimensions. The energy distributes between the two particles according to the ordinary laws of elastic collisions and when the incident energy is very high, greater than a few MeV (and this is the order of energies present in cosmic rays), more than a half of the energy of the incident photon is transferred to the knocked electron. The photon knocks the electron, gives it a rather large energy and comes out in a different direction with a lower energy (that is with a lower frequency). The number of atoms or molecules ionized by the Compton effect in a gas is very small.

The processes through which particles and photons lose their energy are therefore very different. Charged particles lose their energy little by little through a large number of ionization events. If in a beam there are particles possessing the same mass, charge, and kinetic energy, they all slow down in crossing matter at the same rate and stop after having travelled the same distance.

This distance, called *range*, depends of course upon the nature and initial energy of the particles and the material in which absorption takes place. Because the energy losses occur through the collisions with the material atoms and these scale with the density, the energy loss is proportional to the material density and, as a first approximation, is independent from the nature of the atoms. In the case of cosmic rays the absorbing power of the whole vertical atmospheric layer is equivalent to that of about 10 m of water. The range is often measured in g/cm^2. Given the range of a particle in $g \, cm^{-2}$, the path in centimetres is obtained by dividing its value by the material density (g/cm^3).

---

[10]Einstein [32] deduced, from general considerations of statistical thermodynamics, that the entropy of the radiation described by Wien's distribution law has the same form as the entropy of a gas of elementary particles. He used this argument to reason, from a heuristic point of view, that light consists of quanta, each one having an energy content given by the product of Planck's constant and the frequency of the light, and applied this conclusion to explain certain phenomena, among which was the photoelectric effect. The name photon came later and was suggested by Lewis [33].

At high energy, on the contrary, photons travelling in the matter have a low probability to produce an electron by photoelectric effect and the most of their losses are due to the Compton effect. The Compton collision may occur at the beginning of the path or after a considerable distance. Therefore, among the photons that have travelled for some distance, some have still their initial energy while others have lost a large fraction of it in a Compton collision, the latter having been deflected from the beam. Therefore we may predict only the average behaviour of photons, predicting that some fraction of all photons after traversing a given mass of matter is lost from the beam. This fraction depends on the thickness and nature of the material and, if photons have all the same energy, by its value.

The curve which describes the decrease of the fraction of surviving photons as a function of path length is an exponential. The average value of the path travelled by a photon before a collision is called *mean free path*, and, like the particle range, is often measured in g/cm^2.

So, in the 1920s, the popular wisdom was that charged particles lose energy only by ionization and photons only by Compton effect. This view had to be changed later.

Let us now recall that if a magnetic field is applied in a direction perpendicular to that of the particle velocity v, it makes the particle to curve to follow a circular trajectory whose radius R (gyration radius) is given by

$$R = \frac{mv}{qB} = \frac{(2E_{\text{cin}}m)^{1/2}}{qB}, \tag{1.1}$$

where B is the magnetic strength, m is the mass of the particle of charge q, and E_{cin} is its kinetic energy. If the particle moves with a speed close to the light velocity c ($c = 300.000$ km/s), the relativistic version of (1.1) reads

$$R = \frac{m_0 \beta c}{qB}(1 - \beta^2)^{-1/2}, \quad \beta = \frac{v}{c}, \tag{1.2}$$

where m_0 is the rest mass of the particle, that is its mass when at rest or moving with a velocity small compared to the light velocity. The theory of relativity shows indeed that, by increasing the velocity, the mass increases according to the law

$$m = \frac{m_0}{(1 - \beta^2)^{1/2}},$$

A charged particle is therefore curved by the magnetic field and the trajectory, that in general is a helix winding around the lines of the magnetic field, is projected as a circle on a plane perpendicular to the field. A photo taken with the Wilson chamber shows an arc of circle with a radius proportional to the momentum of the particle (mv, or $m_0\beta c/(1-\beta^2)^{1/2}$ if the particle is relativistic). Figure 1.7 shows an example of two low-energy electrons one negative and the other positive (we will talk later of positive electrons) which originate at the centre of the photograph and whose trajectories are curved by the magnetic field, one on one side and the other on the opposite side due to their opposite charges.

Fig. 1.7 Two electrons of opposite charge curved by a magnetic field normal to the paper

A number of data can be derived by the study of the effect of the magnetic field on trajectories. The quantity RB is called *rigidity* and from (1.1) one has

$$RB = \frac{p}{q},\tag{1.3}$$

where p is the momentum of the particle. If the energy E is measured in eV, by giving R in cm and B in gauss (1 gauss $= 10^{-4}$ Wb m^{-2}), (1.3) becomes

$$300RB = \frac{E}{q},\tag{1.4}$$

where the rigidity is expressed in gauss cm. For a given value of the magnetic field and charge, a particle with high rigidity has a larger curvature radius and therefore a higher energy with respect to another particle with lower rigidity.

The total energy of the particle may be expressed as

$$E = mc^2 = m_0(1 - \beta^2)^{-1/2}c^2.\tag{1.5}$$

Because the momentum p in the relativistic case is written as

$$p = \frac{m_0 c \beta}{(1 - \beta^2)^{1/2}},\tag{1.6}$$

one may write for the energy

$$E = mc^2 = \left[(m_0 c^2)^2 + (cp)^2\right]^{1/2}.\tag{1.7}$$

If the particle is very energetic, we may make the approximation $pc \geq m_0 c^2$ and therefore the energy may be written as

$$E \approx E_{\text{cin}} \approx cp,\tag{1.8}$$

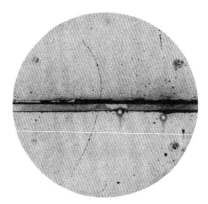

Fig. 1.8 This is a famous photograph of a positron (a positive electron) which shows the effect of absorption in the lead plate. Because the track has a lower curvature in the side below the plate this means that here it has a greater energy. Above the plate the greater curvature indicates that the particle has lost energy in the lead plate and therefore has lower energy. Therefore the particle comes from the bottom (Reprinted with permission from C.D. Anderson, Phys. Rev. **43**, 491 (1933). Copyright 1933 by the American Physical Society)

from which

$$\frac{E}{c} \sim p, \tag{1.9}$$

Because the curvature of a particle in a magnetic field is a function of the momentum, very often the energy is given in MeV/c. We wish to stress that for a given value of the magnetic field, a particle with higher rigidity has a larger curvature radius, and therefore an energy or a momentum higher than a particle with a lower rigidity.

In Fig. 1.8 is shown the trajectory of a particle crossing a plate of lead and curved by an applied magnetic field. Since the curvature radius is larger in the lower part of the photo and smaller in the upper part, the energy of the particle is greater below the plate than above. Therefore, because the particle loses energy in crossing the plate, we may infer that the particle travels from the bottom to the top and not the other way around. The difference of the curvature radii in the upper and lower tracks allows to derive the energy loss suffered in traversing the plate, and also the sign of the electric charge can be derived if the direction of the applied field is known.

However, the method is still imperfect. In fact one may see the trajectory of the particle only in the short fraction of a second in which the expansion is produced and a rather long time is required for the instrument to recover the initial conditions and be ready to work anew.

To be more specific, the particle to be detected should pass through the chamber in a well-defined time-slot in which the expansion takes place. If the particle enters too soon—a few milliseconds before the expansion—the produced ions diffuse away before the gas is cooled (it is the effect used in Fig. 1.4). If the particle arrives

Fig. 1.9 On the left the Bragg draft showing how he figured the probable path of alpha rays. On the right two of the first photos obtained by Wilson (is from C. Chaloner, BJHS **30**, 357 (1997))

too late—a few milliseconds after the expansion—the gases are already warmed before the track forms. Therefore the chamber is sensible only for a time of the order of a hundredth of second each time it is made to expand. In the application to the cosmic ray study this—at the beginning—meant a lot of useless work because the chamber was activated at random and most of the times no particle was passing in that very moment.

Most of Wilson's work was performed in the years 1895–1900 after he was nominated Clerk Maxwell Student at the end of 1894. This allowed him to enjoy 3 years of relative ease during which he was able to dedicate himself entirely to research, including that performed on atmospheric electricity the following year when he was employed at the Meteorological Council. After this period, his new duties—mainly as a teacher—prevented him to continue developing his cloud chamber. His work on it was in fact suspended from 1904 to 1909, during which he investigated atmospheric electric phenomena. However his apparatus was used at the Cavendish Laboratory, by Thomson and Wilson [40], to determine the charge of the electron.

Eventually, in 1911, Wilson [41, 42] with his cloud chamber was the first to see and photograph tracks of individual alpha and beta particles and electrons produced by X-rays. The electron tracks were by him described as "*little wisps and threads of clouds*" [6]. These results arose great interest also because the paths of the alpha particles were just like those Bragg (1862–1942) ([43, 44], see also [45, 46]) had described some years before. Figure 1.9 (from [47]) shows on the left how Bragg drew the probable path of alpha particles [48], and on the right two of the first Wilson's photographs of alpha rays [49]. Bragg's prediction that at the end of their path the rays should show broken branches was fully confirmed by the Wilson photos.

It took some time before the chamber came in common use, but in 1923 it attained a high degree of perfection and Wilson [31] wrote two classical papers, beautifully illustrated on electron tracks.

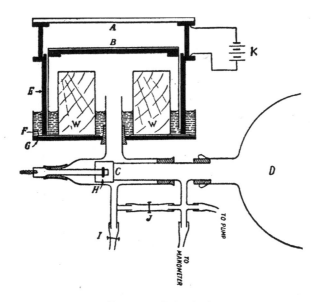

FIG. II-1. Wilson's original cloud chamber.

AB =expansion chamber, cylindrical in shape and completely closed,
B =movable base which slides inside cylinder E and serves as piston,
F =rubber sheet resting on a brass disk G to arrest downward
motion of B,
D =highly evacuated vessel which may be put in communication
with the space below B by opening the valve C,
WW =wooden blocks reducing the air space within the chamber,
I =stopcock on opening which space below B is connected with the
atmosphere and the piston brought back to the original
position,
J =pinch cock for adjusting the initial position of the piston and
hence the expansion ratio,
K =battery providing the electric field to remove stray ions just
before a fresh expansion.

Fig. 1.10 Wilson's original cloud chamber (Reprinted with permission from N.N. Das Gupta, S.K. Gosh, Rev. Mod. Phys. **18**, 225 (1946). Copyright 1946 by the American Physical Society)

The design of the chamber had an evolution since the first experiments. In 1912 Wilson greatly improved it. The essential features of the new chamber [50] are shown in Fig. 1.10. Wilson describes it: *"The cylindrical cloud chamber A is 16.5 cm in diameter and 3.4 cm high; the roof, walls and floor are of glass, coated inside with gelatine, that on the floor being blackened by adding a little Indian ink. The plate glass floor is fixed on the top of a thin-walled brass cylinder (the "plunger"), 10 cm high, open below, and sliding freely within an outer brass cylinder (the "expansion cylinder") of the same height and about 16 cm in internal diameter. The expansion cylinder support the walls of the cloud chamber and rests on a thin sheet of India rubber lying on a thick brass disc, which forms the bottom of a shallow receptacle containing water to a depth of about 2 cm. The water separates completely the air in the cloud chamber from that below the plunger. The base rests on a wooden stand not shown in the diagram.*

The expansion is effected by opening the valve C and so putting the air space below the plunger in communication with the vacuum chamber D... The floor of the cloud chamber in consequence drops suddenly until brought to a sudden stop, when the plunger strikes the India rubber-covered base plate, against which it remains firmly fixed by the pressure of the air in the cloud chamber. To reduce the volume of air passing through the connecting tubes at each expansion the wooden cylinder W was inserted within the air space below the plunger".

An example of how the experiments were performed at the time comes now: *"The valve is opened by the fall of a weight W released by a trigger arrangement* (Fig. 1.11). *On closing the valve and opening communication with the atmosphere through the pinch-cock I, the plunger rises and so reduced the volume of the air in the cloud chamber. By means of the two pinch-cock I and J... the plunger may be adjusted to give any desired initial volume v_1 between the upper limit v_2—the maximum volume of the cloud chamber—and the lower limit reached when the pressure below the plunger is that of the atmosphere...*

In setting up the apparatus, the plunger is placed on the rubber-covered base plate, and the expansion cylinder slipped over it, a hole in the side of the cloud chamber being open at this stage to allow of the imprisoned air escaping. Then, by blowing in air thorough I, momentarily opened for the purpose, the plunger is driven up to a height sufficient to allow of the largest desired expansions being made. The aperture in the wall of the cloud chamber is then closed, and the mass of imprisoned air remains unchanged during subsequent operations."

Many expansions were done to eliminate all the dust particles. Then if in A ions are formed, the expansion will cause condensation upon them. In Fig. 1.10 the plunger is shown at the end of its moving after a complete expansion.

With this chamber Wilson was now able to detect even the fastest beta particles, and X-rays and alpha particles. The illumination used is shown in Fig. 1.11.

After the first photos published by Wilson in 1912, the technique was little by little employed in other laboratories for the study of radioactivity. The following year (1913), Kleeman obtained with it the photographic confirmation that direct ionization of gases by gamma rays is *"comparatively small or non-existent"* and that *"β-rays produce secondary rays of small velocity only, i.e., δ-rays"* [52].[11] In the same year by examination of a large number of alpha particles that in Wilson's photos showed great angular deflections in the last 2 mm of path, M.E. Marsden and T.S. Taylor [53] explained their failure to detect low-velocity alpha particles with the scintillation method.

In his work in 1914, on the structure of the atom, Rutherford [54] saw in the Wilson photographs of alpha rays *"convincing evidence of the correctness of the view that large deflections do occasionally occur as a result of an encounter with a single atom"* [55] (see Fig. 1.12 showing one of these collisions).

---

[11]The name delta ray was introduced by J.J. Thomson to indicate fast electrons emitted in a material medium traversed by ionizing particles. The energy of these electrons is usually sufficiently great to produce further ionization so that the trajectory of delta rays is easily visible, for example, in a Wilson chamber.

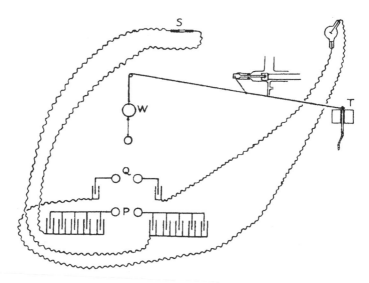

Fig. 1.11 Wilson's arrangement for firing the spark (from C.T.R. Wilson, Proc. Roy. Soc. London A **87**, 277 (1912))

Fig. 1.12 Tracks of α particles. A collision is clearly visible

In 1919, Rutherford [56, 57] by bombarding nitrogen with alpha particles obtained oxygen, realizing the first nuclear reaction, and in 1920, at the Cavendish Laboratory, he established an intense research program on radioactivity, searching evidences for nuclear disintegrations of light nuclei. The experimental methods he employed to investigate the structure of the nucleus were in great part based on scintillation counting using the *spintariscope* in which the particles to be studied hit a fluorescent screen and are detected by the observer as small light points through a microscope. Although recent modifications to the technique had greatly improved it, still with it the different kinds of particles were not clearly identified or distinguished between them. Moreover the method was strongly dependent on the observer. Rutherford hoped that the effect of the human observer fatigue and the measure uncertainty could be eliminated by the objectivity of the photograph made with the Wilson chamber, and asked T. Shimizu [58] to build a Wilson motor-driven chamber to make a great number of photos to see the collisions between alpha particles and gas atoms. However, Shimizu was not able to find any event of this kind over 3,000 photos. Later P.M.S. Blackett [59] (1897–1974), under Rutherford's direction, continued the work of Shimizu, but he too was unable to find any evidence of disintegrations.

With Rutherford's encouragement and generosity, in 1922, the Russian physicist Pyotr Kapitza [60][12] at the time a visitor in the lab, where he earned his PhD in 1923, used a magnetic field to observe the curvature of tracks. The method was used also by P.M.S. Blackett, who in 1948 was awarded the Nobel Prize for his contributions to the further development of the technique and the discoveries made with it (we will tell later about them); in Paris the technique was used by Irene Curie (1897–1956) and Pierre Auger (1899–1993), in Berlin by Bothe (1891–1957), Lise Meitner (1878–1968), and K. Philip; in Leningrad by D. Skobelzyn (1892–1990); in Tokyo by S. Kikuchi.

However Cambridge was the main research centre with the cloud chamber.

In 1913 the Cambridge Scientific Instrument Company, in UK, had produced commercial versions of the chamber, with the assistance from Wilson, and in 1927 the chamber was used for demonstrations to students.

An advertisement on Philosophical Magazine in May 1926 was accompanied with the slogan *"The most wonderful experiment in the world"* as a declaration attributed to an anonymous eminent physicist (see Fig. 1.13) [61].

The Wilson chamber gave a permanent photographic registration that could be carried away and studied with ease, and the examination of the deflections obtained with magnetic fields and the counting of the droplets produced by ionization allowed to have precious information on the nature of the particles that were detected.

A very important result found by using the Wilson chamber was the demonstration of the existence of the recoil electrons in the Compton effect, so establishing without any doubt the reality of the effect. In the following we will have the occasion

---

[12]The Russian physicist Pyotr Leonidovich Kapitza was born in Kronshtadt in 1894 and died in Moscow in 1984. He is mostly known for his work in low-temperature physics to which he gave many important results for which he was awarded the Physics Nobel Prize in 1978.

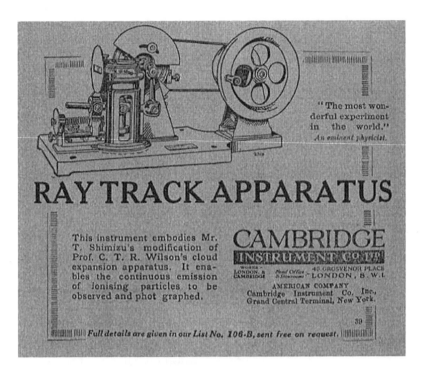

Fig. 1.13 The advertisement of the Cambridge Instruments (is from C. Chaloner, BJHS **30**, 357 (1997))

to describe many other results. Rutherford ([62], quoted in [63]) in a letter written to support Wilson for the Nobel Prize wrote: *"No one will deny the extraordinary interest and importance of this method which showed for the first time and in such minute detail the effects of the passage of ionizing radiations through a gas... I am personally of the opinion that the researches of Mr. Wilson in this field represent one of the most striking and important of the advances in atomic physics made in the last twenty years"*.

In fact it is difficult to exaggerate the importance the Wilson chamber had in the development of modern physics. It allowed to make direct observation of phenomena which otherwise would have remained purely conceptual. Without the evidence given by it, many processes would have been confirmed only through the study of a large number of experiments, slowing the scientific development and with a huge waste of forces.

The technique was eventually extended to cosmic rays in 1927, when D. Skobelzyn casually detected a track of a cosmic ray; we will tell soon of this. After many years two atlantes were published [64, 65] with some of the most beautiful photos made with the Wilson chamber by researchers all around the world.

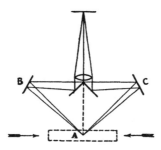

Fig. 1.14 The Shimizu disposition to make stereoscopic photos of tracks in a Wilson chamber. This set-up was used very often. The figure is an integration of the original drawing by Shimizu [58] (Reprinted with permission from N.N. Das Gupta, S.K. Gosh, Rev. Mod. Phys. **18**, 225 (1946). Copyright 1946 by the American Physical Society)

Wilson was awarded the Physics Nobel Prize in 1927 together with Compton. The motivation for Wilson was *"for his method of making the paths of electrically charged particles visible by condensation of vapour"*.

To obtain in three dimensions the effective trajectory of a particle in the chamber, the photo was made placing two mirrors on the sides of the chamber. In this way the scene was taken from two different angles and this allowed to represent it in space and calculate with precision the range, the curvature and the angle among different trajectories in a stereoscopic vision. One disposition used by Shimizu [58] is shown in Fig. 1.14. Two mutually perpendicular images are registered through a single lens on the same photographic plate by means of the two mirrors B and C inclined by 45° with respect to the plane of the chamber A.

This method was later improved by many researchers and in the following we will show many photos taken with this or analogous systems which show a double image.

1.5   The Charge Loss by a Charged Body is Due to the Air Ionization

In the first years of the twentieth century, physicists trying to answer the question of how the production of ions in air occurs, knew that several kinds of radiations could ionize gases and discharge electroscopes: X-rays, radium, and the other radioactive substances.

Elster and Geitel built a device in which the leakage of electricity in the free atmosphere could be measured in an "unobjectionable" way, as they wrote [66]. In this way they proved beyond any doubt that even clear and dust-free air conducts electricity and, based on observations made at different locations arrived at the conclusion that *"the atmospheric air is ionized to a certain degree"* [67]. They also discovered the presence of radioactive substances in the atmosphere [68] and subsequently proved the existence of radioactive substances in that part of the atmosphere that lies below the Earth's surface and which is in permanent exchange by diffusion with the exterior [69]. In this way they established that the radioactivity of the Earth's surface, especially the gaseous emanation observed as the decay product of many radioactive minerals should be considered as the primary source of electrical conduction in the atmosphere.

These results encouraged many other researchers to make experiments. In Vienna, Franz Exner (1849–1926)[13]—a pioneer in the study of atmospheric electricity—created an important centre for the study of atmospheric electricity where particular attention was given to the measurement of the absolute quantity of radioactivity in the atmosphere and to determine if this was sufficient to explain the observed phenomena. Fritz Kohlrausch (1840–1910), at Exner's Institute, invented a new method for the absolute determination of the content of radioactive emanation in the atmosphere [70] and from observations carried out in July 1906 at Gleinstaetten near Graz, he concluded that the emanation and its decay products provide the main reason for the permanent ion content of the atmosphere.

Previously, John Cunningham McLennan (1867–1935) and Eli Franklin Burton [71] (1879–1948) from University of Toronto, Canada, in a paper presented at the 8th meeting of the American Physical Society held on 31st December 1902 in Washington, DC, made a series of observations with atmospheric air contained in sealed vessels of different metals and concluded erroneously that "*the effects observed would seem to indicate that all metals in varying degree are the sources of a marked through feeble radioactive emanation*". This opinion was shared by Robert John Strutt (1875–1947), the elder son of Lord Rayleigh (John William Strutt, third baron Rayleigh 1842–1919) who wrote "*...there are very marked differences in the rate of leak, when different materials constitute the walls of the vessel*" and concluded: "*There can therefore, be little doubt that the greater part—if not the whole—of the observed ionization of air is not spontaneous at all, but due to Becquerel rays from the vessel*" [72]. These conclusions of course are wrong: the ionizing radiations do not come from the walls of the vessel but from outside, crossing the vessel walls.

One may note, however, that McLennan and Burton in the end of their paper reported also of shielding experiments in which they found a decrease of ionization of 37 % by dipping their vessel in a tank covered with 25 cm of water and concluded: "*From these results it is evident that the ordinary air of a room is traversed by an exceedingly penetrating radiation such as that which Rutherford has shown to be emitted by thorium, radium...*" Several weeks later they presented further results at a meeting of the Royal Society of Canada [73].

At the same American Society meeting, Rutherford and H. Lester Cooke [74] reported some of the earliest and more significant shielding experiments observing

[13]Franz Exner was a pioneer in the study of atmospheric electricity and transformed Vienna, where he was appointed a physics professor in 1881, in one of the leading centres of this field. He was director of the Institute of Physics and a dean of the University from 1908.

that, since the atmospheric activity is very similar in character to that of thorium and radium, it would be possible that some penetrating rays might be given off from the surface of the Earth and walls and rooms on which excited activity from the air is distributed. This penetrating radiation could possibly be gamma rays. To test this hypothesis they studied the amount of ionization present in a vessel of about 1 l capacity, placing metal screens outside the vessel. They found little difference on the discharge rate with a lead screen of 2 mm placed around their apparatus, but 5 cm of lead reduced the discharge rate by 30 %. Surprisingly, beyond 5 cm of lead they found no effect, although 5 tons of pig lead was placed around the measuring device. On removing the screens, the discharge came back to the original value. They concluded that their results showed that "*about 30 % of radiation inside a sealed vessel was due to an external radiation of great penetrating power*" and claimed that "*these effects could not be due to the presence of thorium or radium in the laboratory, for similar results were observed in the library which was free from all possible contamination by radioactive substances*". Today we know that in the cosmic radiation a hard component (essentially made by muons) is present which passes easily through those thicknesses, but at that time the muon was yet undiscovered. The work was presented at a Meeting of the American Physical Society in Washington, DC on 31st December 1902.

Notwithstanding these inexplicable facts, the most accepted opinion was, however, that the radiations emitted by the materials present in the Earth's crust and in the atmosphere were responsible for the air ionization. This assumption showed later to be partially correct.

Wilson speculated that some radioactive substance in the atmosphere could be carried down in the rain [75].

Aside from experiments carried out in laboratories, measurements were also made in different kinds of localities such as in caves, on the ocean, on lakes, etc. For example, Elster and Geitel experimented in caves and cellars observing that there the leakage rate was much greater than in the open air. They found that in a cave the loss was seven times faster than in the open air, even when air was clear and free of mist. Also in a cellar, whose windows had been shut for 8 days, the rate of the loss was considerably greater than it was in the outside air [76]. Today we know that this behaviour is due to the increase of radon gas produced by the disintegration of radioactive nuclei contained in the rocks. Radon, which is a heavy gas, in closed rooms tends to accumulate, increasing the radioactivity.

1.6   The Measurements with Height

If the penetrating radiation present on the earth surface would be entirely of terrestrial origin, its effect should be stronger near soil and decrease progressively by increasing height. Of course, this assumption could be checked by comparing observations made at different altitudes. To this purpose, unscreened ionization chambers were brought over the terrestrial crust, but initially contradictory results were obtained.

In May 1900, Elster and Geitel compared the ionization at sea level with that up to 3,000 m on top of Swiss mountains [77], finding that the rate of leakage was greater at high altitudes, than at the low ones.

The adventure in the skies started with Hermann Ebert [78] who made three flights with a balloon in June and November 1900 and in January 1901, reaching a maximum altitude of 3,770 m and finding a change with altitude of the leakage of a charged body taken free in air; however, he observed his results depended on the meteorological conditions. These are probably the first observations made with manned balloons.

Some years later, at the beginning of summer 1906, Heinrich Mache and Travis Rimmer [79] studied the variation of the penetrating radiation intensity in Vienna and found that the radiation changed during the day and depended from the atmospheric conditions. Mache [80] continued the investigations from 1st October 1907 to 15th October 1908 in Innsbruck finding also an annual variation and concluding that the penetrating radiation was made by two parts: one part originated by radioactive substances contained in the upper layers of the Earth and by their decay products, and the other part provided by the decay products created by emanations distributed in the atmosphere. Egon von Schweidler (1873–1948) [81] confirmed these results with observations made in the summers of 1910 and 1911 at the Luftelektrische Station of the Vienna Academy of Sciences in Seeham.

William Walker Strong (1883–1955) of Johns Hopkins University in Baltimore, Maryland, in 1907, also found that the most of the ionization in closed vessels suffers great fluctuations and probably consists of gamma rays that come from radioactive substances on the earth but also come from the ones existing in the atmosphere [82].

In 1909, Karl Bergwitz [83] (1875–1958) in his thesis reported on a balloon ascent in which he found a marked decrease of the total ionization, which at 1,300 m was only about 25 % of its value on the ground. This value agreed with what one could expect assuming the ground was the source of a gamma radiation responsible for the residual ionization. However, his measurements were questioned because his electrometer was damaged during the flight by deformation of its pressure vessel.

In 1910, the German Theodor Wulf made a singular experiment in Paris.

Born at Hamm in Vestfalia in 1868, Theodor Wulf in 1904 became a Jesuit and professor of physics at the Ignatius College of Valkenburg in Holland. He built many electrical instruments among which an electroscope which bears his name, very robust and that was very much employed. He died in Hallenberg in 1946.

The instrument built by Wulf in 1909 was particularly suited as a radiation meter. In it the golden leaves were replaced by two very thin wires held under tension by a light quartz fibre (Fig. 1.15). When charged, the two wires repelled each other and the separation was measured by means of a microscope. The Wulf electroscope was

Fig. 1.15 The Wulf
electroscope

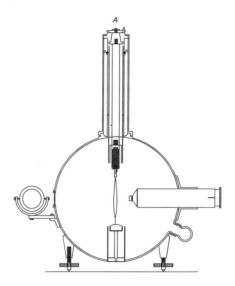

commercialized by the firm Gunther and Tegetmeyer which produced a very robust device that combined portability with accuracy. It is in use also nowadays.

Wulf, in the summer 1909, made extended observations in various localities in Germany, The Netherlands, and Belgium concluding that the penetrating radiation is caused by radioactive substances which are placed in the upper layers of Earth; if part of the radiation came from the atmosphere it was too small to be measured with the methods he had used. He observed also temporal variations that could be explained with the movement of air masses rich of emanation that lay under the Earth surface and were moved by fluctuations in the atmospheric pressure [84].

To test the hypothesis that the whole effect was due to the soil radioactivity, in March and April 1910, Wulf [85] brought one of his electroscope on the top of the Eiffel Tower (300 m) in Paris and experimented for 4 days from 11 a.m. to 5 p.m. On top of the tower he found a reduction of the radiation intensity of 60 % which, however, was less than expected. According to his calculations in fact, at an elevation of 80 m the gamma rays should already reduce to half their value at ground and at 300 m they should be only a few percent of the value at ground. Therefore, Wulf concluded that either another source of gamma rays existed in the upper layer of the atmosphere, or the absorption coefficient of gamma rays in air was smaller than he had assumed.

Wulf's observations were considered of great value because he could take many data in a fixed location and his apparatus did not suffer of any of the difficulties encountered in a balloon ascent, and therefore were considered the most reliable.

Contemporarily, between 1909 and 1911, Albert Gockel (1860–1927),[14] a physics professor at the Fribourg University in Switzerland, undertook several balloon rides. His first flight was made in a special occasion. The meeting of the Society of German Natural Scientists and Physicians, which was held each year in a different German-speaking city, in September 1909 was held in Salzburg, Austria.

Such famous physicists as M. Planck (1858–1947), W.C.W. Wien (1864–1927), A.J.W. Sommerfeld (1868–1951), J. Elster, J. Stark (1874–1957), M. Born (1882–1970), M. von Laue (1879–1955), O. Hahn (1879–1968), L. Meitner (1878–1968), and A. Einstein (1879–1955) were in attendance. At the meeting, Karl Kurz [86] presented a review on the penetrating radiation. Gockel and Wulf [87] discussed their observations made in Zermatt and its surroundings which persuaded them that the source of the radiation was the ground, but they were planning balloon flights to study the variation of the radiation with altitude. In Braunschweig, Germany, in December 1909, an International Balloon Week was taken. During this manifestation, the Swiss Aeroclub put the balloon "Gotthard" at Gockel disposal. Unfavourable weather delayed the flight to the last day of the week: 11 December. The balloon embarked three people, Gockel, dr. de Quervain, who wanted to perform meteorological observations, and Lieutenant Mueller to guide the balloon which reached a maximum height of 4,500 m during a 4 h flight from Zurich to the Jura, west of Bienne. Gockel, used an electrometer of the Wulf type, and concluded that in the free atmosphere, there is in fact a diminution of the penetrating radiation, but by far not to the extent that one could expect if the radiation arose mainly from the ground [88].

Gockel performed several additional balloon ascents; in the first carried out on 15 October 1910 from Zurich to Olten, reaching an altitude of 2,860 m, he found a slight decrease of the penetrating radiation with height, but later he realized that his experimental procedure could be criticized. His last flight was made on 2nd April 1911 with an "Aeroclub Escursion", which carried five people, from Bern to Lake Bienne. Comparing the ionization rate at 2,500 m with an earlier result he had obtained at 2,800 m, and correcting for barometric pressure, he found a weak increase in ionization with increasing height [89]. His measurements were, however, questionable because he used ionization chambers in which the pressure varied with the external pressure and did not make any correction for this effect.

---

[14]Albert Wilhelm Friedrich Eduard Gockel was born in Stockach, Baden, on 27 November 1860. He studied physics at the Universities of Freiburg, Breisgau, and Wuerzburg, at the Technische Hochschule in Karlsruhe and the University of Heidelberg, obtaining his doctorate in 1885. After working as a secondary-school teacher until 1895, he became assistant to Joseph von Kowalski at the University of Fribourg. He was appointed extraordinary professor at this University in 1903 and full professor in 1909. Gockel contributed to various topics in physics and meteorology, natural radioactivity on the earth, storm electricity and atmospheric interference of radiowaves. He died on 4 March 1927 in Fribourg.

As a conclusion neither Wulf nor Gockel found what they expected. The discharge rate did not decrease with height, or at least did not decrease as fast as they expected. The problem was extremely complex because the unquestionable presence of terrestrial radioactivity masked the effects.

One of the founder fathers of quantum mechanics, Erwin Schroedinger [90] (1887–1961)—then an assistant in Wienn—discussed the sources of the penetrating radiation, observing that three possible explanations were suggested for the origin of the radiation, rousing vehement discussions about their relative weight. These were: radioactive substances contained in the ground or precipitated onto the surface of the Earth, radioactive substances suspended in the atmosphere, and eventually an hypothetical extraterrestrial source of radiation. He wrote a paper in which he analyzed the contribution of the substances that could be in suspension in the atmosphere, assuming that this distribution was determined by the up and down moving air currents (circulation), which cause a thorough mixing of the atmosphere in a comparatively short time. His results, very approximate, as he himself wrote, showed that if the contribution of the ground radioactivity was equal to the one of radioactivity suspended in the air, the radiation intensity should be fairly constant from $z = 0$ to $z = 1,000$ m, where z was the altitude; if the radiation from sources suspended in the atmosphere dominates, one could observe an increase in radiation intensity by a factor of about 2 from $z = 0$ to $z = 1,000$ m, and for a dominating ground distribution the radiation intensity at $z = 1,000$ m would be reduced to a fraction of its value at $z = 0$.

He successively, in 1913, performed experimental measurements on the content of RaA in the atmosphere [91, 92].

A new instrument started to be used: the "ionization chamber". A ionization chamber is a vessel filled with some gas in which an electric field is established between two electrodes. When ions are produced in the region between the electrodes, an electric discharge is produced. Studies in which ionization chambers were brought over the Earth's crust were made by J.C. McLennan and E.N. Macallum [93] with inconclusive results, in part because of instrumental difficulties but principally because the effects of the radiation coming from the Earth's crust are superposed to that of the radiation coming from space.

1.7 Domenico Pacini: An Often Forgotten Contribution

To summarize the state of affairs in the first 10 years of the twentieth century, we may say that after having established that air contained ions whose formation was attributed to the effect of a very penetrating radiation, people were inclined to believe that this radiation could be explained in terms of gamma rays emitted by the radioactive substances present on the Earth surface, although an extraterrestrial origin was sporadically discussed (f.e. G.W. Richardson [94] from Trinity College, Cambridge).

The pure radioactive origin was questioned by the Italian physicist Domenico Pacini [95–98] who performed a series of measurements on the ocean and on the Bracciano lake, near Roma.

Domenico Pacini was born on 20th February 1878 in Marino, near Roma, Italy [99]. He took a degree in physics in 1902 at the Roma University and became there an assistant under Professor Pietro Blaserna (1836–1913)[15] in the following 3 years, studying under the guidance of Alfonso Sella phenomena of electrical conductivity in gases. In the same years, Pacini was also aggregated, as an assistant, to the Royal Commission which had the task to examine the presumed efficacy of gunpowder shots against the formation of hail. In 1906, he was appointed assistant to the Ufficio Centrale di Meteorologia e Geodinamica (Central Office of Meteorology and Geodynamics) and the head of the section which studied thunderstorms and electric phenomena in the atmosphere. His position did not change until 1927, when he was promoted to the position of Principal Geophysicist, and the following year he was appointed full professor of Experimental Physics at the Bari University where he was busy with the organization of the Physics Institute for the Medical Faculty and made researches on the light scattering from air. He died in Roma on 23 May 1934.

While working at the Ufficio Centrale di Meteorologia, he was interested in the air ionization. Between 1907 and 1912, Pacini performed measurements of air conductivity on the ground, on the ocean surface, and since 1911, deep in the ocean ([100], see also [101–104]). The purpose of these experiments was to verify if the known causes of ionization (the radioactivity of the terrestrial crust) were sufficient to explain the formation of the measured nearly 13 ions per second per cc. in air at ground level.

Pancini referred to Arthur Steward Eve (1862–1948)—an ex-collaborator of Rutherford—who in 1907 speculated that the ionization over the ocean should be less than over the land because in the seawaters radium was present to a markedly less degree than in the sedimentary rocks on land. And since, radium emanation decays to half value in 4 days, the wind was unable to transport the emanation from land to places in mid-ocean before the activity is decreased. However Eve did not obtain the results he expected [105]. In 1911 he calculated also the effect of RaC [106]. Similar results were obtained also by G.C. Simpson and C.S. Wright [107] who found a considerable rate of ionization on the ocean which could not be explained by radioactivity.

Pacini utilized Eve's calculations to derive the expected ionization on ocean at a distance of more than 800 m from the coast and performing measurements in front of Genoa obtained much greater values than expected on the basis of Eve's considerations.

---

[15]Pietro Blaserna (1836–1918) Italian physicist professor of physics in Palermo (1863) and Roma (1872). Member of the Italian Accademia dei Lincei (since 1873) of which he was the president (1904–1916). Senator of the Kingdom (1890). He worked on electromagnetic problems.

The measurements on the ocean were made on board of the Royal military ship "Folgore", put at his disposal by the Marina Militare. Pacini eventually concluded that ionization on the ocean is of the same order of magnitude as that one observed at land. Starting from 1910, he measured the radiation intensity under the water, three meters deep below the surface in the Golfo Ligure (and successively repeated experiments in the Bracciano lake, near Roma), verifying that the radiation was absorbed by the water layer.

The measurements performed at the Bracciano lake were consistent with the number of ions produced at depth, that he had found in the open ocean.

Discussing the possible origins of the variations, he speculated that a cause of ionization should exist in the air independent from the radioactive substances on ground.

The final results and interpretation are reported in the paper "La radiazione penetrante alla superficie ed in seno alle acque".[16] Pacini writes [98] *"The observations made in the ocean in 1910 led me to conclude that a significant part of the penetrating radiation found in air, had an independent origin from the direct action of active substances contained in the upper layers of the terrestrial crust"*. In fact these observations *"... indicated that on the ocean surface, where the ground action is no more sensible, a ionization cause should exist of such intensity that could not be explained by considering the known distribution of radioactive substances in the water and in air"*.

Pacini added: *"... by dipping the apparatus in the waters, the mean value of the radiation observed at sea level can further be lowered. The apparatus... already used in the previously reported experiments, was sealed in a copper box so to be able to dip it in the waters. The experiments were made at the Accademia navale di Livorno and exactly in the same place where they had been done the previous year. The apparatus was placed on board of the same boat pegged at more than 300 m from the coast, over 8 m of water and observations were made from June 24 to 30 with the apparatus on the sea surface and dipped in the water down to 3 m"*.

He also observed that his results were consistent with an attenuation of an exponential type concluding that the soil radioactivity alone could not explain the results of his experiments, which started submarine physics, and speculated about the possible existence of an extraterrestrial radiation.

References

1. C. De Coloumb, Mem. de l'Acad. des Sciences (Paris), 612 (1875)
2. C.V. Boys, Philos. Mag. **28**(5), 14 (1889) ("Quartz as an insulator" paper read on 13 April 1889, p. 17)
3. G.J. Stoney, Philos. Mag. **38**(5), 418–420 (1894)
4. H.L.F. Helmoltz, J. Chem. Soc. **XXXIX**, 277 (1881)

---

[16]Penetrating radiation at the surface of and in the water.

5. J.J. Thomson, Philos. Mag. **44**(5), 293 (1897)
6. C.T.R. Wilson, *Nobel Lecture* (1927)
7. C.T.R. Wilson, Proc. R. Soc. London **59**, 338 (1896)
8. C.T.R. Wilson, Philos. Trans. **189**, 265 (1897)
9. J.J. Thomson, Proc. R. Soc. Lond. **59**, 274 (1896)
10. J.J. Thomson, J.A. McClelland, Proc. Camb. Phil. Soc. **9**, 126 (1896)
11. W.C. Roentgen, Sitz. Ber. Phys. -Med. Gesell. Wurzburg. **137**, 132 (1895)
12. J.J. Thomson, E. Rutherford, Philos. Mag. **42**(5), 392 (1896)
13. C.T.R. Wilson, Proc. R. Soc. Lond. **64**, 127 (1898)
14. C.T.R. Wilson, Proc. R. Soc. Lond. **65**, 289 (1899)
15. C.T.R. Wilson, Proc. R. Soc. Lond. **59**, 338 (1896)
16. C.T.R. Wilson, Proc. Camb. Phil. Soc. **9**, 337 (1897)
17. J. Elster, H. Geitel, Ann. Phys. **2**, 425 (1900)
18. J. Elster, H. Geitel, Phys. Z. **2**, 560 (1900)
19. A. Pais, Rev. Mod. Phys. **49**, 925 (1977)
20. F. Linss, Meteorol. Z. **4**, 345 (1887)
21. C.T.R. Wilson, Proc. R. Soc. Lond. A **68**, 151 (1901)
22. C.T.R. Wilson, Proc. R. Soc. Lond. A **68**, 151 (1901) (at p. 152)
23. C.T.R. Wilson, Proc. Camb. Phil. Soc. **11**, 32 (1900)
24. C.T.R. Wilson, Nature **63**, 195 (1900)
25. H. Geitel, Phys. Z. **2**, 116 (1901) (published 24 Nov 1900)
26. C. Matteucci, Ann. Chim. Phys. **28**, 385 (1850)
27. E. Warburg, Ann. Phys. Chem. **145**, 578 (1872)
28. C.T.R. Wilson, Proc. R. Soc. Lond. A **68**, 151 (1901) (at p. 159)
29. C.T.R. Wilson, Proc. R. Soc. Lond. A **69**, 277 (1901)
30. R.B. Brode, Rev. Mod. Phys. **11**, 222 (1939)
31. C.T.R. Wilson, Proc. R. Soc. London A **104**, 1 and 192 (1923)
32. A. Einstein, Ann. Phys. **17**, 132 (1905)
33. G.N. Lewis, Nature **118**, 874 (1926)
34. R.A. Millikan, Phys. Rev. **7**, 355 (1916)
35. R.A. Millikan, C.F. Eyring, Phys. Rev. **7**, 18 (1916)
36. R.A. Millikan, C.L. Lauristen, Proc. Nat. Acad. Sci. USA **2**, 78 (1916)
37. R.A. Millikan, Phys. Rev. **18**, 236 (1926)
38. A.H. Compton, Phys. Rev. **21**, 483 (1923)
39. C.D. Anderson, Phys. Rev. **43**, 491 (1933)
40. H.A. Wilson, Philos. Mag. **5**(6), 429 (1909)
41. C.T.R. Wilson, On a method of making visible the path of ionising particles. Proc. R. Soc. Lond. A **85**, 285 (1911)
42. C.T.R. Wilson, On an expansion apparatus for making visible the tracks of ionising particles in gases and some results obtained by its use. Proc. R. Soc. Lond. A **87**, 277 (1912)
43. W.H. Bragg, Philos. Mag. **8**(6), 719 (1904)
44. W.H. Bragg, R. Kleeman, Philos. Mag. **10**(6), 318 (1905)
45. W. Bragg, *Archives of the Roentgen Ray* (1911), p. 402
46. W. Bragg, *Studies in Radioactivity* (Macmillan, London, 1912), p. 34 and 142
47. C. Chaloner, The most wonderful experiment in the world: a history of the cloud chamber. BJHS **30**, 357 (1997)
48. W. Bragg, *Archives of the Roentgen Ray* (1911), p. 405
49. C.T.R. Wilson, Proc. R. Soc. Lond. A **87**, 277 (1912)
50. C.T.R. Wilson, Proc. R. Soc. Lond. A **87**, 277 (1912)
51. N.N. Das Gupta, S.K. Ghosh, Rev. Mod. Phys. **18**, 225 (1946)
52. R.D. Kleeman, Proc. Camb. Phil. Soc. **17**, 314 (1914)
53. E. Marsden, T.S. Taylor, Proc. R. Soc. Lond. A **88**, 443 (1913)
54. E. Rutherford, The structure of the atom. Philos. Mag. **27**(6), 488 (1914)
55. E. Rutherford, Philos. Mag. **27**(6), 488 (1914) (at p. 490)
56. E. Rutherford, Philos. Mag. **37**(6), 537, 562, 571, 581 (1919)

57. E. Rutherford, Science **50**, 467 (1919)
58. T. Shimizu, Proc. R. Soc. Lond. A **99**, 432 (1921)
59. P.M.S. Blackett, Proc. R. Soc. Lond. A **103**, 62 (1923)
60. P. Kapitza, Proc. Camb. Phil. Soc. **21**, 511 (1923)
61. C. Chaloner, BJHS **30**, 357 (1997)
62. E. Rutherford, "Statement of claims of Professor Wilson", 24 Jan 1927
63. V.C. Chaloner, BJHS **30**, 357 (1997)
64. G.D. Rochester, J.G. Wilson, *Cloud Chamber Photographs of the Cosmic Radiation* (Pergamon, London, 1952)
65. W. Gentner, H. Maier-Leibnitz, W. Bothe, *An Atlas of Typical Expansion Chamber Photographs* (Pergamon, London, 1954)
66. J. Elster, H. Geitel, Phys. Z. **1**, 11 (1899)
67. J. Elster, H. Geitel, Phys. Z. **1**, 245 (1900)
68. J. Elster, H. Geitel, Phys. Z. **2**, 560 (1901)
69. J. Elster, H. Geitel, Phys. Z. **3**, 574 (1901)
70. F. Kohlrausch, Sitz. Ber. Akad. Wiss. (Wien) **115**, 1321 (1906)
71. J.C. McLennan, E.F. Burton, Phys. Rev. **16**, 184 (1903)
72. R.J. Strutt, Nature **67**, 369 (1903)
73. J.C. McLennan, E.F. Burton, Phys. Z. **4**, 553 (1903)
74. E. Rutherford, H.L. Cooke, Phys. Rev. **16**, 183 (1903)
75. C.T.R. Wilson, Proc.Camb. Phil. Soc. **11**, 428 (1902)
76. J. Elster, H. Geitel, Phys. Z. **2**, 560 (1901)
77. J. Elster, H. Geitel, Ann. Phys. **2**, 441 (1900)
78. H. Ebert, Ann. Phys. **5**, 718 (1901)
79. H. Mache, T. Rimmer, Phys. Z. **7**, 617 (1906)
80. H. Mache, Sitz. Ber. Akad. Wiss. (Wien) **119**, 55 (1910)
81. E. von Schweidler, Sitz. Ber. Akad. Wiss. (Wien) **121**, 1297 (1912)
82. W.W. Strong, Phys. Z. **9**, 117 (1908)
83. K. Bergwitz, *Habilitationspaper* (Braunschweig, 1910)
84. Th. Wulf, Phys. Z. **10**, 997 (1909)
85. Th. Wulf, Phys. Z. **11**, 811 (1910)
86. K. Kurz, Phys. Z. **10**, 834 (1909)
87. A. Gockel, Th. Wulf, Phys. Z. **10**, 845 (1909)
88. A. Gockel, Phys. Z. **11**, 280 (1910)
89. A. Gockel, Phys. Z. **12**, 595 (1911)
90. E. Schrödinger, Sitz. Ber. Akad. Wiss. (Wien) **121**, 2391 (1912)
91. E. Schrödinger, Sitz. Ber. Akad. Wiss. (Wien) **122**, 2023 (1913)
92. J. Mehra, H. Rechenberg, *The Historical Development of Quantum Theory* (vol. 5, Springer, Heidelberg)
93. J.C. McLennan, E.N. Macallum, Philos. Mag. **22**(6), 639 (1911)
94. G.W. Richardson, Nature Lond. **73**, 607 (1906)
95. D. Pancini, Rend. Acad. Lincei **18**, 123 (1909)
96. D. Pacini, Ann. Dell'Uff. Centr. Meteor. **XXXII**(parte I) (1910)
97. D. Pacini, Le Radium T **VIII**, 307 (1911)
98. D. Pacini, Nuovo Cimento **3**, 93 (1912)
99. A. De Angelis, Riv. Nuovo Cimento **33**, 713 (2010)
100. D. Pacini, Nuovo Cimento **3**, 93 (1912)
101. D. Pacini, Nuovo Cimento **15**, 5 (1908)
102. D. Pacini, Nuovo Cimento **3**, 24 (1912)
103. D. Pacini, Nuovo Cimento **19**, 449 (1910)
104. D. Pacini, Le Radium **VIII**, 307 (1911)
105. A.S. Eve, Philos. Mag. **13**(6), 248 (1907)
106. A.S. Eve, Philos. Mag. **21**(6), 26 (1911)
107. G.C. Simpson, C.S. Wright, Proc. R. Soc. **85**, 175 (1911)

Chapter 2
The Discovery: Victor F. Hess and the Balloon Ascents

2.1 Hess and the Balloon Ascents

The study of the causes of air ionization had reached a state of total uncertainty, when the Austrian physicist Victor Hess became interested in the problem and started a series of balloon ascents that culminated in a memorable flight in August 1912, which convinced him that the origin of a part of the radiation was extraterrestrial.

Victor Franz Hess was born on 24 June 1883 in Schloss Waldstein, near Peggau in Steiermark, Austria. His father, Vinzens Hess, was chief forester on the estate of Prince Oettingen-Wallerstein.

Victor received his entire education in Graz: first at the Gymnasium (1893–1901) and then at Graz University (1901–1905), where he graduated. Intending to do postdoctoral work in optics, he made arrangements to study under Professor Paul Drude (1863–1906) in Berlin. But Drude—a German physicist who performed important research in optics and in the electronic theory of metals and authored the classic textbook *Lehrbuch der Optik* (1900)—committed suicide a few weeks before Hess arrived, and therefore he went to Vienna University to study under Professor Franz Exner.

There, for a short time he worked at the Zweites Physikalisches Institut, under Exner and Professor Egon von Schweidler, who interested him in radioactivity and atmospheric electricity. Hess covered several academic positions and in 1910 earned his PhD and was appointed assistant of Stefan Meyer (1872–1949) at the new Institut fur Radiumforschung of the Austrian Academy of Sciences, a position he held until 1920.

He began his scientific activity by making systematic and quantitative investigations on the radioactive substances contained in the air. In 1919 he received the Lieben Prize of the Austrian Academy of Sciences for his discovery of the "Hohenstrahlung" (radiation that comes from outside), and the following year was nominated extraordinary Professor of Experimental Physics at the Graz University.

M. Bertolotti, *Celestial Messengers: Cosmic Rays*, Astronomers' Universe, DOI 10.1007/978-3-642-28371-0_2, © Springer-Verlag Berlin Heidelberg 2013

From 1921 to 1923, Hess took a leave of absence to make his first trip to the United States. After his return to the University of Graz, in 1925, he was appointed Ordinary Professor of Experimental Physics. He later went to Innsbruck University (1931) and was nominated director of the newly founded Institute of Radiology. In the same year he founded the station at the Hafelekar mountain (2,300 m), near Innsbruck, for the observation and study of cosmic rays.

In 1936 he was awarded the Physics Nobel Prize "*for his discovery of cosmic radiation*". He shared the prize with C.D. Anderson who received the award "*for his discovery of the positron*".

Two months after the Anschluss, in March 1938, he was dismissed from his position, partly because he had a Jewish wife and also because he held a critical position in regard to Nazism. Shortly after he also lost his pension. A sympathetic Gestapo officer warned Hess that he and his wife would be taken to a concentration camp if they remained in Austria, and they escaped to Switzerland 4 weeks before the order came for their arrest. From there they emigrated to the United States where Fordham University in New York offered Hess a full professorship. He became an American citizen in 1944.

In the States he did not continue his research on cosmic rays, but got interested in the effects of atomic radiations, performing many investigations in the field of radioactivity and radiotherapy, in particular on the dosimetry of radiations emitted by radium. In 1946, less than a year after the atomic bomb was dropped on Hiroshima, he and Paul Luger of Seattle University conducted the first tests for radioactive fallout in the United States. Many of these were made from the 87th floor of the Empire State Building. The following year Hess went to the depths of Manhattan, measuring the radioactivity of granite in the 190th Street subway station at the base of Fort Tyron which was covered by 160 ft of rock.

He retired in 1958 and became Professor emeritus, continuing to do research in his laboratory at the school. He passed away on 17 December 1964, in Mont Vernon, New York.

When, in 1936, Hess was awarded the Nobel Prize, Professor H. Pleijel of the Royal Swedish Academy of Sciences, addressed him with the following words at the official ceremony:

> "*By virtue of your purposeful researches into the effects of radioactive radiation carried out with exceptional experimental skill you discovered the surprising presence of radiation coming from the depths of space, i.e., cosmic radiation. As you have proved, this new radiation possesses a penetrating power and an intensity of previously unknown magnitude; it has become a powerful tool of research in physics, and has already given us important new results with respect to matter and its composition. The presence of this cosmic radiation has offered us new, important problems on the formation and destruction of matter, problems which open up new fields of research*" ([1], see also [2]).

In 1911, while reading about earlier experiments by Wulf [3,4] and Gockel [5], Hess speculated as to whether the source of ionization could be located in the sky rather than on the ground. Being a passionate balloonist he made a series of ten ascents (five during the night), two in 1911 [6,7], seven in 1912 [8], and one in 1913 [9] demonstrating that the ionization, after the expected initial decrease with height, started to increase again.

All measurements were done with electroscopes of the Wulf type. The electroscope, once charged, assumes a potential difference with respect to ground and this voltage decreases as the charge disappears. By knowing an electrical characteristic of the instrument (its capacity) it is easy to pass from voltage to charge, and, because the decrease in charge is produced by ions of opposite sign which neutralize it, their number can be easily measured. Therefore, Hess measured the potential drop of his electroscope as a function of time, making approximately a measurement every hour and converting the result to ion pairs created in the gas of the electroscope per cubic centimetre per second.

However the measurements were not very accurate, and measurements performed on different electroscopes presented inconsistent results.

Before making the balloon ascents, Hess compared Wulf's results with the theoretical considerations that Arthur Eve [10] had presented in the paper "*On the Ionization of the Atmosphere due to the Radioactive Matter*", published in the January issue of the journal Philosophical Magazine, and that had been considered also by Pacini. According to these considerations, assuming a uniform distribution of RaC on the Earth surface, an elevation of 10 m should reduce the radiation effects to 83% of the ground value, an elevation of 100 m should reduce it to 36%, and at a height of 1,000 m only 0.1% of the initial value should remain. Such a large disagreement with Wulf's results seemed to Hess of the highest importance and therefore he decided to investigate more deeply the question. Since in Eve's evaluations the gamma ray absorption coefficient played a crucial role, Hess as a first step, measured in the open air the absorption of gamma rays coming from an intense radium source, finding an absorption coefficient of $0.477 \times 10^{-4}\,\mathrm{cm}^{-1}$ of air, in fair agreement with the value extrapolated from the data already known, thus establishing that the gamma rays that came from ground should have been completely absorbed at an height of 500 m [11].

These measurements were very important, in fact, until that moment the gamma ray absorption coefficient had not been determined directly, but had just been obtained by extrapolation from data for solid and fluid substances assuming that the absorption coefficient remained constant at equal density. Hess' result showed that the measurements made by Wulf still had to be explained.

In addition, Hess realized that the gas pressure, in the instruments used by Gockel in his ascents, changed with altitude and that this fact invalidated the measurements. Therefore, he designed an instrument of the kind used by Wulf but with sufficiently thick walls to survive to the pressure changes and to the harshness of an open gondola during the ascent. The instrument was also hermetically sealed so that the pressure of the internal gas remained constant at any altitude, thus making the sensitivity of the instrument independent from height.

The first ascent was made on 28 August 1911 at 8 a.m. leaving from the Prater of Vienna, with the balloon "*Radetzky*" of the Austrian Aeroclub, with Oberleutnant S. Heller as pilot. The balloon reached 1,070 m above ground allowing to make measurements during the 4 h of the flight. A second ride with another balloon ("*Austria*") was made during the night between 12 and 13 October 1911, reaching a height of 360 m. During both flights, Hess found that ionization changed very little

Fig. 2.1 Hess in the balloon
basket after the landing of a
successful flight in 1912

[12] and concluded that, because the radiation at that altitude was not too much different from the one at sea level, some other source of penetrating radiation should exists in addition to the gamma radiation coming from the radioactive substances on the Earth's crust [13]. However he wrote that because neither he nor Gockel had observed the decrease with altitude that Wulf had measured on the Eiffel tower, to obtain reliable values, long flights were necessary, as well as to make observations at modest altitudes. Since better conditions were achieved during night flights, he chose that time to perform many experiments. He also speculated that it was advisable to make measurements with two electroscopes, one with very thin walls (to be able to measure the beta radiation) and one with thicker walls (to allow only gamma ray to pass), and that it also was necessary to extend the observations at higher altitudes, possibly up to 7,000 m (Fig. 2.1).

With these considerations in mind, having obtained a financial support from the Keiserlichen Akademie der Wissenschaften in Vienna, in 1912 he undertook seven more ascents—mostly in the Vienna region—two of which were very important. The first flight in 1912 was made on the occasion of a nearly total Sun eclipse, in the lower Austria on 17 April, between 11 a.m. and 1 p.m., up to an altitude of 2,750 m. During the eclipse, the balloon descended as a result of the gas cooling. Hess found that the radiation at about 2,000 m was greater than at sea level and that the eclipse had no effect. Therefore, he reached the conclusion that a fraction of the radiation is of cosmic origin that hardly comes from the Sun [14].

Table 2.1 Summary of Hess' results

Mean height from ground (m)	Measured radiation (ions per cc per second)		
	Electrosc. 1	Electrosc. 2	Electrosc. 3
0	16.3	11.8	19.6
Up to 200	15.4	11.1	19.1
300–500	15.5	10.4	18.8
500–1,000	15.6	10.3	20.8
1,000–2,000	15.9	12.1	22.2
2,000–3,000	17.3	13.3	31.2
3,000–4,000	19.8	16.5	35.2
4,000–5,200	34.4	27.2	–

He then performed six more ascents: between 26 and 27 of April, at night; between 20 and 21 of May, again at night; on 3 June, 19 June, and 28 June, and the last one on 7 August.

This seventh Hess flight with a hydrogen filled balloon was designed to rise to a very high altitude. On the morning of 7 August 1912 at 6.12 a.m., the balloon ascended from a field near the city of Aussig, Austria. In the gondola were three people: a navigator, a meteorologist, and Hess. The flight lasted 6 h reaching an altitude of 5,350 m. At noon the balloon landed near the German city of Pieskow, 50 km east of Berlin.

In all these flights the Austrian researcher used three apparatuses, two with thick walls and one with very thin walls in order to detect also beta rays.

The results of the last ascent, combined with the previous ones, lead to the conclusion that at 500 m the ionization had approximately decreased by two pairs of ions produced per cubic centimetre per second with respect to the ground value. At 1,500 m the same result as at ground was obtained and, starting from 1,800 m, its increase was evident. At 3,500 m the increase in the produced ion pairs was of the order of 4 and at 5,200 m it was 16. The results did not show any difference between day and night [15]; Hess wrote [16]:

"Immediately above ground the total radiation decreases a little... at altitudes of 1000 to 2000 m there occur again a noticeable growth of penetrating radiation. The increase reaches, at altitudes of 3000 to 4000 m, already 50% of the total radiation observed on the ground. At 4000 to 5200 m the radiation is stronger by (producing) 15 to 18 (more) ions than on the ground".

Table 2.1 shows the pairs of ions per cc per second registered by the three different electroscopes considering all flights [17].

If the difference between the number of ions per cc per second at some height and the one at ground is considered, the variation due to the change of altitude is obtained. Table 2.2 shows the results. While the pairs of ions per cc per second initially decrease with height, after about 2,000 m they start to increase and in the two electroscopes with thick walls there are nearly four pairs more between 3,000 and 4,000 m and from 16 to 18 pairs between 4,000 and 5,200 m. The increase is even stronger in the third electroscope, which was the one with thin walls.

Table 2.2 Differences between altitude and ground measurements

Mean height from ground (m)	Variation between altitude and ground (ions per cc per second)			
	Electrosc. 1	Electrosc. 2	Electrosc. 3	Mean
300–500	−0.8	−1.4	−0.8	−1
500–1,000	−0.7	−1.5	1.2	−0.3
1,000–2,000	−0.4	0.3	2.6	0.8
2,000–3,000	1	1.5	11.6	4.7
3,000–4,000	3.5	4.7	15.6	7.9
4,000–5,200	18.1	15.4	–	11.2

Hess concluded that this ionization should be attributed to a still unknown radiation penetrating from outside the Earth's atmosphere, with an exceptionally high penetration power, still able to ionize air at sea level. On the grounds of the knowledge of the time, this radiation was believed to be very energetic gamma rays.

In the November 1912, issue of the German journal Physikalische Zeitschrift, Hess wrote: "*the first results of my observations are most easily explained by the assumption that radiation of very high penetrating power enters the atmosphere from above and creates, even in the lowest layers a part of the ionization observed in closed vessels*".

Hess called it a very penetrating radiation because it was able to cross the whole atmosphere while still producing ions.

Already at that time people speculated on the origin of the radiation. By using the ascent made on 17 April 1912, at the time of the quasi-total Sun eclipse, Hess observed that the ionization did not show any reduction during the eclipse, and therefore decided that the Sun could not be the source of the rays. Solar cosmic rays—we know today—arrive nearly isotropically and therefore his conclusion was incorrect.

Finally, a new flight of the balloon Astartè, on 1 June 1913, provided to him free of charge by the Austrian Aeroclub and its proprietor E.C. Sigmundt of Trieste, went up to 4,050 m above sea level, thus confirming the previous observations [18].

Therefore, Hess wrote "*the theory developed recently by E. Schrödinger [the theory of 1912 we have discussed in the previous chapter] for the altitude distribution of penetrating radiation fails vis-à-vis the increase of radiation with height*".

In June 1913, Hess submitted to the Vienna Academy of Sciences, a careful analysis of all possible sources of error in his measurements, he again gauged his apparatus and determined as accurately as possible the radiation emerging from its walls. The results of this analysis supported his previous conclusion that an appreciable part of the total penetrating radiation does not emerge from the known radioactive substances of the earth and of the atmosphere.

We may regard this result as establishing definitively the existence of a new unexplained kind of penetrating radiation, and the year 1912 is generally considered the year of the discovery of the cosmic radiation. Hess published a summary of his work in the Physikalische Zeitschrift, which reached out to a larger public than the Wien Sitzungberichte [19].

2.2 The Wien Congress and the Official Birth of Cosmic Rays

Between 21 and 28 September 1913, a very important Congress was held in Vienna: the 85th Versammlung Deutscher Naturforscher und Artze, featuring about 7,000 participants, a reception at the Imperial Court, a banquet hosted by the City of Wien in the town hall, and many other events. In the physical section reports were given by Einstein on gravitation, by Robert Richard Pohl (1884–1976) from Berlin on the photoelectric electron emission, by James Franck (1882–1964) and Gustav Hertz (1887–1975), also from Berlin, on the connection between ionization impact and electron affinity, and by the Viennese Karl Herzfeld on free electrons in metals. A session was devoted to radioactivity and the penetrating radiation in the atmosphere. Hans Geiger—the former E. Rutherford collaborator, now resident in Berlin—presented experiments on methods for counting alpha and beta particles emitted by radioactive substances, and Victor Hess spoke on an improvement in Wulf's apparatus for determining the absolute value of the gamma ray intensity emitted from radium. Theodor Wulf presented a communication in which the results obtained on the ground in Graz (by Benndorf and Veith), in Davos (by Dorno), in Vienna (by Hess and Kofler), in Innsbruck (by Kruse), and in Valkenburg (by Wulf) were compared, concluding that the suggestion of an extraterrestrial source in order to explain the fluctuations of the gamma radiation was not substantiated by these observations. Hess did not reply. He had already presented his research at the 83rd Naturforscherversammlungen in Karlsruhe [20] and at the 84th Naturforscherversammlungen in Muenster [21], however he was excellently replaced by Werner Kolhoerster of Halle, who spoke on balloon measurements that confirmed Hess' results. In the following discussion, also Gockel, mentioning his own measurements on Swiss glaciers, which gave results in agreement with the ones of Kolhoerster, confirmed the hypothesis of a radiation coming from the outer space.

All these presentations agreed with Hess' results and we may say that at the Wien Congress the "Hoehenstrahlung" or "radiation coming from the outside" received its first recognition.

Hess continued his research on cosmic rays in the Austrian Alps. When in 1931 he was appointed a Professor at the University of Innsbruck, he decided to build a small observatory for the study of cosmic rays on the summit of the Hafelekar mountain, 2,300 m in the Alps, a mountain which could be reached by suspended cable railway from Innsbruck in 40 min throughout the whole year. The observatory was named "Station fuer Ultrastrahlungsforschung" and was located in a wooden chalet exactly on the ridge of the Nordkette. It was inaugurated on the summer of 1931. The instruments were set upon large concrete pillars in a room measuring 4.5×4.5 m. The main apparatus was constructed by E. Steinke of Koenigsberg and consisted of a cylindrical ionization chamber of 22.6 l capacity, filled with CO_2 at a pressure of about 9.5 atm, connected with a Lindemann electrometer and a self-registering device.

Within a few years simultaneous observations were organized with similar apparatus in a number of stations on a cooperative plan. The participating stations were the ones located in Abisko (North Norway, researcher A. Corlin), Koenigsberg (E. Steinke), Potsdam (W. Kolhoerster), Dublin (Nolan and C. O'Brolchain), Bandoeng [Java, Jacob Clay (1882–1955) of Amsterdam], and Capetown (South Africa, researcher Schonland) [22].

An interesting correspondence took place between Hess and Domenico Pacini in 1920.[1] On 6 March, Pacini wrote to Hess: "...*I have been able to see some of your publications on the electro-atmospheric phenomena that you sent to the director of the R. Ufficio Centrale di Meteorologia e Geodinamica. Someones were already known to me in the summaries which could arrive during the war. One I did not know, the one titled 'Die Frage der durchdring Strahlung ausserterrestrischen Ursprungers'. While I must make you my compliments for the clarity with which you expound in a simple form the status of this important question, I regret you did not quote the Italian works on this argument, works that have without any doubt the priority, for what pertains to the forecast of the very important conclusions to which successively came Gockel, You yourself, Mr. Hess, and Kolhoerster; and I am so much displeased, because in my publications I never forgot to quote the pertaining people...*" Hess reply is from 17 March 1920: "*Very estimable Mr. Professor, your very esteemed letter of 6 was particularly welcome to me, because it renews our relations so long interrupted during the unhappy war: I had been pleased to write to you before, but unfortunately I did not know where you were. The short note on the problem of the penetrating radiation of extraterrestrial origin is the publication of a popular conference, and therefore the references do not claim to be exhaustive. Since it deals essentially of measurements made by balloons, I did not extend to speak of your measurements on the ocean, that I well know. I apologise for this omission that was far from any intentions of mine...*"

2.3   The First Confirmations

Werner Heinrich Gustav Kohloerster was born in Schwiebus, Germany, on 28 December 1887 and died in Munich in 1946 as a consequence of a car accident. He studied physics at Halle University under Professor Friedrich Ernst Dorn (1848–1916) (co-discoverer of radon with M. Curie). He then joined the Aerophysikalische Forschungsfonds in Halle, devoting himself to the observation of the penetrating radiation both at high altitudes in a balloon and under water. In 1914 he continued his studies in physics in the Hans Geiger's laboratory in the Physikalisch-Technische Reichsanstalt, Berlin. During World War I he performed measurements of atmospheric electricity on Bosporus in Turkey (1916–1918) and after the war he taught

---

[1] Quoted in the funeral commemoration of Pacini by Prof. Rizzo, University of Napoli, in Bari in 1935 and reproduced in ([23] see also [24]).

in school for several years. In 1922 he came back at Physikalisch-Technische Reichsanstalt. Eight years later the Prussian Academy supported the establishment of a laboratory in Potsdam, in which Kolhoerster carried research on cosmic rays. Eventually, in 1935, he was appointed a full Professor of geophysics and director of the Institut fur Hoehenstrahlungforschung in Berlin-Dalhem.

Kolhoerster made five balloon ascents, at very dangerous altitudes, in 1913 and 1914, using more sophisticated techniques than the ones used by Hess, reaching a maximum height of 9,300 m so extending by far Hess' measurements [25–27].

He constructed an apparatus similar to the one of Wulf, which was particularly suited for balloon flights, because it could be sealed airtight so that the local air pressure would not affect the measurements [28] and, before the rides, he tested the effects of the low temperature on it, a particular to which Hess did not pay attention.

Kolhoerster performed three ascents in 1913, all starting from Bitterfeld (about 25 km northeast of Halle), and going to Schuttenhoven in Boemia, Neutomischl in Posen and close to Halle, reaching altitudes of 3,600, 4,000, and 6,300 m, respectively.

These were the rides he presented to the Vienna Conference reporting results in agreement with Hess' ones.

In 1914 he ascended to 9,300 m and found an ionizing radiation considerably stronger than the one detected by Hess, with an increase of the ionization up to 50 times the value at sea level. The absorption coefficient was evaluated to be 10^{-5} per cm of air in standard conditions (i.e. 20° C of temperature and 760 mm of mercury pressure). This value caused a great surprise being eight times smaller than the absorption of air for the most penetrating gamma rays known at that time.

The ionization–height curves were extrapolated from the altitudes where the ionization was certainly given by the sole cosmic radiation down to the sea level. It was estimated that the cosmic rays coming from outside were responsible for the production of about 1–2 ion pairs per cc per second of the ionization at sea level, while most of the ionization which amounted to 6–10 ion pair per cc per second should be attributed to the soil radioactivity.

Therefore only about 20% of the ion pairs present on the Earth surface at sea level is of cosmic origin, while the remaining part is due to natural radioactivity. This result well explains the original difficulties encountered in distinguishing the small extraterrestrial fraction from the bigger bias due to the environment radioactivity.

2.4   However a Great Confusion and Scepticism Survive

The newly found radiation had not yet received a generally accepted name. Egon von Schweidler introduced the term "*Hess rays*", but Hess himself used the term "*ultragamma radiation*" and Kolhoerster called them "*Hoehenstrahlung*".

During World War I and immediately afterwards, there was not much activity. Gockel, Hess, and Martin Kofler continued their researches in mountain ascensions and balloon flights [29, 30].

Kohloerster performed observations in his meteorological observatory [31]. However there was no major progress, and in spite of the measurements made by Hess and Kohloester, many people were not yet convinced. As an example, C.T.R. Wilson [32, 33], the inventor of the cloud chamber and one of the world expert of ionization phenomena, speculated that the radiation could be produced in thunderstorms in the high atmosphere. Other people continued to maintain that the atmosphere could contain small traces of radioactive gases (it was known that some radioactive substances as the element 86, radon, exist at the gaseous state). If for some reason, these radioactive gases had the tendency to concentrate in the high atmosphere, this could explain the increase of ionization at high altitudes.

We may observe, just as it could have been observed at the time, that in both cases the intensity of the unknown radiation should have changed depending on the meteorological conditions, on the hour, on the day, and the season. Obviously thunderstorms do not take place every moment, so it is hard to believe that the distribution of the hypothetical radioactive gases in the atmosphere can be constant at any moments of the day, or of the year, and for all weather changes.

Indeed, notwithstanding disagreements and doubts due to the limited accuracy of the instruments used at that time, the radiation as a whole was remarkably uniform. It came during the day and at night, in the summer and in the winter, rain or shine, with little changes from day to day, and from one location to another if they were at the same altitude.

It was speculated that the penetrating radiation of cosmic origin consisted of gamma rays of very high energy, because the gamma radiation from natural radioactive substances was much more penetrating than any other kind of corpuscular radiation known at the time. This assumption persisted until 1929, when the research on cosmic rays received a strong turn with the discovery that the majority of rays are corpuscles.

Be as it may, after the publication of the results of Hess and Kolhoerster, a violent controversy arose on the existence of the radiation and on its provenance to which took part, among others, Millikan and Bowen [34], Hoffmann [35], and Behounek [36]. However, Hess [37] confirmed his original results.

Hoffmann and Behounek, at first, did not find confirmation of the existence of a radiation more penetrating than gamma rays from radioactive substances. Their observations were made with ionization chambers shielded with lead absorbers. The rate of ionization in the chamber was measured as a function of the led thickness surrounding the chamber and the decrease observed for the first 10 cm was of the same order of what could be expected for gamma rays.

The reason why these observations did not allow to detect the existence of cosmic rays was found later. Due to secondary effects, the ionization rate under the lead is lower than under an equivalent mass of air and therefore the total intensity of cosmic rays drops abruptly when the measuring instrument is screened with an absorber only a few centimetres thick. This strong decrease was confused with the absorption of radioactive rays from the surroundings.

As we shall see, the cosmic radiation becomes "hard" passing through absorbing bodies. That is, while a fraction of the radiation is absorbed after traversing a few lead centimetres, the remaining part is able to pass through much greater thicknesses, up to a metre and more. This characteristic is expressed by saying that the radiation is harder. What effectively happens is that the few initial lead centimetres filter out a part of the radiation, and the remaining part is very penetrating. Therefore the radiation, after traversing 10 cm of lead, is much less attenuated by additional absorbers than one would expect. This effect was neglected and the measured ionization with more than 10 cm lead was ascribed to the residual ionization of the chamber.

The cosmic ray absorption in lead was observed by Hoffmann [38–40] a year after these experiments were performed, and was studied also by other researchers [41, 42].

At first, Millikan did not believe that the radiation had an extraterrestrial origin and assumed that it was all produced by radioactive substances present in the ground or in the air. He later changed his mind.

References

1. Nobel Lectures, *Physics* (Elsevier, Amsterdam, 1965), p. 356 (1922–1941)
2. Q. Xu, L.M. Brown, Am. J. Phys. **55**, 23 (1987)
3. Th. Wulf, Phys. Z. **10**, 997 (1909)
4. Th. Wulf, Phys. Z. **11**, 811 (1910)
5. A. Gockel, Phys. Z. **12**, 595 (1911)
6. V.F. Hess, Phys. Z. **12**, 998 (1911)
7. V.F. Hess, Wien. Sitz.-Ber. **120**, 1575 (1911)
8. V.F. Hess, Phys. Z. **13**, 1084 (1912)
9. V.F. Hess, Phys. Z. **14**, 610 (1913)
10. A. Eve, Philos. Mag. **21**(6), 26 (1911)
11. V.F. Hess, Sitz. Ber. Akad. Wiss. (Wien) **120**, 1205 (1911)
12. V.F. Hess, Sitz. Ber. Akad. Wiss. (Wien) **120**, 1575 (1911)
13. V.F. Hess, Phys. Z. **12**, 998 (1911)
14. V.F. Hess, Phys. Z. **13**, 1084 (1912)
15. V.F. Hess, Sitz. Ber. Akad. Wiss. (Wien) **121**, 2001 (1912)
16. V.F. Hess, Sitz. Ber. Akad. Wiss. (Wien) **121**, 2001 (1912) (Presented at the Munster meeting on 17 Oct. 1912)
17. V.F. Hess, Phys. Z. **13**, 1084 (1912)
18. V.F. Hess, Sitz. Ber. Akad. Wiss. (Wien) **122**, 1481 (1913)
19. V.F. Hess, Phys. Z. **14**, 610 (1913)
20. V.F. Hess, Phys. Z. **12**, 998 (1911)
21. V.F. Hess, Sitz. Ber. Akad. Wiss. (Wien) **121**, 2001 (1912)
22. V.F. Hess, Terr. Magn. Atmos. Electr. **37**, 399 (1932)
23. A. De Angelis et al., Il Nuovo Sagg. **24**, 70 (2008)
24. A. De Angelis, Riv. Nuovo Cimento. **33**, 713 (2010)
25. W. Kolhoerster, Phys. Z **14**, 1153 (1913) (Presented on 23 Sept. 1913 at the 85th Natur-forcherversammlung in Vienna, published in issue of 15 Nov. 1913)
26. W. Kolhoerster, Naturforschung Gesell. Zu Halle, Abhandl. Neue Folge 4 (1914)

27. W. Kolhoerster, Dtsch. Phys. Ges. Verh. **16**, 719 (1914)
28. W. Kolhoerster, Phys. Z. **14**, 1066 (1913)
29. A. Gockel, Dtsch. Schweiz. Natur. Ges. **54**, 1 (1917)
30. V.F. Hess, M. Kofler, Phys. Z. **18**, 585 (1917)
31. W. Kolhoerster, Phys. Z. **11**, 379 (1922)
32. C.T.R. Wilson, Proc. Camb. Phil. Soc. **22**, 534 (1925)
33. C.T.R. Wilson, Proc. R. Soc. **31**, 320 (1925)
34. R.A. Millikan, I.S. Bowen, Phys. Rev. **22**, (1923) (198 minutes of the Am. Phys. Soc. meeting, Pasadena, 5 May 1923)
35. G. Hoffmann, Phys. Z. **26**, 669 (1925)
36. F. Behounek, Phys. Z. **27**(8), 536 (1926)
37. V.F. Hess, Phys. Z. **27**, 159 (1926)
38. G. Hoffmann, Ann. Phys. **80**, 779 (1926)
39. G. Hoffmann, Ann. Phys. **82**, 413 (1927)
40. G. Hoffmann, Z. Phys. **42**, 565 (1927)
41. E. Steinke, Z. Phys. **42**, 570 (1927)
42. L. Myssowsky, L. Tuwim, Z. Phys. **50**, 273 (1928)

Chapter 3
The Confirmation: Millikan and the "Birth Cry" of Created Atoms

3.1 The Research Moves to USA: Millikan Provides the First Unambiguous Confirmations

The final confirmation of the existence of a penetrating radiation coming from outer space and hitting our Earth was given by a series of measurements made in the United States by R.A. Millikan (Fig. 3.1), one of the most influential American physicists of his time. He—who at the beginning was very sceptical about the extraterrestrial origin of the radiation—played a pivotal role in its study for more than 20 years.

Robert Andrew Millikan (1868–1953) was born on 22 March 1868 in Morrison,[1] Illinois, the son of a preacher. After getting his master degree, he went to Columbia University, where at the time was the sole graduate student in physics, receiving his PhD in 1895. He then travelled in Europe—as customary for the young American graduates—visiting the Universities of Berlin, Gottingen and Paris and had meeting with Max Planck, Walther Nernst (1864–1941), and Henri Poincarè (1854–1912). In 1896 Albert A. Michelson (1852–1931)—who had been his teacher at Chicago University during the summer of 1894—offered him an assistantship in his department which Robert—then still in Europe—promptly accepted. For the next ten years his main task was teaching and writing textbooks, which were adopted in many schools and colleges in the USA, for six hours a day; other six he devoted to research. When he decided to "*get busy on some more serious research*" he started developing the—later famous—technique to study the falling of small oil drops in order to measure the electric charge and thus determine the electron charge, eventually finding it to be the smallest existing charge [1–3].

The following research project was to verify Einstein's photoelectric theory, studying its characteristics, and eventually finding the best confirmations of

---

[1]His family moved to Maquoteca, Iowa, where the young Robert attended schools. In 1886, he entered the Oberlin College where he graduated in 1891 and obtained a Master degree in 1893.

M. Bertolotti, *Celestial Messengers: Cosmic Rays*, Astronomers' Universe,
DOI 10.1007/978-3-642-28371-0_3, © Springer-Verlag Berlin Heidelberg 2013

Fig. 3.1 R.A. Millikan

Einstein's photon theory—which he did not completely trust—in a series of measurements performed between 1907 and 1917 [4–7]. In 1910 he became professor at Chicago University and in 1916 he was nominated president of the American Physical Society and a member of the National Research Council. In 1917 he moved to Washington to organize the mobilization of civilian scientists and their collaboration with the military agencies. Later in 1921 he accepted a very advantageous offer from the new and almost unknown California Institute of Technology (CalTech). His researches were fully recognized in 1923 with the award of the Physics Nobel Prize for his "*precise measurements of the electronic charge and the Planck constant*".

He retired in 1945 and passed away on 19 December 1953 in San Marino in California.

His great contribution to American education was the developing of the California Institute of Technology. His life and scientific results were described by himself in a biography [8]. The main characteristics of Millikan's research consisted in a thorough study of the work of his predecessors, to find all possible sources of errors or weak points that he then tried to eliminate.

The first reports in America on the presence of a penetrating radiation in the atmosphere had been presented to the Congress of the American Physical Society in Washington, DC, on 31 December 1902 by E. Rutherford and H.L. Cooke [9], then at Montreal, and by J.C. McLennan and E.F. Burton [10] from Toronto. The following researches by A. Gockel, V. Hess, and W. Kolhoerster in the 1910s proved the existence of an extraterrestrial radiation, but the quantitative part of the measurements was still rather deficient. Millikan, in particular, was rather sceptical about its extraterrestrial origin and, therefore, decided to see if the experimental data given by the European researchers were correct.

We have already said that when an electromagnetic radiation travels through a medium it is absorbed, its intensity decreasing accordingly with the path length. The mathematical law (absorption law) is usually described by an universal exponential decrease

$$I(x) = I_0 e^{-\mu x}, \tag{3.1}$$

where $I(x)$ is the intensity after a path length x, I_0 is the initial intensity (for $x = 0$) and μ is called the *absorption coefficient*. The absorption coefficient of the penetrating radiation in the atmosphere, as derived from Kolhoerster's measurements was about 10^{-5} cm^{-1} of air, a value much smaller than that of the gamma rays emitted by natural radioactive materials (Hess had measured a value nearly 4.5 times larger for the gamma rays emitted by radium). This meant that the path of the mysterious radiation, before being completely attenuated, was much longer than that of the radiations emitted by the radioactive bodies known at the time.

In 1914, immediately after the measurements by Gockel, Hess, and Kolhoerster, discussed in Chap. 2, Millikan decided to understand why such a low absorption coefficient was found and to verify it. To this aim, he started to design special electroscopes to be sent in the high atmosphere, carried by sounding balloons. That was a revolution: no more men carried together with instrumentation by a big balloon, but small and cheap unmanned balloons barely carrying the instrumentation. The project started in 1915–1916, but the necessary instrumentation was not completed because the war put a stop to further activity in this direction. After the war, about 2 years were spent in trying to build the right kind of sounding balloons. Eventually, immediately after coming to Pasadena, Millikan was able to send his sounding balloons and the instrumentation in the high atmosphere. The first four electroscopes were built with the aid of Mr. Julius Pearson at Norman Bridge Laboratory of Physics at CalTech, Pasadena, in the winter 1921–1922. Millikan, with the scientific collaboration of his assistant Ira Sprague Bowen (1898–1973)—a skilful spectroscopist later professor of physics at CalTech (1946–1964) and director of the astronomical observatories of Mount Wilson (1946–1964) and Palomar (1948–1964)—had them launched from Kelly Field, near San Antonio, Texas, in March and April 1922. The electroscopes reached altitudes up to 15.5 km. The first communication of these experiments was made on Saturday 5 May 1923 by Millikan and I.S. Bowen [11] at the 121th Meeting of the American Physical Society which was held in the Norman Bridge Laboratory of Physics, Pasadena. In another communication during the same meeting, Russel M. Otis [12], from Millikan's group, reported on measurements done with a specially designed Wulf electroscope, very similar to the one used by Kolhoerster, except that the insulation losses were counteracted by an electrical method. Measurements in airplanes and balloons, up to an altitude of 5,340 m, indicated first a decrease, then an increase in radiation, the increase being however not so large as that found by Kolhoerster. The American physicist C.H. Kunsman [13] from the University of California suggested that the increase measured by the Europeans was due to neglected additional charge losses due to the low temperature present at high altitude. This possibility was ruled out by Otis through accurate checks.

At the same meeting, in a second communication, Otis [14] also reported measurements made near the top of Mt. Whitney, California, at an elevation of 13,552 ft (4,130 m), which had shown no sensible difference between night and day values. The mean value of the ionization at that level was 19.6 ions cc^{-1} s^{-1}, as compared to 9.33 ions cc^{-1} s^{-1} at Pasadena (245 m above the sea level).

The results of these and other measurements were later fully published in three papers on Physical Review, in 1926. In the first paper [15], submitted to Physical Review on 24 December 1925, were described the measurements made with the auto-recording electroscopes sent up with sounding balloons at altitudes from 5 to 15.5 km. The electroscopes were provided with an original system which registered on a photographic film the divergence of the electroscope fibres, the temperature and the barometric pressure. Thanks to the good relations established by Millikan with the Army, he obtained that Lieutenant McNeal would be sent from Washington to Kelly Field to assist in measuring by the two-theodolite method the altitude reached by the balloons in their ascent.

Four launches were organized, each one equipped with the electroscope, barometer, thermometer, and relative registration mechanism. The total weight of each pack with all its contents was but 190 g. Each of these electroscopes was carried up by two balloons 18 in. across when deflated and weighting about 300 g apiece. At some altitude one of the two balloons burst and the other brought gently the instruments safely to earth. During the flight the data were registered on the photographic film that was later developed. Four flights were performed. Three of the four balloons were found at distances about 80 miles away from the starting point and two of them reached the altitudes of 11.2 and 15.5 km, respectively, and produced satisfactory registrations. A third flight reached an altitude of 11.4 km and also yielded similar results.

These were the first stratospheric flights of electroscopes showing that the increase of ionization does not continue constantly with altitude; this circumstance seemed to prove that some form of radioactivity conveniently distributed in the high atmosphere was sufficient to explain the results and seemed to exclude an extraterrestrial radiation more penetrating than those which originate in the radioactive transformations. Millikan however had also found that after some altitude the ionization increases with height and the report on the Kelly Field measurements stressed two circumstances: *"This shows quite unambiguously, in agreement with the findings of Gockel, Hess, and Kolhoerster, that the discharge rates at high altitudes are larger than those found at the surface. Quantitatively, however, there is complete disagreement between the Hess–Kolhoerster data and our own, the total loss of charge of our electroscope in the two hours spent between the altitudes of 5 km and 15.5 km having been but about 25 percent of that computed from the Hess–Kolhoerster curve"*.

H. von Schweidler [16] reckoned that in order to fit the Hess and Kolhoerster data, the rays had to have an absorption coefficient of 0.57 m^{-1} of water, and an ionizing power within a closed vessel sent to the top of our atmosphere of at least 500 ions cc^{-1} s^{-1}, in place of the 10 or 12 ions found in ordinary electroscopes at the surface.

Fig. 3.2 The electroscope used by Millikan and Otis [17]. The external cage was in brass thick about 1.7 mm. The air in the sailed vessel was maintained dry by a drying material put in D at the bottom of the instrument. The two very thin fibres of the electroscope start from A and are sealed to an arch B that takes them in tension. Their divergence is read by a microscope which is not shown in the figure (Reprinted with permission from R.A. Millikan and R.M. Otis, Phys. Rev. **27**, 645 (1926). Copyright 1926 by the American Physical Society)

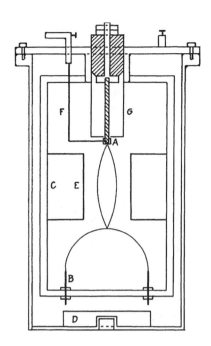

Millikan concluded that his results *"constitute definite proof that there exists no radiation of cosmic origin having such characteristics as we had assumed. They show that the ionization increased much less rapidly with altitude than would be the case if it were due to rays from outside the earth having an absorption coefficient of 0.57 per meter of water. Taken in conjunction, however, with experiments on absorption coefficients to be reported in the successive articles II and III— experiments which present unambiguous evidence for the existence of a cosmic radiation of extraordinary penetrating power, μ calculated as above being as low as 0.18 per meter of water, these experiments at very high altitudes have important bearings upon the distribution in wave-length of these hard rays as they enter the atmosphere"*.

It was the beginning of the conviction, lasting for many years, that the radiation was composed of high energy gamma rays.

In the second paper, sent 1 March 1926, Millikan [17] reported measurements that his student Russel M. Otis had done on Mt. Whitney (4,130 m), which were repeated by him and Otis the following summer on Pike's Peak (4,300 m) and were completed with measurements done with airplanes and aerostatic balloons. Using a fibre electroscope of the Wulf type shown in Fig. 2 of the paper (here is Fig. 3.2) and improved with respect to the ones previously used, Millikan found that the results obtained in a long series of airplane flights made by Otis in 1922 at Marsh Field (near Riverside) and in 1923 at Rockwell Field (near San Diego)—which had reached altitudes of more than 5,000 m—agreed with the measurements by Millikan and Bowen, as reported in the first paper of the mentioned trilogy. They showed a charge loss at the highest altitudes reached markedly lower than those reported by

Hess and Kolhoerster. However, as a whole, all the measurements agreed with those of the European observers in [18] *"showing that the intensity of the penetrating radiation first decreases to a minimum after which it increases continuously with altitude"*.

Moreover, the results proved that, within the limits of the experimental error, the penetrating radiation did not change from day to night and did not depend on the position of any celestial body. Two observations performed in Bolivia *"at a time when the Milky Way was wholly beneath the horizon gave no indication of a rate of discharge lower than that found when the Milky Way was overhead [19]"*. The Bolivia measurements were described later [20]. They were done, at an altitude of 4,700 m, in a deep valley in the Andes in which the Milky Way was for a few hours completely screened by the mountains.

Further, a comparison with the number of ions per cc per second in different places [in the campus at Pasadena (11.6 ions cc^{-1} s^{-1}), on Mt. Whitney at 4,130 m (19.6) and at 3,660 m (17.3)], showed that the variation among the measurements at different altitudes was about the same of that obtained with flights made at the same altitudes. The evidence of the whole research, made both in the mountain and in numerous flights, showed that there was *"a definite variation with altitude alone and that mountain and airplane observations can be brought into approximate agreement by suitable precautions for eliminating the activity of the adjacent rocks"* [19].

Also, experiments performed after screening the electroscope with lead were reported. The absorption experiments with lead screens 4.8 cm thick on Pikes Peak, together with the results obtained during a snow storm which had produced an attenuation of the ionization, brought to the conclusion that on the mountains there was a copious radiation of local origin, of a hardness not greater than the one of gamma rays from radium or thorium.

The final conclusion was than that if cosmic rays existed they should be responsible only of a small part of the ionization measured on earth and should be very penetrating. All was however still very hypothetic.

Eventually, Kolhoerster [21–23] on one side, and Millikan and Cameron [24] on the other side, made direct measurements of the ability of the mysterious radiation to penetrate the atmosphere (the so-called *penetrating power*). The German researcher first made measurements under water and then, together with G. V. Salis, performed a series of measurements on Jungfrau, reaching with a funicular an altitude between 2,300 and 3,500 m, and using instruments developed and tested at Berlino-Charlottenburg from the Physikalisch-Technische Reichsanstalt. In the experiments the absorption of the radiation by several thicknesses of ice was measured. The first trials, made in the Eiger glacier at 2,300 m, in an ice cave, proved that the rays may pass ice thicknesses from one and a half to 3 m. Following experiments, in a crevasse on Jungfrau at 3,550 m, were made putting the instruments at depths of 2.5, 4.5, and 9.7 m below the ice surface and showed that the radiation, after travelling through the maximum thickness is still able to sensibly ionize air.

These measurements allowed a better estimate of the absorption coefficient, which was found to be 0.5×10^{-5} per air cm or 0.25×10^{-2} per water cm. The new

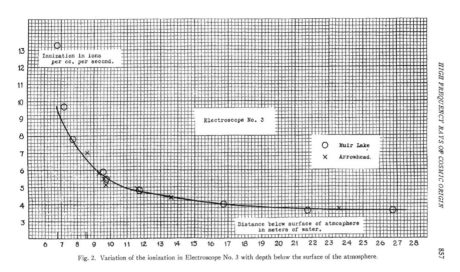

Fig. 2. Variation of the ionization in Electroscope No. 3 with depth below the surface of the atmosphere.

Fig. 3.3 This is Millikan's Fig. 2 (from Millikan and Cameron [24]). On the ordinates the readings in ions per cc per second and on abscissas the depth in meter of water equivalent measured from the top of the atmosphere (Reprinted with permission from R.A. Millikan and G.H. Cameron, Phys. Rev. **28**, 851 (1926). Copyright 1926 by the American Physical Society)

value was recognized by Millikan [25]: "*now so low as to be no longer incompatible with the Kelly Field sounding balloons experiments*".

In August 1925, Millikan and G. Harvey Cameron sank their electroscopes at increasing depths down to 27 m beneath the surface of the waters of Muir Lake, California, 11.800 ft. (3,640 m) above sea level, just under the brow of Mount Whitney, the highest peak in the USA, a beautiful snow-fed lake (therefore without radioactivity) 100 ft. deep and some 2,000 ft. (700 m) in diameter, and of the Arrowhead Lake in the San Bernardino mountains, 300 miles farther south and 6,700 ft. lower in altitude (1,530 m above sea level). By increasing the depth in water, they found a continuous decrease of the discharge rate of electroscopes.

For each radiation travelling vertically through the atmosphere, the thickness of about 2,000 m of air over the Arrowhead Lake represents an absorbing mass per unit area of the surface of the lake equivalent to 1.8 m of water. The two researchers wrote: [26]

"*Within the limits of observational error, every reading in Arrowhead Lake corresponded to a reading 6 feet (1.8 m) farther down in Muir Lake, thus showing that the rays do come in definitely from above, and that their origin is entirely outside the layer of atmosphere between the levels of the two lakes. This, taken together with the sounding balloon data, appears to eliminate completely the idea that the penetrating rays may have their origin in thunder-storms, a possibility recently suggested by C.T.R. Wilson and repeated by Eddington [27]*".

Figure 2 from Millikan's paper (here is Fig. 3.3) shows the curve obtained by plotting all the readings taken in the two lakes as ordinates, and as abscissas the depths in meters beneath the top surface of the atmosphere, reduced to the equivalent depth beneath water. On these graphs the depth beneath the top of the atmosphere

Table 3.1 Atmospheric pressure at different altitudes measured in atmospheres cm of mercury, or meters of water

Altitude (m)	Pressure (atm.)	Pressure (cm Hg)	Pressure (m H_2O)
0	1.0000	76.00	10.333
1,000	0.8870	67.41	9.165
2,000	0.7845	59.62	8.106
3,000	0.6918	52.58	7.148
4,000	0.6082	46.23	6.284
5,000	0.5330	40.51	5.507
10,000	0.2606	19.816	2.693
20,000	0.05449	4.141	0.5630
30,000	0.01137	0.864	0.1175
40,000	0.00274	0.208	0.0283
50,000	0.00092	0.070	0.0095
60,000	0.00035	0.027	0.0036
70,000	0.00012	0.0090	0.0012
80,000	0.000032	0.0024	0.00033

of the surface of Muir Lake is 6.75 m, that of Arrowhead Lake 8.6 m, and that of Pasadena 9.98 m. The figure shows that the readings obtained in the two lakes, expressed as a function of the equivalent thickness of water that the rays have to traverse to reach the instrument, superpose exactly.

As we already said, the energy carried by a beam of electromagnetic radiation decreases with the length of the path according to an exponential law. The absorbing coefficient appearing in the law is proportional to the number of atoms per unit volume (density). In the case of cosmic rays, the absorbing power of the whole vertical path through the atmosphere is equivalent to that of about 10 m of water.

Table 3.1 shows the water equivalent thicknesses corresponding to various heights in the atmosphere.

Millikan assumed that equal masses per unit surface of air or water absorb the cosmic rays in the same way. As a matter of fact, as we will see later, equal masses per unit area of air and water do not absorb exactly the same fraction of cosmic rays. If the experimental data would have been more precise than what the state of the art allowed at the time, Millikan would have found a difference between the readings of Arrowhead Lake and those of the Muir Lake, a water depth of 1.8 m below. Probably this could have prevented him to reach the correct conclusion that all the radiation came from over the higher lake and there was no radiation originated in the air between the levels of the two lakes. He also found that the radiation was composed by a *harder* (i.e. more energetic, so as to penetrate notable thicknesses) part and a *softer* (less energetic) one, with absorption coefficients of 0.30 and 0.18 m^{-1} of water, respectively.

Measurements were also made in Bolivia in the Miguilla lake at 4,570 m and in lake Titicaca at 3,820 m to verify if in the southern hemisphere the ionization–altitude curve coincided with the results found in the two American lakes in the northern hemisphere (the Muir and Arrowhead lakes), finding a perfect agreement [20].

The great penetration power of these rays convinced Millikan that the radiation discovered by Hess came effectively from outer the earth atmosphere, and induced him to call them **cosmic rays** [28].

Millikan and his collaborators' work was important not only for the precision of the results obtained, but also because of the great innovation of using sounding balloons and employing a new and ingenious self-registration technique. Hess and Kolhoertser had to accompany personally their electroscopes in the flights. The use of unmanned balloons eliminated the danger and the high price of man flights, but took away also the flavour of the adventure.

After Millikan obtained the Nobel prize in 1923 for his work on the elementary electrical charge and the photoelectric effect, he became the most popular physicist in America. The New York Times published [29] an editorial entitled "Millikan Rays": *"Dr. R.A. Millikan has gone out beyond our highest atmosphere in search for the cause of a radiation mysteriously disturbing the electroscopes of the physicists. His patient adventuring observations through twenty years have at least been rewarded. He found wild rays more powerful and penetrating than any that have been domesticated or terrestrialized, travelling toward the earth. The mere discovery of these rays is a triumph of the human mind that should be acclaimed among the capital events of these days. The proposal that they should bear the name of their discoverer is one upon which his brother-scientists should insist. "Millikan rays" ought to find a place in our planetary scientific directory all the more because they would be associated with a man of such fine and modest personality"*.[2] TIME, Science [30–32] and other journals echoed. Millikan did not make anything to deny it, and the term *"Millikan Rays"* was used quite often; he enjoyed being designated as the "discoverer". Hess and many other researchers expressed their chagrin about this, and Millikan cunningly solved the problem writing to Hess that he had done no claim of the discovery, the really important thing being that everybody agreed that cosmic rays existed. When in 1936 Hess was awarded the Nobel Prize for the discovery, Millikan sent him a congratulation letter.

However, neither the evidences given by Kolhoerster nor the ones by Millikan and Cameron were sufficient to convince all physicists. As late as 1925, C.T.R. Wilson [33] maintained that the origin of the penetrating radiation was to be found in the upper layers of the atmosphere, and these ideas continued to be discussed even in 1937 [34].

For many years the electroscopes developed by Millikan and his collaborators were more precise than the ones employed by other researchers.

In the introduction of a work written with Cameron [35], the two authors observed that, until about a year before, they were so busy with the demonstration of cosmic rays existence, that they had had little time to improve the precision of the instrument. As an example they reported that with two different electroscopes they had obtained, under the same conditions 1.4 and 1.6 ions cc^{-1} s^{-1}, respectively, as the value of the sea-level ionization due to the cosmic rays. *"Such differences—they*

---

[2]The New York Times, 23 November 1925.

wrote—*whatever their cause, are tolerable only in the initial phase of work on any given physical quantity*" and continued observing that "*at high altitudes Millikan and Bowen got a total discharge of an electroscope about one-fourth that computed from the curves of Kolhoerster and Hess*". They reported some other examples and concluded "*Such differences are quite like those found, for example, in the early determination of e/m which showed fluctuations of 100 %, but they cannot long be permitted*". Then they described a new method leading to greatly increased precision in the measurements.

A new absorption curve was obtained in Gem Lake[3] (9,080 ft) and Arrowhead Lake which showed the existence of more penetrating cosmic rays than those previously found and—again under the assumption they were gamma rays— furnished "*indications that the cosmic rays consist chiefly of three bands for which the mean absorption coefficients are approximately 0.35, 0.08, and 0.04*" [36] affording definite evidence for the existence of bands in the spectrum of cosmic rays. At the end of the paper, an evaluation of the total energy in cosmic rays was given: "*the total energy coming into the earth in the form of cosmic rays is very close to one-tenth the total energy of starlight*".

Due to the superior precision of his measurements, Millikan could so prove that the statements, made in several cases, of large day variations in the intensity of cosmic rays depending from the Sun and stars position were groundless. In fact since 1923, his school maintained that diurnal variations were very small or nonexistent.

For the same reason, Millikan and his collaborators obtained good curves of the intensity of cosmic rays at high altitudes.

In the first measurements the variation of ionization produced by cosmic rays up to moderate altitudes was measured, to establish the extraterrestrial origin of the radiation. The flights with aeroplanes gave accurate results since 1933; the sounding balloons whose data at the beginning were little reproducible, gradually became highly reliable. It was so possible to measure the variation near the top of the atmosphere, approximately in the last 1% of the atmospheric pressure.

To these researches are associated the names of A. Piccard (1884–1962) [37, 38] M. Cosyns [39], G. Suckstorff [40], E. Regener [41], I.S. Bowen and R.A. Millikan [42]. Auguste Piccard ascended to the altitude of 16,196 m in a pressurized cabin. Many of these researchers used ionization chambers and measured the radiation coming from all directions.

G. Pfotzer, on 25 July 1935, at Stuttgard, used a vertical coincidence system carried by balloons, which we will describe later, in order to select only the rays which arrived from the vertical direction. The intensity was found to increase with altitude up to some height, to reach a maximum and then to decrease. Pfotzer with his precise measurements showed that the flux reaches the maximum at about 15 km and then steeply decreases [43]. Similar measurements which confirmed these results were done in Groenland by H. Carmichael and E.G. Dymond [44, 45].

---

[3]Gem Lake was 225 ft. deep and was 250 miles north of Arrowhead lake.

E. Regener [46,47] made very elaborated measurements also at great depths with an ionization chamber. In the Constance Lake, he closed an ionization chamber with a photographic registration system in a sealed vessel sank at various depths down to 230 m. During these measurements, the batteries were discovered to be radioactive so that the measurements were invalidated; the inconvenient was then eliminated and good curves were obtained [48]. Later many other researchers measured the absorption under water in the Constance lake [49], in mines down to depths equivalent to 1,380 m of water [50,51], or in the sea down to 1,420 m [52].

3.2 The Theory of Atom-Building

Most researchers working on cosmic rays were convinced that the rays were simply photons of higher energies than the ones observed until then. Millikan firmly believed they were gamma rays entering the atmosphere equally from all directions (that is they were isotropically distributed). Generalizing the researches on X-rays, he thought that he could measure the rays' energy by studying their absorption. At that time, only two absorption processes were known: Compton scattering and ionization. Millikan assumed that both processes depended only on the photon energy and the density of matter. The absorption law was therefore of an exponential type, as we already said.

To measure the energy of the cosmic ray, Millikan plotted the ionization measured with an electroscope against the thickness of the absorber and fitted the results with an exponential curve. The μ value that better fitted the experimental data would allow to determine the energy of the photons, by using suitable theoretical models for the absorption due to Compton scattering and ionization. However a single μ did not fit the curve, and it was necessary to fit different parts of it with different μ values.

From the energy values obtained in this way, Millikan began to speculate on the origin of cosmic rays. Eddington and Jeans [53, 54][4] had already proposed that protons and electrons could annihilate each other in stars, producing very hard gamma rays which would radiate outward. In 1926, as soon as they confirmed that cosmic rays existed, in the third paper of the trilogy, Millikan and Cameron objected to this mechanism because the rays would be too hard (that is they would have too high energies) and suggested instead that nuclear processes took place "*not in the stars but in the nebulous matter in space, i.e., throughout the depths of the universe*" [55]. The processes they had in mind were "*(1)the capture of an electron by the nucleus of a light atom,(2)the formation of helium out of hydrogen, or (3)some new type of nuclear change, such as the condensation of radiation into atoms*".

---

[4]A.S. Eddington (1882–1944) was an English astronomer, mathematician, and physicist who gave important contributions in astrophysics and relativity. J.H. Jeans (1877–1946) was an English astronomer, mathematician, and physicist who gave important contributions.

With their specially designed electroscopes, Millikan and Cameron [56] measured the ionization rate in several lakes as a function of water depth and, applying the same model used in 1926, they managed to determine a detailed spectrum of the absorption constants. By fitting different parts of the experimental curve with exponential curves, they choose three coefficients representative of three exponential curves that in their opinion better reproduced it; they were 0.35, 0.08, and 0.04 m^{-1} of water.

Although the method employed was open to criticism, because by suitably choosing the portions of the curve to be approximated one may find whatever value of μ, the two researchers published their results on the prestigious journal Physical Review, under the title *"The origin of the cosmic rays"*. Considering Einstein's equation between mass (m) and energy $E = mc^2$, where c is the light speed, the authors calculated the energy released by creating atoms from protons. This energy in their view was emitted as a photon by using the other Einstein's equation which connects the energy to the frequency v of the radiation through the Planck constant ($E = hv$), and these photons were the primary cosmic rays. Any other particles Millikan contended were only the secondary product of the interaction of primary photons in the earth's atmosphere.

By using a theory developed by Dirac on the energy loss due to Compton scattering, Millikan [57] guessed that there should exist three groups of photons (he called them bands) with energy 26, 110, and 220 MeV. He then observed that the interstellar space is filled with very diluted hydrogen and assumed that the atoms of the heavier elements were built in a continuous way by the sudden union of these hydrogen atoms. For example, a helium atom was created by the sudden union of four hydrogen atoms. Because the helium atom weighs a little less than four hydrogen atoms, the mass difference, according to the Einstein's relativity theory, should be emitted as energy. This energy was 27 MeV and corresponded perfectly to that of the first group of gamma rays. Therefore Millikan concluded that this photon group originated by the union of four hydrogen atoms to produce a helium atom.

Similarly a nitrogen atom could be thought as made by 14 hydrogen atoms, and the energy release in this case would be 100 MeV. In the same way, the making of oxygen would release 120 MeV. Finally, if 28 hydrogen atoms would be grouped to make a silicon atom, the released energy would have been nearly equal to that of the third group of photons. Millikan concluded therefore that the cosmic rays were the "birth cry" of atoms continuously created in space [58].

Millikan and Cameron found in this way an extraordinary agreement for the production of oxygen, nitrogen, helium, and silicon (the most abundant elements on earth). Table 3.2 shows the absorption coefficient they obtained and the one experimentally measured with their method for the various processes.

We may agree with Gallison that the two physicists, in order to explain the origin and the nature of cosmic rays, used the wrong particle (the photon) produced in a process which does not occur (building up of nitrogen, oxygen, etc. by the direct union of hydrogen atoms), invoking an absorption law which is incorrect (it misses pair production, the effects of electron binding, etc.).

Table 3.2 Millikan's calculation for the cosmic ray origin

	Absorption coefficient per m of water	
Atom building process	Theory	Experiment
O and N produced by H fusion	0.08	0.08
He produced by H fusion	0.30	0.35 esp. in lakes and atm.
		0.30 lakes only
Si produced by H fusion	0.041	0.04
Fe produced by H fusion	0.019	Not inconsistent with experiment

Soon after the publication of the paper, Millikan received by Oppenheimer a letter in which he draw attention to a work by Klein and Nishina which showed that the gamma ray absorption in the region of interest to Millikan had to be corrected. Millikan, however, did not change his mind, and even tried to match the data with the new formula [59]. He continued to support vigorously the "birth cry" theory.

He liked his theory also because of his beliefs according to which the scheme of Nature must be simple and understandable to everyone. Millikan also draw an analogy between the results he obtained with the energies of the cosmic rays—which according to his results were grouped into bands—and the results he had found making experiments on the electron charge, according to which the charge on the oil drops changed by discontinuous steps. At the time, to obtain that right result, he discarded a number of measurements that did not agree with his hypothesis that Nature had a granular representation. And surely, had he taken into consideration all data, he would have found a mass of undifferentiated values of charge. Now, plucking up from the previous success, he thought that also the energies of cosmic rays distributed in discrete bands. This forcing the experimental results to confirm his idea agreed with his way to interpret natural phenomena as something which one should easily understand and visualize.

For some time his theory was accepted and publicised in every way [60–62][5] and Millikan maintained that the creation of elements was going on continuously throughout the Universe saving it from the heat death, as someone had proposed on the grounds of thermodynamics. In December 1930, Millikan presented anew his ideas at a meeting of the American Association for the Advancement of Science. The New York Times reporter who was covering the event, William L. Laurence had already publicized Millikan's views on the Sunday supplement and now wrote a six column paper on the journal the 1st January 1931 [63], reporting Millikan claim that the Creator was "still on the job".

Later, the same year, at a meeting on nuclear physics in Rome, Millikan described his theory of the "birth cry".

---

[5]The analysis of the curve into three spectral bands was presented at the Physics Seminar of the California Institute on 16 February 1928. The evidence that these bands were the signals of the atom building was presented to the Association of the California Institute on 16 March and was reported by the Associated Press on 17 March 1928.

References

1. R.A. Millikan, L. Begeman, Phys. Rev. **26**, 197 (1908)
2. R.A. Millikan, Philos. Mag. **19**, 209 (1910)
3. R.A. Millikan, L. Begeman, Phys. Rev. **32**, 349 (1911)
4. R.A. Millikan, Phys. Rev. **2**, 109 (1913)
5. R.A. Millikan, Phys. Rev. **4**, 73 (1914)
6. R.A. Millikan, Phys. Rev. 7, 355 (1916)
7. R.A. Millikan, Philos. Mag. **34**, 1 (1917)
8. R.A. Millikan, *Autobiography* (Prentice-Hall, New York, 1950)
9. E. Rutherford, H.L. Cooke, Phys. Rev. **16**, 183 (1903)
10. J.C. McLennan, E.F. Burton, Phys. Rev. **16**, 184 (1903)
11. A. Millikan, I.S. Bowen, Phys. Rev. **22**, 198 (1923)
12. R.M. Otis, Phys. Rev. **22**, 198 (1923)
13. C.H. Kunsman, Phys. Rev. **10**, 349 (1920)
14. R.M. Otis, Phys. Rev. **22**, 199 (1923)
15. R.A. Millikan. I.S. Bowen, High frequency rays of cosmic origin I. Sounding balloon observations at extreme altitudes. Phys. Rev. **27**, 353 (1926) (this was the first paper of the trilogy)
16. H. von Schweidler, Elster u. Geitel Festschrift. 411 (1915)
17. A. Millikan, R.M. Otis, High frequency rays of cosmic origin II. Mountain peak and airplane observations. Phys. Rev. **27**, 645 (1926) (this is the second paper of the trilogy)
18. A. Millikan, R.M. Otis, Phys. Rev. **27**, 645 (1926) (at p. 651)
19. A. Millikan, R.M. Otis, Phys. Rev. **27**, 645 (1926) (at p. 653)
20. R.A. Millikan, G.H. Cameron, Phys. Rev. **31**, 163 (1928)
21. W. Kolhoerster, Proc. Berl. Acc. 366 (1923)
22. W. Kolhoerster, Phys. Zs. **27**, 62 (1926)
23. W. Kolhoerster, Die Naturwiss. **15**, 31 (1923)
24. R.A. Millikan, G.H. Cameron, High frequency rays of cosmic origin III Measurements in snow-fed lakes at high altitudes. Phys. Rev. **28**, 851 (1926) (this is the third paper of the trilogy submitted on 7 Aug 1926)
25. R.A. Millikan, G.H. Cameron, Phys. Rev. **28**, 851 (1926) (at p. 853)
26. R.A. Millikan, G.H. Cameron, Phys. Rev. **28**, 851 (1926) (at p. 856)
27. A.S. Eddington, Nature **107**(25 suppl.), 32 (1926)
28. R.A. Millikan, Nature **116**, 823 (1925)
29. Q. Xu, L.M. Brown, Am. J. Phys. **55**, 33 (1987)
30. TIME **6**, 26 (1925) (Nov. 23)
31. TIME **9**, Cover, and p. 16 (1927) (Apr. 25)
32. Science **62**, 461 (1925)
33. C.T.R. Wilson, Proc. Camb. Phil. Soc. **22**, 534 (1925)
34. M.C. Holmes, J. Franklin Inst. **223**, 495 (1937)
35. R.A. Millikan, G.H. Cameron, New precision in cosmic ray measurements; yielding extension of spectrum and indications of bands. Phys. Rev. **31**, 921 (1928)
36. R.A. Millikan, G.H. Cameron, Phys. Rev. **31**, 921 (1928) (at p. 928)
37. A. Piccard, E. Stachel, P. Kipfer, Naturwissenschaften **20**, 592 (1932)
38. D. Devorkin, *Race to the Stratosphere: Manned Scientific Ballooning in America* (Springer, Berlin, 1980)
39. A. Piccard, M. Cosyns, CR Acad. Sci. **195**, 604 (1932)
40. G. Suckstorff, Naturwissenschaften **20**, 506 (1932)
41. E. Regener, Phys. Z. **34**, 306 (1933)
42. I.S. Bowen, A.R. Millikan, Phys. Rev. **43**, 695 (1933)
43. G. Pfotzer, Z. Phys. **102**, 23 and 41 (1936)
44. H. Carmichael, E.G. Dymond, Nature **141**, 910 (1938)

45. H. Carmichael, E.G. Dymond. Proc. R. Soc. London A **171**, 321 (1939)
46. E. Regener, Phys. Z. **31**, 1018 (1930)
47. E. Regener, Phys. Z. 34, **306**, (1933)
48. E. Regener, Z. Phys. **74**, 433 (1932)
49. A. Ehmert, Z. Phys. **106**, 751 (1937)
50. J. Clay, A. van Gemert, Proc. Acad. Amst. **42**, 672 (1939)
51. J. Clay. Rev. Mod. Phys. **11**, 128 (1939)
52. V.C. Wilson, Phys. Rev. **53**, 337 (1938)
53. J.H. Jeans, Nature **116**, 861 (1923)
54. A.S. Eddington, Nature **107**, 25 (1926)
55. R.A. Millikan, G.H. Cameron, Phys. Rev. **28**, 851 (1926) (at p. 868)
56. R.A. Millikan, G.H. Cameron, Phys. Rev. **31**, 921 (1928)
57. R.A. Millikan, G.H. Cameron, Phys. Rev. **32**, 533 (1928)
58. P. Gallison, Centaurus **26**, 262 (1983) (for a bibliography on this subject)
59. P. Gallison, Centaurus **26**, 262 (1983)
60. R.A. Millikan, G.H. Cameron, Nature **121**, 19 (1928)
61. R.A. Millikan, Science **67**, 401 (1928)
62. R.A. Millikan, Phys. Rev. **31**, 921 (1928)
63. V.D.J. Kevles, Phys. Today **31**, 23 (1978)

Chapter 4
A Turn: Things are not as they are Assumed to be

4.1 The Current Belief is that Cosmic Rays are Gamma Rays

The current belief was that cosmic rays were simply photons of greater energy than the ones observed in radioactive transformations. In fact, still in 1929, the German literature usually referred to cosmic rays as *Ultragammastrahlung* (ultra-gamma radiation). The reason is that the most penetrating radiations known at those times were the gamma rays from radioactive substances. Indeed, the mean free path of gamma ray photons emitted by these substances in air were hundreds of metres, while beta rays of similar energies had ranges of a few metres only, and the ranges of alpha particles were even shorter.

Moreover, theoretical calculations based on the assumption that high energy photons were absorbed prevalently through Compton collisions predicted a mean free path which increased by boosting up the photon energy. Therefore it was sufficient to think that cosmic rays were gamma photons with enough energy to explain their long range. Millikan and many others had this belief. However, while everybody was using this assumption as a working hypothesis, and Millikan was toying with it, offering a simple and appealing explanation for the cosmic origin of the rays, a great revolution was starting.

4.2 Dimitrij V. Skobelzyn: For the First Time a Cloud Chamber "Sees" a Cosmic Ray

The revolution concerning the nature of cosmic rays developed through experimental observations made by using two completely different techniques. The Russian physicist Dimitrij V. Skobelzyn [1], in 1927, while studying the gamma rays emitted by RaC in a small horizontal cloud chamber with a magnetic field of 1,500 gauss, found tracks of particles that did not show a measurable curvature and that he could ascribe to cosmic rays (see Fig. 4.1). Repeating the measurements, the following

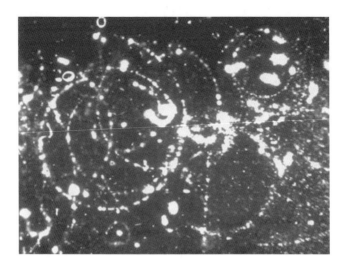

Fig. 4.1 This is one of the photos obtained by Skobelzyn. In a sea of curved tracks due to low energy particles emitted by the radioactive sample under examination, on the *right* in the lower part of the photo one may distinguish a rectilinear track which crosses the photo and was too energetic to be curved by the applied magnetic field (is from D. Skobelzyn, Z. Phys. 43, 354 (1927))

year [2], by examining 613 photos, he found 32 more tracks which originated outside the Wilson chamber and were not affected noticeably by the magnetic field. Each one of these tracks was due to a particle with a momentum >15 MeV/c. Also these tracks were ascribed by Skobelzyn to cosmic rays. Their ionization was of the same order of magnitude of the one needed to explain the ionization one could expect from the whole beam of cosmic rays incident in the chamber volume, and their energy was at least ten times higher than that of any known radioactive radiation.

These tracks were not associated in direction with the radioactive source under investigation, and indeed were predominantly in directions well removed from the horizontal plane. Skobelzyn identified them with particles arising from cosmic rays: *"Diese β-Strahlen sind als von Hessenchen Ultra -γ-Strahlen erzeugte sekundaere Elektronen zu deuten"*[1] and estimated the flux of particles across a horizontal surface to be $1.2/cm^2/min$, a rather accurate value when compared with later estimates $(1.07/cm^2/min)$ [3].

Some photos also showed two or three associated rays that crossed simultaneously the chamber (they were member of a *shower* as we will see later).

Dimitrij Vladimirovic Skobelzyn was born on 24 November 1892 in St. Petersburg, the son of a professor at the St. Petersburg Polytechnic Institute. After graduating from the University of that city, which in the meanwhile had changed its name to Leningrad, in 1925, he became a research fellow of the Leningrad

---

[1]One should assign these beta rays to the secondary electrons created by Hess ultra gamma rays.

Physicotechnical Institute. Here he started to study the electrons scattered by the Compton effect and, later the electrons of cosmic rays. From 1929 to 1931 he worked in the laboratory of Marie Curie, in Paris.

A short time before World War II, he went to work at the Lebedev Physical Institute of the Soviet Academy of Sciences in Moscow (FIAN). There he started a series of investigations about cosmic ray showers. He established the Institute of Nuclear Physics at Moscow State University of which he was the director from 1946 to 1960, being also the director of FIAN from 1951 to 1973. He died on 16 November 1990 in Moscow.

Skobelzyn was the first in Russia to built and use the Wilson cloud chamber, and was the first to interpret some tracks he found with it as the ones of secondary high velocity electrons produced by an energetic primary cosmic ray of electromagnetic radiation.

4.3   The Geiger–Müller Counter and the Coincidence Method Step In

Contemporaneously to Skobelzyn's results, around 1928, W. Bothe and W. Kolhoerster [4, 5] started a novel way to perform research, employing a new electronic instrument: the Geiger–Müller counter.

Of course, there was the Wilson chamber which allowed to "see" tracks and there were the point counters, but the Wilson chamber was difficult to operate and was sensitive only for a short while after the expansion, and the point counter was not very stable. Therefore, when in 1928 Hans Geiger with his student W. Müller, improving the old counter he had developed with Rutherford, invented what we today call the Geiger–Müller counter [6, 7] (see Fig. 4.2), physicists had in their hands a new and powerful instrument able to "count" the cosmic rays.

In 1907, Rutherford went to Manchester where he found a young graduate from Erlangen, Hans Geiger (1882–1945) and with him set up an electric method to count directly alpha particles from radioactive substances, the ancestor of the *Geiger counter*, as it was later called [8]. The device consisted of a cylindrical tube filled with a gas along whose axis a conducting wire insulated from the walls was placed. A potential difference was applied between the walls of the tube and the wire, and the current was measured with an electrometer.

The alpha particles passing through the gas produced some ions that under the action of the electric field were accelerated and started a discharge that could be registered. This possibility to count the particles one at the time was a great improvement. It was found that a gram of radium emits 3.4×10^{10} particles per second and on 18 June 1908, Rutherford communicated to the London Royal Society that the charge of an alpha particle is positive and has a value two times the charge of the electron [9]. The following year, he demonstrated [10], with one of the most beautiful experiment in the radioactivity history, that they are helium nuclei.

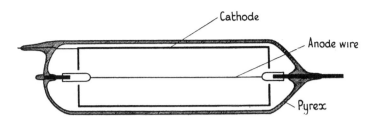

Fig. 4.2 A sketch of a Geiger–Müller counter

With the earlier version of his counter, Geiger in 1911, together with E. Marsden (1889–1970) [11, 12], under the direction of Rutherford, studied the deviation suffered by charged particles (alpha rays) traversing matter, giving the experimental basis on which Rutherford built his "planetary" model of the atom [13], according to which the atom is formed by a central nucleus, charged of positive electricity in which practically the whole atom mass is concentrated, around which orbit a certain number of electrons such as to balance the positive nuclear charge and make the atom neutral.

In 1913, Geiger developed a counter similar to the former, but with the central electrode made by a metallic point (*point counter*), able to detect ionizing particles near the point itself. Eventually, in 1928 in Berlin (where in meantime he had moved being appointed to the physics chair) Geiger, by sharpening the axial wire and applying a greater potential, in collaboration with his student W. Müller, created the *Geiger–Müller counter*. This counter was an essential tool in nuclear physics and was one of the most powerful instruments in the research on cosmic rays.

The final version of the Geiger–Müller counter is a metallic cylinder filled with a suitable gaseous mixture with a central metallic wire insulated from the cylinder (Fig. 4.2). The central wire (anode) is maintained at a positive potential with respect to the outer cylinder (cathode). The passage of an ionizing particle in the counter is accompanied by a migration of ions between the electrodes and, if the counter is powered through a resistance, a voltage pulse originates. In the counter the applied voltage is made sufficiently high so that the passage of the ionizing particle in the device is accompanied by a strong discharge throughout the gas which can be suitably detected. In earlier times, the detection was made by observing the discharge of an electroscope with a circuit as the one shown in Fig. 4.3.

By using Geiger–Müller counters in a suitable configuration (configuration with *coincidence counters*), Bothe and Kolhoerster proved the presence of charged particles in the cosmic rays in the atmosphere.

Walther Wilhelm Bothe was born on 8 January 1891 at Oranienburg, near Berlin. From 1908 to 1912 he studied physics at the University of Berlin, where he was a pupil of Max Planck, earning his doctorate before the outbreak of World War I. During the war, he was taken prisoner in Russia on his way back to Germany in 1920. Here he began a collaboration with H. Geiger at the Physikalisch-Technische Reichsanstalt, in Berlin. Together with Geiger, whose influence determined much

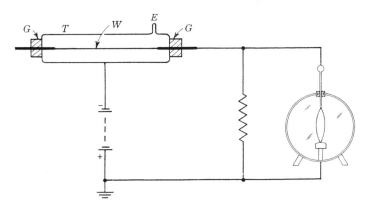

Fig. 4.3 Mounting of a Geiger–Müller counter whose discharge is visualized through an electroscope. T is the metallic tube, G are insulating rings in glass, and W is the metallic wire

of his scientific work, he published, in 1924 [14–16], his method of coincidence, which he applied to many physical problems.

This is how the story went: the newly discovered Compton effect was showing an irrefutable particle behaviour of the radiation that seemed incompatible with the Maxwell wave treatment. To overcome this difficulty, N. Bohr et al. [17] suggested that energy was only conserved in a statistical way and not in the single atomic processes. Bothe and Geiger wanted to see if the Compton effect could really be interpreted as a collision between two particles—the photon and the electron—following the laws of conservation of energy and momentum, and designed an experiment that was employing two of the primitive Geiger counters to detect the coincidence between the recoil electron and the scattered photon. The result of the experiment demonstrated that in every collision process between a photon of the radiation and an electron, the laws of conservation of energy and momentum were fulfilled, disproving Bohr's hypothesis.

After the studies we discuss below, Bothe focused his efforts on radiation scattering problems and continued to study the Compton scattering. He also studied several nuclear transformations. He remained at the Physikalish-Technische Reichsanstalt, Berlin, until 1930, when he was appointed professor of physics and director of the Institute of Physics at the University of Giessen. In 1932 he was appointed director of the Institute of Physics at the University of Heidelberg, as successor to Philipp Lenard (1862–1947), at the Kaiser Wilhelm (now Max Planck) Institut fur mediz. Forschung. At the end of the Second World War, Bothe returned to the Department of Physics in the University of Heidelberg. He died on 8 February 1957.

For his discovery of the method of coincidence and the discoveries subsequently made with it, which laid the foundations of nuclear spectroscopy, Bothe was awarded, jointly with Max Born (1882–1970) the Nobel Prize for Physics in 1954 on the following grounds: *"for his coincidence method and his discoveries made*

Fig. 4.4 Two counters
placed one over the other and
crossed by a particle. The
counters are represented with
the double circles and the line
is the particle trajectory

therewith". Max Born, who may be considered one of the founders of quantum
mechanics, received the Prize with the following motivation: "*for his fundamental
research in quantum mechanics, especially for his statistical interpretation of the
wave-function*".

In 1928, in collaboration with Kolhoerster, Bothe introduced the method of
coincidence for the study of cosmic rays.

Bothe and Kolhoerster's work was titled "Das Wesen der Hoehenstrahlung"
(Nature of radiation coming from outside). In the work, the authors reported
experiments and conclusions that marked a turning point in the history of cosmic
ray research. Using Geiger–Müller counters to study cosmic rays, they observed
that when two counters were placed one over the other, some distance apart, they
often discharged simultaneously [18].

In the experiment each Geiger–Müller counter was connected with a fibre
electroscope and the two electroscopes side-by-side were photographed with a
continuously moving film. The pulses from the two counters which resulted visually
coincident were measured with a temporal resolution of about one-hundredth of a
second. These simultaneous discharges or *coincidences* could not be due to chance,
or at least not entirely, since they were always less frequent when the distance
between the two counters was increased. The two researchers interpreted what was
happening by assuming that the same particle passed through both counters simultaneously. Therefore if two or more counters are aligned along a line (see Fig. 4.4)
and by a suitable method one finds that they are simultaneously discharged (we will
say they are in *coincidence*), one may deduce that they were all crossed by the same
particle of which the trajectory is so determined. Of course, there is always the possibility that both counters are simultaneously discharged by particles that come from
different directions and by chance are passing in the same moment in one counter
or in the other one, but this possibility may be greatly lowered by stacking, for
example, three counters instead of two. This geometry would surely allow to select
the particles which are moving along the line of the counters, measuring in this way
their direction of provenance and therefore, for example, their angular distribution.

In principle, as someone proposed, a photon could produce a coincidence also
by a double Compton effect. However the probability of a Compton collision in
the walls or in the gas of the counter is very small, and the probability to have
simultaneously two collisions, one in each counter, is fully negligible. As Bothe
and Kolhoerster correctly concluded, the observed coincidences should be due to
the passage of individual charged particles through both counters. Moreover, the

Fig. 4.5 Two counters with a
thickness of gold between
them

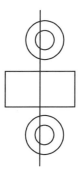

particles could not be alpha or ordinary beta rays, because the counter's walls (thick
1 mm and made of zinc) had stopped them.

By itself this result did not contradict the hypothesis that the primary radiation,
that is the cosmic rays hitting the atmosphere from space, could consist of high
energy photons. Since photons suffer Compton collisions in the atmosphere, the
ionizing particles could be the recoil electrons originated from these collisions.
People believed, in fact, that the energies of the hypothetic primary photons would
be in the range from 20–30 MeV up to several hundreds of MeV. The recoil electrons
from these photons would have more than enough energy to cross the counter walls.

To verify this possibility, Bothe and Kolhoerster placed (see Fig. 4.5) a gold
block of 4.1 cm thickness between the two counters (they choose gold because of
its high density and consequently high stopping power). Now only particles with
a range >4.1 cm of gold could cross both counters and produce coincidences. The
researchers found that the rate of coincidences was still 76 % of the ones without the
gold block. In other words, 76 % of charged particles present in cosmic radiation at
sea level could penetrate 4.1 cm of gold!

This was a surprising result because according to the most generous estimates
only a small fraction of recoil electrons in any point of the atmosphere could
have that range. Therefore Bothe and Kolhoerster concluded that the current ideas
concerning the nature of cosmic rays were probably wrong.

A first obvious result of their findings was that cosmic rays at sea level consist
principally of ionizing particles. However the authors [19] went forward and
affirmed that even the primary radiation consisted of charged particles rather than
photons. In their opinion the results showed that the corpuscular radiation is just
as strongly absorbed as the high altitude radiation itself. The conclusion was that
the high-altitude radiation, as far as it had manifested itself in the phenomena
so observed until then, was of corpuscular nature, and to sustain this affirmation
they invoked the recently discovered geomagnetic effect, which we will discuss in
Chap. 5. Although at that time this statement could not be proved, the work of the
two Germans was a milestone in the cosmic ray story.

In their experiments, the two researchers observed the discharge of the counters
using electroscopes. Even though the discharge was rather fast (about one-hundredth
of second), the temporal resolution was still low and could not guarantee that the

counters discharged exactly at the same time. A counter could have discharged a while before the other one. However Bothe and Kolhoerster were convinced that their conclusions about the existence of charged particles in cosmic rays were correct.

At this point Bruno Rossi, then 24 years old, enters the stage.

Bruno Benedetto Rossi (Fig. 4.6) was born in Venice in 1905, graduated in Bologna in physics and began his carrier in 1928 at the Physics Institute of the University of Florence, as an assistant under professor Antonio Garbasso (1871–1933), a senator of the Kingdom (1924), fellow of the Accademia dei Lincei (1921), mayor from 1920 to 1928, and afterwards podestà of the town. In 1932, Rossi was nominated professor of physics at the University of Padova, where he remained until the fall of 1938, when he was expelled because he was a Jew, just as Jewish was his wife, Nora Lombroso, the daughter of the famous anthropologist Cesare Lombroso (1835–1909). After a short stay in Copenhagen and Manchester, he emigrated in the United States and arrived at the University of Chicago in 1939. There he obtained a temporary position as associate researcher. Three years later (1942), he was offered a position as associate professor at Cornell University, and in 1945 he obtained the American citizenship. During the Second World War he was a consultant first in the radar development at the Radiation Laboratory, MIT, and then at Los Alamos as a co-director of the detection group, responsible for the development of the instrumentation for the experiments on the atomic bomb. Eventually in 1946 he was appointed physics professor at MIT, where he created a group on cosmic rays. In the late 1950s, when experiments with accelerating machines became dominant in the study of particles, he started experiments with rockets discovering the first extra-solar source of X rays—the Scorpius X-1—and, in 1961, he designed the first space experiment on board of Explorer X that demonstrated the existence of the solar wind. He retired from MIT in 1970 and came back to Italy, teaching in the University of Palermo, Sicily, from 1974 to 1980. He died at home in Cambridge, Massachussetts in 1993 [20].

Rossi improved in a substantial way the Bothe and Kolhoerster experiment. In the summer 1930, Rossi went to Charlottenburg, Germany, working several months in Bothe's laboratory, at Physikalische-Technische Reichsanstalt. After coming back to Arcetri, in a few months he built his first Geiger–Müller counters, conceived a new method to detect coincidences and began some experiments, improving [21, 22] the technique used by Bothe and Kolhoerster. His ingenious contribution to the coincidence technique was the invention, using electronic valves, of a circuit that emitted a pulse only when two or more pulses were transmitted to it at the same moment [22]. A device of this kind was exactly what was needed for the work on cosmic rays. In a modified form, the "*coincidence circuit*" became one of the most used systems in experimental physics. On top of that, it was the first logic circuit able to give a positive answer only if two signals arrived contemporaneously in input (such an element is called AND, and is a fundamental element in today's electronic computers).

Fig. 4.6 B. Rossi (on the *left*) with Millikan (*centre*) and Compton (*right*)

Eventually, Rossi's results, together with the ones of other researchers, showed in an unquestionable way that the penetrating part of cosmic rays is made of ionizing particles and not gamma rays.

In the fall of 1931 the Accademia d'Italia—whose president was Guglielmo Marconi (1874–1937)–organized in Rome, Italy, an International Scientific Conference, and Enrico Fermi[2]—already an Academy member—invited Bruno Rossi to hold the opening speech on the then very discussed subject of cosmic rays. All the most influential personalities in the field of nuclear physics and cosmic rays were present, among them Millikan and Arthur Compton. Rossi rejected Millikan's "*birth cry*" theory and in his talk demolished it by using very convincing arguments that he expounded with great clarity. He showed that the last experiments proved that the cosmic radiation consisted principally of charged particles and not gamma rays. It is not surprising that Millikan got so enraged that, in Rossi's words [23], since that moment refused to consider his existence.

---

[2]Enrico Fermi was born in 1901 in Rome. He studied at the Scuola Normale di Pisa earning his PhD in physics in 1924. After spending some time abroad in Gottingen and Leiden he returned in Italy where he was appointed Professor of Physics at the University of Roma. He performed outstanding work both in theoretical and experimental physics and created at the Physics Institute in via Panisperna in Roma a very famous group of young researchers (later called "I ragazzi di via Panisperna"). His main contributions are the beta decay theory, the discovery of slow neutrons and of the reactions produced by them. In 1938, he was awarded the Nobel Prize for physics and in the same year migrated in the United States, first at Columbia University (1939–1942) and then at Chicago University. There he constructed the first atomic reactor in 1942. He died in 1954 in Chicago.

Fig. 4.7 The experimental
disposition used by Rossi to
measure the penetrating
power of cosmic rays. The
number of lead layers
between counters may be
varied up to a thickness of
more than 1 m. Only the
particles able to traverse the
absorber produce
coincidences

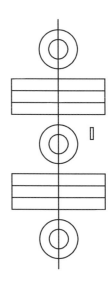

Bothe and Kolhoerster had demonstrated that a large fraction of cosmic ray particles were able to traverse 4.1 cm of gold, and Rossi's work in the Bothe's laboratory confirmed this result. Then Rossi [24, 25] placed three counters along a vertical line[3] (Fig. 4.7) in a configuration that was later called "*counter telescope*" and interposed some lead between them. Initially he chose a thickness of 25 cm, which he then changed. This way, he found that 60 % of the particles able to cross 25 cm of lead could also get across 1 m. Because 25 cm of lead absorbed about 45 % of the radiation, this result means that about 35 % of the particles found at sea level had in lead a range >1 m. The fact that most of the coincidences observed after 1 m of lead were due to single particles that traversed the whole thickness, was afterwards proved by Rossi. By displacing the central counter in Fig. 4.7 of the geometric line that passed through all the three counters, the coincidences number dropped to about one-sixth of the original value. This conclusion was confirmed using cloud chambers [28–30].

This was a very important result if one considers that beta rays in lead have a maximum range of the order of 1 mm.

In conclusion the experiments with both Geiger–Müller counters and the Wilson chamber gave the conclusive proofs of the corpuscular nature of the cosmic radiation near the ground. Moreover, with only minor exceptions, all particles produced nearly the same ionization density than that found along the tracks of fast electrons and, because for a given velocity the ionization density should be proportional to the electric charge squared, this meant that presumably all the particles had the same absolute value of charge as the electron [31, 32].

---

[3]The use of three aligned counters was already proposed by M.A. Tuve [26] and L.M. Mott-Smith [27].

Still, there was someone who disbelieved the hypothesis of charged particles and was suggesting that the majority of cosmic rays was electrically neutral (e.g. photons) ([33–35], see also [36]). A.H. Compton replied to all of them with a frantic activity [37–41] and the paper *"Cosmic rays as electric particles"* [42]. The proofs of the charged nature of the particles of the primary radiation were discussed, along with the latitude effect and the east–west asymmetry, the coincidence experiments of Bothe and Kolhoerster, the experiments of P. Auger and Ehrenfest [43] on showers, and the ones of J.C. Street et al. [28] in which a Wilson chamber interposed between two or more aligned counters showed that any time a coincidence occurred, the chamber was traversed by a track. Moreover, all the particles which produced coincidences showed the latitude effect.

4.4 What about the "Birth-Cry"?

The 1928 experiments of Bothe and Kolhoerster had been a deathblow to the Millikan theory of the *"birth-cry"*.

The theory that cosmic rays were particles and counters were used to detect them was really not accepted by Millikan who, on December 1932 at a Symposium on cosmic rays held in Atlantic City, later published on Physical Review [44], declared *"I have been pointing out for two years in Pasadena seminars, in the Roma congress on nuclear physics in October 1931, in New Orleans last Christmas at the A.A.A.S. meeting, and in the report for the Paris Electrical Congress, that these counter experiments never in my judgement actually measure the absorption coefficients of anything"*. So, when later the Rossi's results appeared, Millikan was ready to reply. Anderson collaborated with him in the composition of the paper with which in December 1933, they rejected the program counter/particle [45].

In his experiments Rossi had also observed that in conjunction with the passage of a primary particle, secondary particles appeared that at variance with the primary had a mean penetration in lead of about 1 cm.

Also Anderson and Millikan observed the presence of these showers of secondary particles and with S. Neddermeyer (a post doc student) and W. Pickering (a graduate) used this circumstance to criticize Rossi's interpretation, using also an important experimental observation: the number of particles in showers increases by increasing the lead thickness. This is true; however, they assumed, without any further measurement, that the number of particles in the shower augmented indefinitely with the lead thickness, which is not true. Therefore, they maintained that the Rossi coincidences could not be due to the passage of a single charged particle through the counters and the interposed lead, but should be due to some mechanism through which many different particles simultaneously but separately trigger the counter [45].

The paper starts with a strong statement: *"As pointed out by Millikan [44, 46] in 1932, coincidences in the response of two counters placed on a vertical line and separated by considerable thicknesses of heavy matter, such, for example as a metre of lead, cannot in general be due to the passage of one charged particle through both counters and the intervening lead, but must[4] rather be due to some mechanism by which a photon can release successively along, or in the general neighbourhood of, its path a number of different particles whose separate but practically simultaneous action on the two or more counters is responsible for the observed coincidences".*

To support this position, the authors reiterated a vision of the nucleus by their own.

When the latitude effect, showing that the primary rays are charged particles (we will describe it in Chap. 5) was definitely accepted in 1934, Millikan [36] continued to speak of photons produced by atom-building. Then, in 1942 he changed the model [47] and assumed that [48] *"...an atom out in interstellar space has the capacity not possessed by an atom in the stars or in any other region in which it is continuously subjected to bombardment from neighbouring atoms, of occasionally transforming the whole of its rest mass energy into a charged-particle pair which for the present may be called an 'electron pair'. The kinetic energy of such an electron pair can then easily be computed from the known rest mass of the atom m with the aid of the equation $E = mc^2$, c denoting the velocity of light and E energy in ergs".*

Eventually, in 1949, on the Reviews of Modern Physics, Millikan [49] presented anew this modified version of his theory. Now, in *"The present status of the evidence for the atom-annihilation hypothesis"*,[5] he dropped the atom-building theory from hydrogen atoms, suggesting instead that the origin of cosmic radiation could occur in the opposite process of annihilation of H, He, C, N, O, and Si atoms. However, nuclear physics was too much advanced at this point for this suggestion to be considered.

Once Millikan was summoned as an expert in a trial against a man charged of having sold bottles containing a liquid that he claimed to posses healing properties due to irradiation with cosmic rays. Asked to give his opinion Millikan said: *"Who did not make incorrect assertions on cosmic rays?"* The anecdote may not be true, but, as an Italian way of saying goes, "se non è vero è ben trovato".

References

1. D. Skobelzyn, Die Intensitatsverteilung in dem Spektrum der γ-Strahlen von RaC. Z. Phys. **43**, 354 (1927)
2. D. Skobelzyn, Uber eine neue Art sehr schneller β-Strahlen. Z. Phys. **54**, 686 (1929)
3. B. Rossi, Rev. Mod. Phys. **20**, 539 (1948)
4. W. Bothe, W. Kolhoerster, Naturwissenschaften **16**, 1044, 1045 (1928)
5. W. Bothe, W. Kolhoerster, Z. Phys. **56**, 751 (1929)
6. H. Geiger, W. Müller, Phys. Z. **29**, 839 (1928)

---

[4]The underlined text is in italics in the paper.

[5]The theory was discussed also in Millikan [50].

7. H. Geiger, W. Müller, Phys. Z. **30**, 489 (1929)
8. E. Rutherford, H. Geiger, Proc. R. Soc. **A81**, 141 (1908)
9. E. Rutherford, H. Geiger, Proc. R. Soc. Lond. A **81**, 162 (1908)
10. E. Rutherford, T. Royds, Philos. Mag. **17**(6), 281 (1909)
11. H. Geiger, E. Marsden, Proc. R. Soc. London A **82**, 495 (1909)
12. H. Geiger, E. Marsden, Philos. Mag. **25**(6), 604 (1913)
13. E. Rutherford, Philos. Mag. **21**(6), 669 (1911)
14. W. Bothe, H. Geiger, Z. Phys. **25**, 44 (1924)
15. W. Bothe, H. Geiger, Naturwissenschaften **13**, 440 (1925)
16. W. Bothe, H. Geiger, Z. Phys. **32**, 639 (1925)
17. N. Bohr, H.A. Kramers, J.C. Slater, Philos. Mag. **47**(6), 785 (1924)
18. W. Bothe, W. Kolhoerster, Z. Phys. **56**, 751 (1929)
19. W. Bothe, W. Kolhoer Die Natur der Hohenstrahlung Naturw **17**, 17 (1929)
20. A. Pascolini (ed.), *The Scientific Legacy of B. Rossi* (University of Padova, Padova, 2006)
21. B. Rossi, Linc. Rend. **11**, 831 (1931)
22. B. Rossi, Nature **125**, 636 (1930)
23. B. Rossi, *Moments in the Life of a Scientist* (Cambridge University Press, Cambridge, UK, 1990)
24. B. Rossi, Naturwissenschaften **20**, 65 (1932)
25. B. Rossi, Z. Phys. **82**, 151 (1933)
26. M.A. Tuve, Phys. Rev. **35**, 651 (1930)
27. L.M. Mott-Smith, Phys. Rev. 35, 1125 (1930)
28. J.C. Street, R.H. Woodward, E.C. Stevenson, Phys. Rev. **47**, 891 (1935)
29. P. Auger, P. Ehrenfest, J. Phys. Rad. **7**, 65 (1936)
30. L. Leprince-Ringuet, J. Phys. Rad. **7**, 67 (1936)
31. R.A. Millikan, D. Anderson, Phys. Rev. **40**, 325 (1932)
32. D. Anderson, Phys. Rev. **41**, 405 (1932)
33. H. Kulenkampff, Phys. Z. **30**, 561 (1929)
34. E. Regener, W. Kramer, E. Lenz, Z. Phys. **85**, 411 (1933)
35. I.S. Bowen, R.A. Millikan, H.V. Neher, in *Address before the London Conference on Physics*, October 1934, London, 1935, p. 206
36. I.S. Bowen, R.A. Millikan, H.V. Neher, Phys. Rev. **46**, 641 (1934)
37. A.H. Compton, Guthrie Lecture of the Physical Society, 1 Feb. 1935
38. A.H. Compton, Proc. Phys. Soc. Lond. **47**, 747 (1935)
39. A.H. Compton, Proc. Am. Philos. Soc. **75**, 251 (1935)
40. A.H. Compton, Am. Phys. Soc. (Address before 6 Jan. 1936)
41. A.H. Compton, Rev. Sci. Instrum. **7**, 71 (1936)
42. A.H. Compton, Phys. Rev. **50**, 1119 (1936)
43. P. Auger, P. Ehrenfest, CR **199**, 1609 (1934)
44. R.A. Millikan, Phys. Rev. **43**, 661 (1933)
45. C. Anderson, R.A. Millikan, S. Neddermeyer, W. Pickering, Phys. Rev. **45**, 352 (1934)
46. R.A. Millikan, in *Institut Poincarè Proceedings* (1933)
47. R.A. Millikan, H.V. Neher, W.H. Pickering, Phys. Rev. **61**, 397 (1942)
48. R.A. Millikan, H.V. Neher, W.H. Pickering, Phys. Rev. **63**, 234 (1943)
49. R.A. Millikan, Rev. Mod. Phys. **21**, 1 (1949)
50. R.A. Millikan, *Electrons (+ and −), Protons, Photons, Neutrons, Mesotrons and Cosmic Rays* (University of Chicago Press, Chicago, 1947)

Chapter 5
The Earth's Magnetic Field and the Geomagnetic Effects

5.1 The Earth's Magnetic Field

The researches described until now regarded the nature of cosmic rays on the Earth, but had given no direct information on the nature of the "primary" cosmic rays that are falling on the Earth from outside. This information came from the study of new effects, due to the presence of the Earth's magnetic field.

The Earth, like the Sun and many planets and stars, is surrounded by a magnetic field.

The magnetic field surrounding the Earth resembles the field of a magnetic dipole—the field produced by a magnetized bar of small length in comparison with the Earth's radius—placed near the Earth's centre with its north pole towards the geographic south and its south pole towards the geographic north.[1] More precisely, the centre of the terrestrial magnetic dipole is displaced by a few 100 km (about 342 km) from the centre of the Earth, and its axis intersect the Earth's surface in two points, one at 78.3° S latitude and 142° E longitude, and the other at 78.3° N latitude and 100° W longitude, forming an angle of about 11.5° with respect to the rotation axis of the Earth (see Fig. 5.1). The two magnetic poles are therefore one in front of the Adelie Land in Antarctica and the other one offshore the Bathurst island in the Arctic circle. The maximum circle passing from the centre of the dipole and normal to its axis is the *geomagnetic equator*. In analogy with the geographic coordinates, one defines geomagnetic coordinates with reference to the magnetic poles and equator. The distance of a point on the Earth from the geomagnetic equator is called *geomagnetic latitude*.

The position of the magnetic poles is not fixed in time. The inclination angle of the magnetic dipole oscillates sensibly (in 400 years has suffered a total variation of about 40°), so that the position of the magnetic poles moves (see Fig. 5.2) and,

---

[1] The dipole moment is 8.1×10^{22} A m^2 and it produces a field that on the Earth's surface is about 0.5 Gauss.

M. Bertolotti, *Celestial Messengers: Cosmic Rays*, Astronomers' Universe, DOI 10.1007/978-3-642-28371-0_5, © Springer-Verlag Berlin Heidelberg 2013

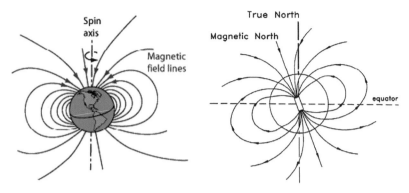

Fig. 5.1 The Earth's magnetic field

Fig. 5.2 Position of the
South geomagnetic pole in
the northern hemisphere at
various times during the past
five millions of years. The
white cross (barely visible
near the central *left side*) is
the present position
(Reprinted with permission
from P.H. Roberts and G.A.
Glatzmaier, Rev. Mod. Phys.
72, 1081 (2000). Copyright
2000 by the American
Physical Society)

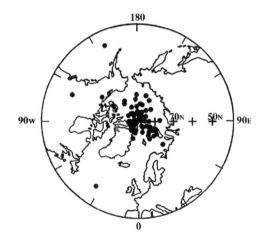

indeed, the north and south poles may invert, as it happened nine times in four
millions of years, with the last inversion 780,000 years ago.

The produced field extends well beyond our atmosphere in the space surrounding
the Earth up to about 60.000 km and although it is weak, it has an important
influence because it may act on particle path lengths of thousands of kilometres. At
the sea level its intensity decreases, moving from the poles to the magnetic equator.
In the space surrounding the Earth it decreases in a complex way, approximately as
the reciprocal of the cube of the distance from the centre, and at great distances it
becomes asymmetric. Measurements with spacecrafts have shown that the situation
outside the Earth is indeed very complex because the field is strongly perturbed by
the solar field and by currents of particles, principally protons and electrons coming
from the sun (*solar wind*) (see Fig. 5.3). Therefore it does not extend indefinitely
into the space, as it would be if the Earth was in vacuum, but is confined in a region
surrounding our planet, called *magnetosphere*. We will come back on this point in

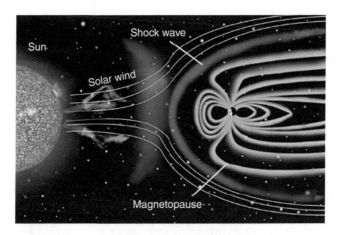

Fig. 5.3 The Earth's magnetic field is modified by the solar wind

Chap. 12; for the moment we may consider the field as if it was approximately that of a dipole, also because this was the representation used in the researches described here.

At a distance from the Earth of about 2,000 km, its intensity is still about half the value of that on surface. By contrast, the remaining air mass, beyond about 40 km, constitutes <1 % of the total atmosphere.

A precise theory of the origin of the Earth's magnetic field does not exist but the mechanism of its production is now understood in its general lines and ascribed to a process of a dynamo self-excited by convective motions at the Earth's interior ([2–4], see also [5]). We must remember that the Earth's interior may be schematized as a solid central inner core of a radius of about 1,270 km, surrounded by a shell of liquid metals (external nucleus) that reaches a radius of about 3,490 km. Over this liquid outer core is placed the mantle, covered by the crust. In the mechanism— called *geodynamic*—that is sought to originate the magnetic field, the fluid of the outer core, a nickel–iron alloy too hot to be magnetized but that is however an excellent conductor of electricity, originates a self-excited dynamo. While this fluid rotates relatively to the Earth, due to the convective currents generated by the heat released by the inner core that is cooling, it crosses some magnetic force lines already existing (for example, the weak Sun's magnetic field) thus generating the geomagnetic field at the expenses of its kinetic energy. The treatment of the phenomenon is very complex because the axial symmetry of the rotation induces columnar vertical motions with an axis parallel to the rotation axis. Figure 5.4 shows a section with the solid inner core, the outer liquid core with the convective currents that surrounds it and the solid mantle (Fig. 5.4a), and the axial shape of the convective currents in the fluid core (Fig. 5.4b). Today, some simulations are able to account for the principal characteristics of the model with considerable success.

Fig. 5.4 (**a**) Schematic section of the Earth's interior with the solid nucleus at the centre, the liquid nucleus with the convective currents and the surrounding solid mantle. (**b**) The axial symmetry of the terrestrial rotation induces columnar vortex motions with axis parallel to the Earth rotation axis that in the figure is coplanar with the paper and directed upward

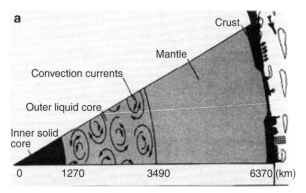

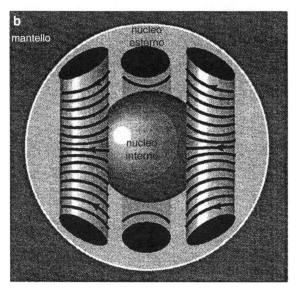

5.2 Stoermer and the Aurora Borealis

The effect of the Earth's magnetic field on charged particles coming from outside was first discovered and studied in connection to the *aurora borealis*.

The *aurora* is a distinct luminosity, most often visible at high latitudes, that comes from definite regions in the high atmosphere (see Fig. 5.5). The luminous phenomenon manifests itself under different shapes: arcs, bands (ribbon like), rays, curtains, diffuse lights, etc. Some auroras are standing and of invariant shape, while others move in the sky and change their shape, as if they were dancing. They may be white or red, yellow, or greenish. The upper limits of the aurora light range from nearly 100 to 1,000 km, while the inferior limits are between about 85 and

Fig. 5.5 An aurora borealis

170 km. The auroras are very common both in the northern and southern regions.[2] In the northern regions they are called *aurora borealis*—a name given by Pietro Gassendi (1592–1655) to the general phenomenon, when the *aurora australis* was not yet known—and are more frequent in a zone limited by a line that starting from Cape North, nearly grazes the northern coasts of New Zemlya, the extreme northeast point of Siberia, Alaska, the Hudson bay, the Labrador (Canada), the southern lands of Greenland, and Iceland to close the loop back at the starting point. Outside this belt, towards south, the phenomenon becomes less frequent but, although very seldom, aurora at lower latitudes may occur. In Italy, for example, they may appear one or two times in a year.

The aurora was discussed as far back as Aristotle (384–322 BC). He surmised in his *Meteorologica* (about 330 BC) that it was some kind of vapour that rose from the ground when heated by the Sun.

Among the inhabited countries, Norway has the better conditions for the study of aurora borealis and in fact it has long traditions and an ample literature on this phenomenon. One of the first realistic descriptions of aurora is found in the Norwegian chronicle *The King's Mirror*, dating back to 1,230. It was originally written as a textbook, probably for the young King Magnus Lagabote, by his father. Sophus Tromholt (1851–1896) organized a network of northern lights observation

---

[2]They are called aurora borealis or northern lights, in the North, and aurora australis or southern lights in the South.

stations and enumerated a great number of auroras, from the most remote times until 1878. Anders Angstrom (1814–1874) discovered that the spectrum of aurora light consists of lines and bands, not of a continuum. He rejected Descartes' idea that the aurora was caused when tiny ice crystals suspended in the air reflected sunlight.

Today we know that the aurora light is initiated by the passage, through the upper atmosphere, of energetic charged particles, coming from outside. The particles include both electrons and protons. Most of the visible aurora light is due to the ionization or excitation of the upper atmospheric constituents (nitrogen and oxygen) by incoming electrons and also by secondary electrons produced thereby. Protons, after having a number of collisions, recombine with electrons emitting on some lines of hydrogen. Therefore, lines characteristics of atomic nitrogen, oxygen, and hydrogen are present in the spectrum of aurora light. A line at 5,577 Å in the green, for long time considered of mysterious origin, is today attributed to a particular transition of oxygen.[3]

The Italian astronomer Giovanni Battista Donati (1826–1873), the successor of G.B. Amici (1786–1863) in the direction of the Florence Sun observatory—that he moved onto the Arcetri hill in 1872—discovered the relation between aurora and solar phenomena and realized that the phenomenon should derive from the action of electrified particles emitted by the Sun. The Norway physicist Kristian Birkeland [6] (1867–1917) suggested that the Sun emits beams of electrons, sensible to the action of the Earth's magnetic field that, by bombarding the upper layers of the atmosphere, excite the atoms of the gas present there.

Birkeland [7] projected a beam of cathode rays in an evacuated chamber towards a globe of a shape similar to the Earth—that he named "*terrella*"—coated with a fluorescent paint, with a magnet placed in its interior, showing that the cathode rays made luminescent the polar regions of the "*terrella*", which are exactly the regions where usually auroras are formed. That was the demonstration of the fact that a magnetic field of dipolar kind is able to deflect charged particles (in this case the electrons of the cathode rays), conveying them essentially on the polar regions where they could interact with the atmospheric gases and produce the light emission. The Birkeland studies have been described in ref. [89].

The Norwegian geophysicist Fredrik Carl Muelertz Stoermer[4] (1874–1957), a Professor at the Oslo University, stimulated by Birkeland's ideas, undertook the study of the motion of charged particles projected from a distance into the Earth's magnetic field. Early in the summer 1903, he succeeded to obtain elements for a

---

[3]This line is due to a forbidden line of the oxygen atom. The production of forbidden lines in the upper atmosphere is favoured by the low density. Some atoms can remain in a metastable excited state for seconds or even hours, rather then radiating in 10^{-6}–10^{-8} s. In laboratory experiments, an atom in a metastable state is usually de-excited by collisions with particles or with the wall of the vessel. At the low density prevailing above 100 km, however, the atom has a larger probability of remaining undisturbed long enough to radiate a forbidden quantum.

[4]The most important C. Stoermer works on the motion of charged particles are: "Sur les trajectoires des corpuscules électrisés dans l'espace sous l'action du magnetisme terrestre" [8, 9]. "Ein Fundamentalproblem der Bewegung einer elektrisch geladenen Korpuskel im kosmischen Raume" [10–12]; The book [13].

Fig. 5.6 A few electron trajectories reaching the Earth coming from two points on the Sun. Each trajectory is represented by means of copper wires

more quantitative explanation of the appearance of auroras. He thought that they were produced by charged particles coming from the Sun at the time of an increase of the solar activity. To improve the study, he had to derive singularly the different trajectories via complex calculations that allowed him to determine some of their general characteristics. At the time computers were not available, so each trajectory had to be calculated by hand, starting from the Sun and arriving on the Earth (while taking into account the relative positions of the two points), and for each different direction and initial velocity of the particle. In the period between 1904 and 1907 he, together with his collaborators, calculated numerically nearly 120 trajectories.

Initially the calculation was made starting from an emission point on the Sun and trying to find the trajectory that allowed the electron to reach the Earth. Eventually, Stoermer found out it was better to calculate the trajectory in the reverse way, choosing a starting point on the Earth and looking for the particles that arrived on the Sun. The trajectories calculated in this way were represented with copper wires in a very impressive vision (see Fig. 5.6).

Stoermer's work dealt with the motion of a single charged particle, subjected only to the influence of the Earth's magnetic field. The research is summarized in his book *"The Polar Aurora"*. Although Stoermer's basic hypothesis was correct, his theory was only a first approximation. One of the reason is that the particles of the aurora have relatively low energies and are therefore strongly influenced also by the weak magnetic and electric fields present in the interplanetary space, whose existence was unknown when Stoermer was developing his theory. Aurora is in fact a very complex phenomenon we will not discuss any further.

By studying the mathematical equations of motion of charged particles in the geomagnetic field, Stoermer[5] found a general expression for the solutions that connected the Earth to the Sun. He also found solutions that are closed, that is with no connection between the Earth and the Sun, that he decided (mistakenly) did not contribute to the production of aurora, and lacked to suggest that the regions occupied by these solutions could be filled with trapped particles, which in fact are the source of aurora in conditions of extreme geomagnetic disturbance. The trapped particles were later observed directly with the first American satellite by van Allen, as we will discuss later, and are largely confined to the "forbidden" regions of Stoermer's theory.

5.3   The Theory for Cosmic Rays

The primary cosmic rays that arrive from outside the Earth encounter the terrestrial magnetic field well before having a chance to collide with the molecules of the atmosphere. Travelling in a region of space where a magnetic field exists, a charged particle is subjected to the Lorentz force which diverts it from its original trajectory in the same way as it does for charged particles in the Wilson chamber. This field acts only on charged particles and its action depends on their mass, velocity, charge, intensity of the field, and on its orientation with respect to the velocity of the particle. If the field is parallel to the trajectory, its action is null. If it is normal, its action is at a maximum and the trajectory is an arc of a circle. If the trajectory is not directed in one of these two particular directions, the resulting trajectory is an helix. Because the Lorentz force is always orthogonal to the particle velocity, it does not change the particle energy that therefore is only deflected by the field in one direction or the other depending on whether its charge is positive or negative. Moreover, although the terrestrial field is much weaker than the fields produced in our terrestrial laboratories, its enormous extent allows it to act over a very large length of the primary ray trajectory outside the terrestrial atmosphere, making its effect important. Therefore even if the charged particles have a great energy, the deflections may be very large.

If they are photons, they will pass through the field without being deflected.

At the time of the studies we are describing, people assumed that the field acting on the primary particles was only that of the Earth that was assimilated in a first approximation with that of a dipole.

It is mostly thanks to Stoermer, Paul S. Epstein [14] (1883–1966), the Belgian priest G. Lemaitre (1894–1966) and to the Mexican Manuel S. Vallarta [15–17] (1899–1977) who all studied the problem theoretically, that the effect of the Earth's magnetic field on cosmic rays, was understood.

---

[5]A treatment of these first Stoermer's studies was by him made in the book [13].

The study of the spatial variations of the cosmic rays intensity is simplified by the circumstance that temporal variations are small. The "primary" cosmic rays have a notable constancy and isotropy. Most of the variations of the intensity may be attributed to changes of atmospheric pressure and temperature. The *barometric effect*, as it is called, that is the decrease of the cosmic ray intensity with the increase of the atmospheric pressure, was discovered by L. Myssowsky and L. Tuwim [18] and confirmed by E. Steinke [19]. It is a small effect due to the absorption of cosmic rays in the air. After correcting for the atmospheric absorption, variations associated with magnetic storms remain. Because the variations due to magnetic storms are produced near to the Earth or, in any case, inside the solar system, the radiation in interstellar space seems to remain constant.

To understand what is happening to a charged particle moving in the geomagnetic field, the classic work of Stoermer [13, 20–22] was fundamental. Although the cosmic ray particles have an energy much greater than the aurora particles, there is no reason why their trajectories should not be of the kind calculated by the Norwegian scientist. Of course, calculations are very complex because the cosmic rays come from all directions. However some characteristics are easy to see.

Stoermer himself, in 1930, applied his theory to cosmic rays. The theory was applied also by P. Epstein [23] and later developed by Lemaitre and Vallarta in 1933, and summarized by the latter ([24] see also [25]). Fundamental contributions were given also by E. Fermi and B. Rossi [26], and many others [27].

As shown by the theory of aurora, the low energy particles are deflected by the Earth's magnetic field near the geomagnetic equator where the field is perpendicular to their trajectories, but may reach the Earth near the geomagnetic poles because their direction is parallel to the field.

If the primary radiation is made of photons, they will pass through the field without being deflected; therefore, they will arrive in the same number in polar as well as in equatorial regions. When they arrive in the high atmosphere at some 15 km over the sea level they could interact with the air molecules and produce charged particles, but these particles would not be influenced by the magnetic field because it is too weak and the length of the path in the atmosphere is too short.

If we assume that the primary radiation is made of charged particles uniformly distributed at great distances from the Earth, the ones arriving along the polar direction would suffer no action and could reach the polar regions whichever their energy, while the particles arriving in the equatorial plane would suffer the strongest deviation. The slower particles would be deviated so much as to be unable to reach the Earth's surface and would be rejected into the space. The ones with a greater energy would be less deviated, and it is possible to calculate which is the minimum energy a particle should have to arrive vertically on the equator. Moreover, if the rays are made of positively charged particles, at low latitudes more particles would arrive from the west than from the east, and the reverse should be true for negative particles.

Therefore if a part of the primary cosmic ray is charged and has sufficiently low energy, the radiation reaching the equatorial regions should be less than the one reaching the polar regions, that is it should depend on the latitude; such an effect was called *latitude effect*. I repeat that this applies to primary cosmic rays, because the deflection of the particles in the relatively short extension of the terrestrial atmosphere is negligible.

At the geomagnetic poles particles with any value of the momentum (energy) may arrive, while only particles with a sufficiently great momentum may arrive at lower latitudes or at equator. In order to reach the equator from a vertical direction, a particle with unitary charge (for example, an electron or a proton) should have an energy larger than 15.000 MeV/c, and only particles with a momentum larger than 60.000 MeV/c may arrive coming from all directions.

These results show that for charged particles below some energy, there is an unapproachable equatorial zone. At some critical latitude the rays start to arrive inside a narrow cone around the western horizon, if they are positive particles, and increasing the latitude this cone enlarges until eventually, at a second critical latitude, it embraces the whole hemisphere. Although in this analysis the unapproachable regions were well defined, the intensity inside the accessible cone was not derived. This problem was solved by Lemaitre and Vallarta [15], and Swann [28] who showed that when an isotropic radiation enters a magnetic field, this field may forbid the radiation to reach some point in some direction, but in the allowed directions the intensity does not change. Otherwise stated, an observer finds the same intensity (for a given energy) he would find in the absence of the magnetic field (allowed cone), while he finds zero intensity in all other directions (forbidden cone).

Considering particles with some definite value of energy, one finds that they cannot arrive on the Earth for latitudes lower than some value λ_1. For latitudes larger than some other value λ_2, all particles of that energy may reach the Earth's surface independently of their arrival direction. Between λ_1 and λ_2, only particles whose arrival directions are inside a cone (centred towards west for positive particles and towards east for the negative ones) may arrive. This cone enlarges gradually with increasing energy and embraces the whole hemisphere when the energy reaches 60 GeV. Because the effect is strictly linked to the Earth's magnetic field, the latitude to be considered is the geomagnetic one and not the geographic latitude. These two latitudes may be very different each from the other due to the inclination of the axis of the terrestrial magnetic dipole and because of its eccentricity.

The existing asymmetry in the intensity distribution due to the fact that positive particles should arrive less numerous from east than from west while negative particles should have the opposite behaviour, was named *east–west effect*.

The study of the latitude effect has allowed to determine the momentum spectrum of primary rays up to some 6×10^{10} eV/c, an energy above which the rays may enter the Earth's atmosphere even at equator. The result has been that many incident particles have momenta of this order of magnitude.

Further studies by Lemaitre and Vallarta [16, 17] showed that, when the problem is considered more deeply, there are forbidden directions for some particles that were instead allowed by the simple Stoermer theory.

In summary, and speaking rather approximately, for an observer on the Earth, the sky may be divided into the following regions:

(a) The Stoermer cone, within which particles with a momentum below a threshold value cannot enter.

(b) The main cone, within which all particles with a given momentum may enter.

(c) The so called "penumbra", that lies between the Stoermer and the mean cone, in which there are alternating bands of allowed and forbidden directions, and then,

(d) The shadow cone within which lie only orbits which have passed one or more times through the Earth before having arrived at the point in question. This last region finds itself close to the horizon adjacent to the nearer pole and is the more important at large zenith angles.

The penumbra region is difficult to calculate. It has the greater importance at low latitudes. At high latitudes the simple Stoermer theory may be applied. At the equator, the principal and the Stoermer cone in the east–west plane coincide and therefore the simple Stoermer theory may be applied.

The great importance of these geomagnetic effects was that they demonstrated the presence of charged particles in the primary radiation. The treatment we have described does not take into account the effects produced by the Sun, effects that were unknown at the earlier times, although Epstein [14] tried to make some estimates. It was however by the study of the geomagnetic effects that the first information on the primary radiation was obtained. The status of the field after the first researches was well described in a paper by T.H. Johnson [27].

5.4 The Latitude Effect

Although practically all interpretations on the nature of cosmic radiation before 1929 were based on the hypothesis that the primary rays were photons, the evidence supporting this assumption was of course, very incomplete, and this point of view was justified mainly because gamma rays from radioactive substances were known to be more penetrating than electrons of the same energy. The nearly exponential shape of the ionization curve as a function of altitude also allowed an interpretation based on the photon hypothesis.

Between 1927 and 1930, the Dutch physicist Jacob Clay (1882–1955)—then a physics Professor at Bandoeng in the Java island—measured the intensity of cosmic rays by using a ionization chamber in some trips between Java and The Netherlands and found a diminution of nearly 11% near the Suez Canal, giving thus the first prove that the intensity of cosmic rays was varying with latitude.

On 17 December 1927, Clay qualitatively reported a decrease between Amsterdam and Batavia [29]. In November 1928 [30], he measured a very large (about 45%) variation between Leyden and Singapore, and in a last paper [31] on 30 September 1930, he quoted a somehow lower, but still large variation of 25% between Amsterdam and Singapore. Later, he gave a variation of about 6% between The Netherlands and the equator.

However, the first Clay results were generally not accepted until later, because of a number of negative results obtained by other researchers.

In fact Millikan and Cameron [32], in 1928, published that they had not found any significant change in measurements done in the summer 1926 between the Andes in Bolivia (17° S latitude) and Pasadena, CA (34° N latitude). Actually the points on the ionization curve obtained in Bolivia were 5 or 6 % below the ones obtained in California, but the accuracy of the measurements was not good enough to say the difference was significant. In the paper the authors reported also measurements made on board of a ship in a voyage from Los Angeles to Mollendo, Peru, which did not show any effect. In the report published on Nature [33] they wrote: "*We found no variation in sea-level reading with geographical position*".

Later Millikan [34] wrote that when in the summer 1926 embarked to Mollendo, the measurements could start to be taken only in the second day of travel because the electroscopes were not working properly, and therefore the measurements began to be taken only when the ship was 900 km more south, so much south, we may say now, that more than one half of the diminution observable between Los Angeles and the equator had already been produced.

Again in 1930, Millikan [35] found practically no change between Pasadena (34° N) and Churchill, Canada (59° N) 730 miles from the north pole, on the west side of the Hudson Bay, with measurements performed during that summer. That same year, Bothe and Kolhoerster [36] did not detect any variation with latitude in the north sea between Hamburg (53° N) and Spitzbergen (81° N), and also Grant [37] travelling from Adelaide (Australia) in the Antarctica, did not detect anything, while the Swedish physicist Axel Corlin [38] believed he had found a small latitude effect in the Baltic sea while travelling from 50° N to 70° N in Scandinavia. Again we may say that they were travelling in regions where the effect was not anymore appreciable. At the time, however, they did not understand this.

Therefore, the existence of the effect seemed to be very doubtful [39]. In any case the effect, even if present, was very small and could not be attributed with certainty to the Earth's magnetic field. People observed, and with good reasons, that the different conditions of the Earth's atmosphere in different geographic locations could by themselves produce significant changes in the cosmic ray intensity at the sea level.

For Millikan and his collaborators, the absence of a latitude effect was the proof that the primary cosmic rays were not charged particles and this fact corroborated his "*birth cry*" theory.

Millikan's insistence in maintaining that primary cosmic rays were photons was probably the main reason why his group failed to discover the latitude effect. Still in 1931, they had found "*not a shred of evidence*" for the latitude effect and Millikan celebrated the success of his "*birth cry*" theory. The reason for these negative results is that Millikan's travel between Bolivia and California developed along regions all having nearly the same geomagnetic latitude.

A.H. Compton—who had ceased to trust the "*birth cry*"—was no longer convinced that the primary rays were all photons and did not share Millikan's certitude that the effect did not exist. The relation between Millikan and Compton, started as an amicable relationship, became one of the most acrimonious and publicized scientific dispute of the time. Consulting Millikan's archives one may see that both men became personally involved, even casting aspersions on each other's scientific integrity [40]. Obviously amused at the spectacle of two Nobel laureates engaged in such a "dogfight", as Millikan called it, the press raised the issue to front-page news.[6]

Arthur Holly Compton was born on 10 September 1892 in Wooster, Ohio and died on 15 March 1962 in Berkeley, California. His father Elias was a philosophy Professor and Karl (1887–1954), Arthur's brother, was also a physicist who became president of MIT. Arthur was educated at Wooster College, where his father was a Professor, and in Princeton where he earned his PhD in 1916. He then started teaching at the University of Minnesota and after a 2-year period at Westinghouse Corporation in Pittsburgh he was appointed Physics Professor at the Washington University in St. Louis, Missouri. Most of his career was spent at the University of Chicago, from 1923 to 1945, after which he came back to Washington University. The principal contribute of Compton was the discovery and the explanation in 1923 of the effect which bears his name, for which he was awarded the Physics Nobel Prize in 1927 together with C.T.R. Wilson, who, among other things, proved experimentally Compton prediction by using the cloud chamber. In the 1930s, Compton focused on the study of cosmic rays. During the Second World War, he had an important role in the Manhattan project.

Compton [41–47], together with a group of researchers, at the end demonstrated in an unquestionable way the existence of the effect of the Earth's magnetic field, with studies that started in 1930 and were conducted in many different places in the world. In the summer of 1930, Compton, R.D. Bennett, then both at Chicago University (Bennett successively moved to MIT), and J.C. Stearns—of Denver University—decided to initiate a coordinated study of the geographic distribution of cosmic rays. They associated other colleagues in America, South Africa, and India and began a series of measurements in different places in the world via a meticulous organized research. Seventeen physicists, equipped with large ionization chambers filled with argon at 30 atmospheres, all of similar characteristics, measured the intensity of cosmic rays in 69 stations distributed around the world. Seven similar sets of apparatus were constructed, and seven expeditions were organized that made measurements contemporarily following the same procedure. At each station, the ionization due to cosmic rays was compared with that due to a standard radium capsule so to have a uniform calibration of the instrumentation and to be able to compare among them measurements taken in different places by different research groups.

Measurements were made not only at the sea level, but also at different altitudes. The study of all measurements showed clearly the dependence of the intensity of cosmic rays on the geomagnetic latitude.

---

[6]New York Times, 5 February 1933.

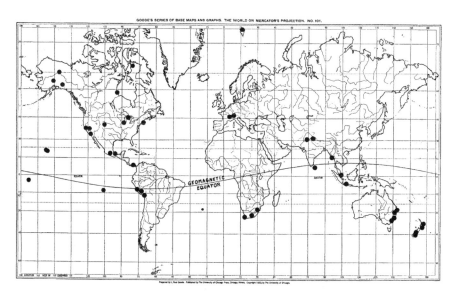

Fig. 5.7 Map showing location of major stations for studying the latitude effect (from Compton [42]). In the figure the position of the geomagnetic equator is marked (Reprinted with permission from A.H. Compton, Phys. Rev. **43**, 387 (1933). Copyright 1933 by the American Physical Society)

Figure 5.7 shows the geographic location of the major stations where measurements were made.

Data were corrected, among other things, for barometric pressure, and in order to make a precise comparison of the results for different latitudes, were reduced to three standard barometric pressures (see Fig. 5.8 [42]). All the data concurred in showing a marked difference between the intensities found within 20° of the magnetic equator as compared with those at latitudes higher than 50° north or south. At the equator (0° latitude) the ionization has the lower value, while for latitudes higher than 50°, up to the pole, the ionization stabilizes at a maximum value. The increase with higher latitude averaged 14 % at sea level, 22 % at 2,000 m (barometer 60 cm), and 33 % at 4,360 m (barometer 45 cm).

Figure 5.9 shows the values obtained at the sea level on a larger scale. In this figure, Compton added also the data taken by Millikan in 1930 between Pasadena and Churchill [35] and other ones of Clay and Berlage [48], obtained in a trip between Genoa and Singapore. All data practically show the existence of the effect. In his paper, Compton writes "... *the latitude effect is so large as to be unquestionable...*".

At the end for his campaign of measurements, Compton had travelled for nearly 100,000 km and visited five continents.

The fast increase in the intensity of cosmic rays between geomagnetic latitudes from 20° to 50° is exactly what one could expect from the calculations made by Lemaitre and Vallarta [15, 49] who considered the effect of the Earth's magnetic field on the motion of electrons approaching the Earth from remote space. These

Fig. 5.8 Intensity vs. geomagnetic latitude for different elevations (Reprinted with permission from A.H. Compton, Phys. Rev. **43**, 387 (1933). Copyright 1933 by the American Physical Society)

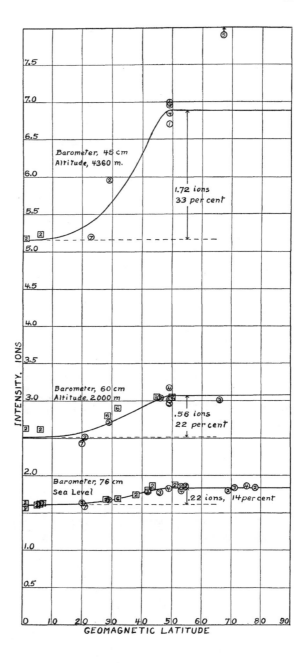

calculations are shown as smooth curves drawn in Figs. 5.8 and 5.9, and Compton was the first to point out that the effects correlated better with geomagnetic latitude rather than with the geographic one.

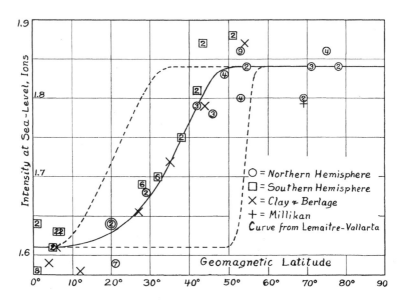

Fig. 5.9 Intensity vs. geomagnetic latitude at sea level, including Data of Clay and Millikan. (Reprinted with permission from A.H. Compton, Phys. Rev. **43**, 387 (1933). Copyright 1933 by the American Physical Society)

At high altitudes the effect is more evident, but the zone where the intensity varies with latitude is the same as at sea level. Compton's group continued the measurements [43] and later A.H. Compton and R.N. Turner [50], in a series of trips between Vancouver, British Columbia and Sydney, Australia, performed very accurate measurements at sea level. Other expeditions were made the following years but the effect was already ascertained.

Millikan continued to take measurements, maybe hoping to demonstrate the absence of the effect, but eventually got convinced it really existed.

A peculiar tract of his character, in my opinion, was that, even if he did not believe in some effect or phenomenon, if he was convinced that it deserved attention, he would undertake its study with great diligence. And so it happened that, even if still in February 1933 he was excluding any latitude effect could exist [51], somehow after the dispute with Compton, he, with his pupil V.H. Neher, made an extensive investigation on the intensity of cosmic rays on the earth's surface at sea level published in 1936 [52] with the aim "*to attain as high an order of accuracy in comparing sea-level ionization all over the world as had been attained in 1930 in comparing the mean cosmic-ray intensity at Pasadena with that at Churchill, Manitoba*". They sent their instrument to many parts of the globe on 12 different voyages, as shown in Fig. 5.10. Their results are shown in the successive Fig. 5.11, which shows the intensity variation as a function of geomagnetic latitude.

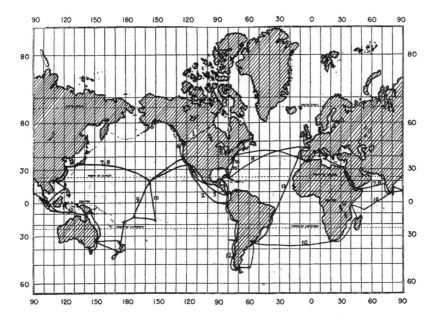

Fig. 5.10 The result of 12 different trips which included most of the chief commercial routes on the oceans at the time (Reprinted with permission from R.A. Millikan and H.V. Neher, Phys. Rev. **50**, 15 (1936). Copyright 1936 by the American Physical Society)

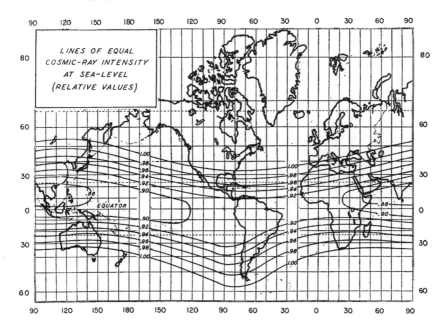

Fig. 5.11 The results obtained by Millikan and Neher [52] (Reprinted with permission from R.A. Millikan and H.V. Neher, Phys. Rev. **50**, 15 (1936). Copyright 1936 by the American Physical Society)

In September 1936, Millikan [53] presented a detailed report on high altitudes measurements. All measurements taken until then—he noted—suffered from the limitation that the electroscopes used, if designed to discharge slowly at low altitudes, discharged suddenly at high altitudes because of the great increase of ionization. If they were designed to give good measurements at high altitudes they were not suitable at sea level and, because they could not be recharged during the measurements, the measuring range was limited.

Therefore, Neher had designed an electroscope which charged up automatically from a battery, say every 5 min during a 3- or 4-h flight, so that the difficulty was eliminated. Successively Neher and S.K. Haynes at Norman Bridge Laboratory succeeded in making condensers of very high capacity that could be used to recharge the electroscope and weighted much less than the battery. These condensers were tried out in August, 1935, in flights made by Millikan and Neher in collaboration with Colonel Prosser and the staff of the Signal Corps of the US Army at Fort Sam Houston, San Antonio, Texas, but because of some imperfections no records higher than 45.000 ft. (13.5 km) were then obtained.

By July 1936, however, the condensers were perfected and the instrumentation could make measurements up to 3 h. On July 6th, 7th, and 8th, 1936 five different Neher self-recording electroscopes, which were automatically charged up every 4 min during the flight, were sent up into the stratosphere. The total weight of each instrument with photographic film, barometer, thermometer, and electroscope was 1,200 g. With the parachute, the steel-wire basket for breaking the fall, the insulating covering for keeping the temperature nearly constant, and the supporting tape and cord, without counting the balloons themselves, the weight raised to 1,400 g. A string of five balloons of 1-m diameter, arranged in tandem 20 ft. (6 m) apart, were inflated as to provide a lift which assured a rate of ascent from 150 to 200 m min^{-1}. The whole string, with parachute and basket attached, was about 130 ft. (39 m) long (see Fig. 5.12). As this convoy approached the top of its flight, the bursting of one balloon slowed down the rate of rise, the burst of a second balloon enabled the instrument to remain at nearly the same level for a long time or until a third balloon broke. There was thus an automatic "levelling off" of the altitude that was ideal for measurements.

A timer was incorporated with a special contact to detach the parachute at the end of three and a half hours, and the instrumentation could be brought safely to the Earth. The parachute was of red silk so as to attract the attention of farmers or passers by, and an envelope offering a reward for the return of the instrument was attached.

Three of the five instruments came back within 2 days of the flight, two having been found within 12 h of their fall. All of these had good records. A fourth instrument was found some time later. It laid for 2 months underneath a tree in which the parachute got caught, and was recovered by a horseman. It fell within 30 miles from San Antonio, while the other three fell some 80 miles distant from the starting point. The most interesting result of these flights is presented in Fig. 5.13, which shows that the ionization increases from the sea level value, reaches a maximum, and then decreases.

Fig. 5.12 Launching electroscopes with balloons (Reprinted with permission from R.A. Millikan, H.V. Neher, S.K. Haynes, Phys. Rev. **50**, 992 (1936). Copyright 1936 by the American Physical Society)

The measurements continued also at different altitudes and the following year Millikan [56] published new measurements taken at different latitudes all showing the presence of a maximum of about $360\,\text{ion}\,\text{cm}^{-3}\,\text{s}^{-1}$ at somehow less than half a meter of water at 60° latitude which decreased to nearly $100\,\text{ion}\,\text{cm}^{-3}\,\text{s}^{-1}$ at 1 m of water at Madras at 3° latitude. Millikan with his collaborators [57] continued to measure the latitude effect up to high altitudes (2 cm of Hg of pressure), confirming that the effect increases considerably with height.

In 1937, Compton and Turner [50] found evidence of a small diminution of the cosmic ray intensity at sea level by increasing the atmospheric temperature. Because

Fig. 5.13 Ionization as a function of height given in meters of water equivalent, below the top of the atmosphere (Reprinted with permission from I.S. Bowen, R.A. Millikan, H.V. Neher, Phys. Rev. **52**, 80 (1937). Copyright 1937 by the American Physical Society)

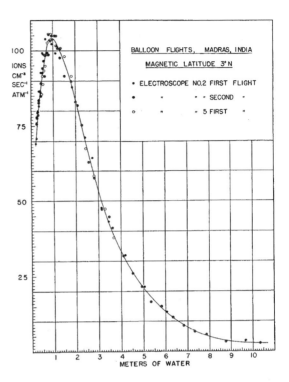

the mean atmospheric temperature is greater at the equator than at the poles, this variation of temperature superposed on the latitude effect. Precise measurements showed that about one-third of the latitude effect found at sea level is in fact due to the temperature variation and the Earth's magnetic field is responsible only for two-thirds of the total effect (see Fig. 5.14).

Figure 5.14 [50] shows the effect at sea level emphasizing the part due to temperature and the one due to the true geomagnetic effect. From the data of Fig. 5.14 and the theory of the geomagnetic effect by Lemaitre and Vallarta, one may estimate that only 1 % of the total energy of the hard component reaching the sea level is carried by primaries of energy between 2 and 7 GeV, about 7 % by primary of energy between 7 and 15 GeV, and about 92 % by primaries of energy larger than 15 GeV.

Millikan [57] now who had accepted the existence of a latitude effect wanted to turn the results in favour of his atom annihilation hypothesis and his old idea of bands of energy of the primary rays. In the paper he presented results that in his opinion showed that the intensity of incoming radiation had a step behaviour as a function of latitude which resulted from entrance of electrons coming up from the annihilation of silicon (which produced particles of 13.2 GeV energy), oxygen (7.5 GeV), nitrogen (6.6 GeV), carbon (5.6 GeV), and helium atoms (1.9 GeV). The final (wrong) result was: "…*namely, that from the standpoint of the atom annihilation hypothesis as to the origin of cosmic rays the incoming cosmic-ray*

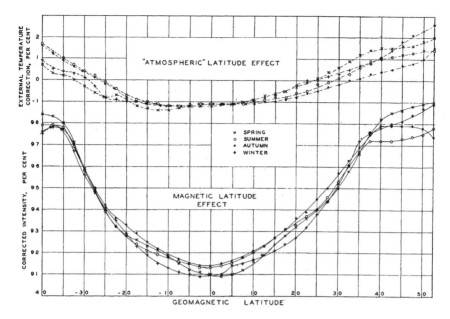

Fig. 5.14 The upper curve shows the effect due to the temperature and the lower one the true latitude effect. Sum of these two effects gives the observed total effect of Fig. 5.9 (Reprinted with permission from A.H. Compton and R.N. Turner, Phys. Rev. **52**, 799 (1937). Copyright 1937 by the American Physical Society)

charged particles must be electrons rather than protons or any other particles essentially more massive than electrons".

Poor Millikan missed the target twice using an inconsistent hypothesis of annihilation and assuming energy was given to the wrong particles (electrons).

In 1945, S. Kusawa [59] showed that within the experimental error, the points of the Millikan curves could be fitted with an inverse cube spectral energy distribution down to about 6 GeV, at which point the Sun's magnetic field causes a decrease in the spectral intensity. Therefore no need of steps in the energy distribution.

5.5 The East–West Effect

In 1930, Bruno Rossi, in Berlin, heard about Stoermer's theory of the motion of charged particles in the Earth's magnetic field and made immediately a simple application of it. In a letter [60] sent to Physical Review on 3 July 1930, he predicted that an asymmetry should exist in the direction of arrival of cosmic rays depending on the sign of their electric charge. The effect is more pronounced near the equator and is called the *east–west effect*.

Rossi himself, in his book on cosmic rays [61], explains the reasons for his interest, by noting that the effect provided a surer test of the magnetic field influence on cosmic particles than the latitude effect, because it did not involve a comparison of cosmic ray intensities at different locations, but it only implied making measurements towards east and west in a same place and with the same instrument pointed once in one direction and then in the other direction. Variations in the conditions of the atmosphere from one region of the Earth to another, which could simulate or mask a latitude effect, could not produce an east–west asymmetry. The second reason was that if cosmic rays were electrically charged, the effect would also reveal the sign of the charge.

A first attempt, made in the fall of 1930, near Florence in the laboratory of the Physics Institute of Arcetri, where he was an assistant, gave inconsistent result [62,63]. This negative result was later justified [26].

To measure the effect, Rossi studied the coincidences of two or more Geiger–Müller counters arranged with their centres on a straight line. Obviously these could take place only for particles coming from directions near the axis of the system which, if inclined with respect to the vertical, could indicate their provenance. This configuration became known as *"cosmic ray telescope"*. In the experiment, the telescope was pointed towards east and then towards west and a comparison was made between the number of counts found in the two inclinations.

Rossi was determined to demonstrate the existence of the effect and started to plan an expedition in Eritrea, an Italian colony in Africa. However, in 1932, he was appointed Professor in Padua, where he was busy to create the Physics Institute, and the expedition had to be postponed until the fall of 1933. The lower geomagnetic latitude and the higher elevation of the selected place, near Asmara, Eritrea, provided better conditions for the measurements that were performed in collaboration with Sergio Benedetti, on October and November 1933, and this time showed the presence of the east–west effect [65,66].

In the same expedition, while testing the instrumentation, Rossi and his collaborators discovered also the presence of extensive showers of particles produced by interactions of cosmic rays in the atmosphere, a phenomenon subsequently studied by Pierre Auger whose name became associated with this discovery.

The delay in making the measurements proved fatal because meanwhile, a few months earlier, two separate experiments in Mexico, one by T.H. Johnson [67] and the other by L. Alvarez and A.H. Compton [68] were able to detect the effect before Rossi.

Johnson and Street [69] had previously made experiments on mount Washington, geomagnetic latitude 55° N, elevation 1,920 m, an altitude corresponding to an atmospheric thickness of 8 m water equivalent, with a disposition of three counters in coincidence, that allowed a better precision and angular resolution than that of Rossi, and found that the east–west distribution was practically symmetric, even if, at a closer examination, some effect was there. On April 1933, Johnson— using calculations by Lemaitre and Vallarta [15] who, on the basis of the intensity measurements of Compton, predicted that an asymmetry should be detectable in geomagnetic latitudes between the equator and 34° with the better situation being between 20° and 30°—carried out measurements in Mexico City on the roof of the Hotel Genève and this time found an asymmetry ([67] see also [70]).

Also Luis Alvarez and Compton were interested to the effect and grounding on the same paper by Lemaitre and Vallarta planned a campaign of measurements in Mexico City. The two groups were therefore together in the same town and their papers were published one after the other in the same issue of Physical Review.

The results showed that the west intensity is greater than that of the east at angles between 30° and 65° from the zenith reaching a maximum at about 45°. In the equatorial region, the excess, at sea level, amounts to about 15 %. At higher elevations the effect is larger. The asymmetry decreased for larger latitudes and at 50° N it was only of the order of 2 or 3 %.

None of the two papers was referencing the work of Bruno Rossi of 1930, quoting only the 1933 paper by Lemaitre and Vallarta.

The asymmetry confirmed the corpuscular nature of the primary radiation and showed that the primaries that produced the effects at the sea level were mostly positively charged.

Independently from the sign of the incident particles no east–west effect may be detected at the polar calotte even if the particles had all the same sign. The Earth's magnetic field acts on the charged particles at the interior of the polar calotte as much as to the exterior, but because the particles are uniformly distributed as many are deviated in one direction as in the other.

The effect is maximum at the magnetic equator. Johnson performed also measurements at Cerro de Pasco (Peru, magnetic equator, elevation 4,300 m) [71].

5.6 The Longitude Effect

A result obtained by Millikan with his electroscopes, sent on a number of voyages from Los Angeles to Mollendo and back and also around the world on ships of a Dollar line, was that in going from the north temperate zone into the equatorial belt the decrease in the sea level cosmic ray intensity was lower between San Francisco and Mollendo, while it was greater when the passage from the temperate zone into the equatorial belt takes place on the other side of the world, that is, in the region of Singapore and Batavia.

In August 1933, Millikan asked the captains of some ships to place in their room a self-recording electroscope which was carried on voyages around the world for several months. The reading of the data was completed in January 1934 and the measurements showed that while the equatorial decrease in going from Los Angeles to Mollendo was only 8 %, going to Singapore via Kobe and Shanghai it was 12 %. The estimated error was 1 %. The effect may be seen clearly in Fig. 5.15 [54, 72] and was present both with screened and unscreened electroscopes.

Millikan [73] reported that also Clay had observed a similar qualitative difference during two voyages which he made at essentially the same time as those made by Millikan, between Amsterdam and Batavia: one around the Cape of Good Hope and the other one through the Suez Canal [74, 75].

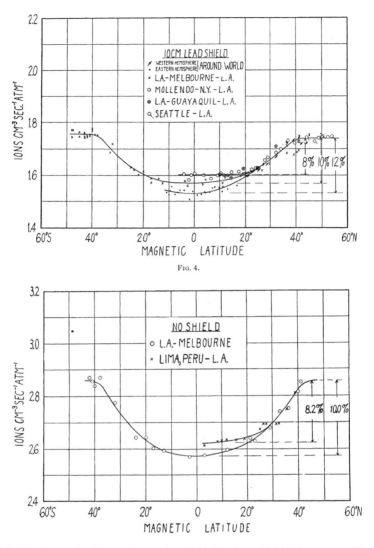

Fig. 5.15 The longitude effect as showed by shielded and not shielded instruments (Reprinted with permission from R.A. Millikan and H.V. Neher, Phys. Rev. **47**, 205 (1935). Copyright 1935 by the American Physical Society)

These results showed that there is also a variation of the cosmic radiation intensity with longitude at a fixed altitude and geomagnetic latitude. The effect was named *longitude effect*. It demonstrated that the Earth's magnetic field, even at distances of thousands of kilometres out in space is essentially dissymmetrical with respect to any line passing through the centre of the Earth. Later, when satellites came into the space, it was possible to have more information.

5.7 And Eventually the Sun Steps in the Stage

Before the Second World War, during the 1930s, the intensity of cosmic rays was measured on a world scale from several groups, and Hess in Austria studied the temporal variations associated with the solar activity cycle on his observatory on mount Hafelaker from 1931 to 1937. The world network, mentioned in Chap. 2, started also to study variations on short term of the global ionization simultaneously in the north and south hemisphere correlating with Sun disturbances [76].

In America, Scott Ellsworth Forbush (1904–1984) began a detailed study of temporal variations of the cosmic ray intensity.

Forbush was born on a farm near Hudson, Ohio. After majoring in physics in 1925, he entered Ohio State University to study physics but soon decided that observational geophysics was much more appealing than pure physics and sought employment in that field. After a year's employment by the National Bureau of Standards in Washington, in September 1927, he joined the staff of the Department of Terrestrial Magnetism of the Carnegie Institution of Washington. His first job was as an observer at the magnetic observatory at Huancayo, Peru, in the Andes about 250 km east of Lima. Two years later he joined the staff of the *Carnegie ship*, a famous vessel built so to be nonmagnetic to allow a worldwide survey of the geomagnetic field without influencing it. On 29 November 1929 an explosion and consequent fire destroyed the ship while she was anchored about a mile offshore in the harbour of Apia, Western Samoa. Forbush was on board at the time, but by a quirk of fate, he escaped unscathed, because he had decided to take a daytime nap rather than return to work in the ship's photographic darkroom in a compartment near the explosion. On his return to Washington, in 1931 he took graduate courses in physics and mathematics at Johns Hopkins University. Eventually in 1932, in response to a recommendation by Millikan and Compton, the Carnegie Institution sponsored the development of a network of detectors for the continuous recording of cosmic ray intensity. The measurements were made with a chamber especially designed by Compton of which Forbush had studied the characteristics and was aware of all of its secrets. The first detector of this kind was operated at Cheltenham, Maryland. The second and third meters of the network were put into operation in June 1936 in Huancayo and Christchurch, New Zealand. Others were added later. Forbush began a patient and meticulous study of the recordings obtained by the network, utilizing also sophisticated statistical methods for the treatment of data. In 1937 he published a short paper [77] *"On the Effects in Cosmic-Ray Intensity Observed During a Recent Magnetic Storm"* in which he reported changes in cosmic ray intensity observed simultaneously at two stations during the magnetic storm of 25 April to 30 April 1937. Figure 5.16[7] gives, for two stations [Cheltenham (Maryland) and Huancayo (Peru)], in Greenwich mean time, the departures, in percent of the absolute value, of each bi-hourly mean of cosmic ray intensity,

---

[7]Fig. 5.1 of [77].

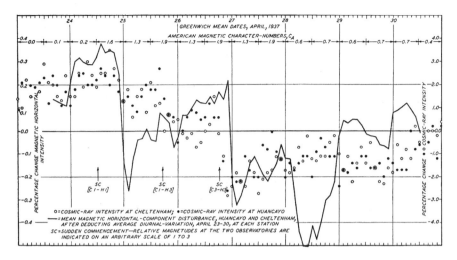

Fig. 5.16 Bi-hourly departures expressed in percentage of absolute values for cosmic ray intensity and for disturbance of horizontal magnetic component April 23–30, 1937, Huancayo and Cheltenham magnetic observatories. (Reprinted with permission from S.E. Forbush, Phys. Rev. **51**, 1108 (1937). Copyright 1937 by the American Physical Society)

showing a striking similarity of the simultaneous changes at the two stations. The solid curve indicates the departures in percent of absolute value of the bi-hourly means of the horizontal component of the Earth's field recorded at the two stations. The major changes in cosmic ray intensity were by Forbush [78] found to be worldwide and correlated with the magnetic changes. Such impulsive decreases of cosmic ray intensity in a few hours at the time of the sudden commencements of the magnetic storms and the subsequent slow (typically a few days to a few weeks) recovery of intensity thereafter received the designation of "*Forbush decreases*".

In successive papers [78–80], in 1938 and 1939 Forbush used cosmic ray and magnetic storm data from Cheltenham, Huancayo, Hafelaker, Christchurch, and Teoloyucan (Mexico) to establish the occurrence of worldwide decreases in cosmic ray intensity in association with magnetic storms, also considering results obtained by Hess [81, 82].

The magnetic storm of 1 March 1942 produced a great decrease that attained the value of about 11 %. After a first rapid recovery for about one-half the initial drop that coincided approximately with the end of the storm, the recovery continued very slowly [83].

During the war, Forbush was busy with the support activity to the war, but immediately after that, studying again the records of the three high- and mid-latitude stations he found large impulsive increases in the cosmic ray intensity on 28 February 1942 and 25 July 1946, each following an exceptionally large solar *flare.*[8] These large increases of cosmic ray intensity started indeed to be observed in association with solar flares from 1942 [84, 85]. The brief (hours) increases were precursors to large Forbush decreases at all stations [86]. He identified the solar origin of the impulsive emission of energetic particles that he surmised had energies up to at least 3 GeV, but less than the geomagnetic cutoff at Huancayo, about 15 GeV. The study of the sporadic solar emission of energetic particles subsequently became a major field of research by spacecrafts.

Eventually in the paper *"Worldwide Cosmic-Ray Variations"* published in 1954 [87], Forbush demonstrated that the intensity of galactic cosmic rays varied with the 11 years solar cycle and was negatively correlated with the number of sunspots, that is the intensity of cosmic rays was greater when the solar activity was minimal.

To these findings added also the discovery of diurnal variations of cosmic ray intensity and their periodicity of 22 years. Of course the Sun is responsible for these variations and the geomagnetic effects cannot be explained plainly by assuming that the Earth's field is simply that of a dipole. Discrepancies between the measured intensity of cosmic rays and the one expected by the simple dipole representation of the geomagnetic field had already been noticed by Johnson [27] and were later demonstrated by Simpson [88].

Twenty years passed before the physical causes of Forbush decreases were convincingly identified and we will discuss them in Chap. 12.

References

1. M.W. McElhinny, *Paleomagnetism and Plate Tectonics* (Cambridge University Press, Cambridge, 1973)
2. P.H. Roberts, G.A. Glatzmaier, Rev. Mod. Phys. **72**, 1081 (2000)
3. G.A. Glatzmaier, P.H. Roberts, Phys. Earth Planet. Inter. **91**, 63 (1995)
4. G.A. Glatzmaier, P.H. Roberts, Nature **377**, 203 (1995)
5. R. Ladbury, Phys. Today **49**, 17 (1996)
6. K. Birkeland, Arch. Sci. Phys. **1**, 497 (1896)
7. K. Birkeland, Vid. selsk. Skr. **1** (1901)
8. C. Stoermer, "Sur les trajectoires des corpuscules électrisés dans l'espace sous l'action du magnetisme terrestre". Arch. Sci. Phys. Genève **24**, 5, 113, 221, 317 (1907)
9. C. Stoermer, "Sur les trajectoires des corpuscules électrisés dans l'espace sous l'action du magnetisme terrestre". Arch. Sci. Phys. Genève **32**, 117 (1911) (190, 277, 415, 501)
10. C. Stoermer, "Ein Fundamentalproblem der Bewegung einer elektrisch geladenen Korpuskel im kosmischen Raume". Z. Astrophys. **3**, 31 (1931) (227)
11. C. Stoermer, "Ein Fundamentalproblem der Bewegung einer elektrisch geladenen Korpuskel im kosmischen Raume". Z. Astrophys. **4**, 290 (1932)

---

[8]A solar flare is a sudden liberation of energy from a localized region on Sun, in the form of electromagnetic radiation and usually also of energetic particles. Flares occur in active Sun regions, especially at the boundary zones between solar spots of opposite magnetic polarity; their frequency changes during the solar cycle in a similar way as the spots.

12. C. Stoermer, "Ein Fundamentalproblem der Bewegung einer elektrisch geladenen Korpuskel im kosmischen Raume". Z. Astrophys. **6**, 333 (1933)
13. C. Stoermer, *The Polar Aurora* (Oxford University Press, UK, 1955)
14. P.S. Epstein, Phys. Rev. **53**, 862 (1938)
15. G. Lemaitre, M.S. Vallarta, Phys. Rev. **43**, 87 (1933)
16. G. Lemaitre, M.S. Vallarta, Phys. Rev. **49**, 719 (1936)
17. G. Lemaitre, M.S. Vallarta, Phys. Rev. **50**, 493 (1936)
18. L. Myssowsky, L. Tuwim, Z. Phys. **39**, 146 (1926)
19. E. Steinke, Z. Phys. **42**, 570 (1927)
20. C. Stoermer, Astrophys. Norveg. **1**, 115 (1934)
21. C. Stoermer, Astrophys. Norveg. **2**, 193 (1936)
22. C. Stoermer, Z. Astro. Phys. **1**, 237 (1934)
23. P. Epstein, Proc. Nat. Acad. Sci. **16**, 658 (1930)
24. M.S. Vallarta, *On the Allowed Cone of Cosmic Radiation* (Toronto University Press, Toronto, Canada, 1938)
25. M.S. Vallarta, *Handbuch der Physik*, vol. XLVI/1 (Springer, Berlin, 1961), p. 88
26. E. Fermi, B. Rossi, Acc. Linc. Att. **17**, 346 (1933)
27. T.H. Johnson, Rev. Mod. Phys. **10**, 193 (1938)
28. W.F.G. Swann, Phys. Rev. **44**, 224 (1933)
29. J. Clay, Proc. Ned. Akad.v.Wet. **30**, 1115 (1927)
30. J. Clay, Proc. Ned. Akad.v.Wet. **31**, 1091 (1928)
31. J. Clay, Proc. Ned. Akad.v.Wet. **33**, 711 (1930)
32. R.A. Millikan, G.H. Cameron, Phys. Rev. **31**, 163 (1928)
33. R.A. Millikan, G.H. Cameron, Nature **121** (19 Suppl.) (1928)
34. R.A. Millikan, Rev. Mod. Phys. **21**, 1 (1949) (at p. 4)
35. R.A. Millikan, Phys. Rev. **36**, 1595 (1930)
36. W. Bothe, W. Kolhoerster, Ber. Ber. 450 (1930)
37. K. Grant, Nature **127**, 924 (1931)
38. A. Corlin, Lund. Medd. 121 (1930)
39. G. Hoffmann, Phys. Z. **32**, 633 (1932)
40. P. Galison, Centaurus **26**, 262 (1983)
41. A.H. Compton, Phys. Rev. **39**, 190 (1932)
42. A.H. Compton, Phys. Rev. **43**, 387 (1933)
43. A.H. Compton, J.M. Benade, P.G. Ledig, Phys. Rev. **45**, 294 (1934)
44. A.H. Compton, Phys. Rev. **41**, 111 (1932)
45. A.H. Compton, Phys. Rev. **41**, 681 (1932)
46. A.H. Compton, Phys. Rev. **42**, 904 (1932)
47. R.D. Bennett, J.L. Dunham, E.H. Bramhall, P.K. Allen, Phys. Rev. **42**, 446 (1932)
48. J. Clay, H.P. Berlage, Naturwissenschaften **37**, 687 (1932)
49. G. Lemaitre, M.S. Vallarta, Phys. Rev. **42**, 914 (1932)
50. A.H. Compton, R.N. Turner, Phys. Rev. **52**, 799 (1937)
51. R.A. Millikan, Phys. Rev. **43**, 661 (1933)
52. R.A. Millikan, H.V. Neher, Phys. Rev. **50**, 15 (1936)
53. R.A. Millikan, H.V. Neher, S.K. Haynes, Phys. Rev. **50**, 992 (1936)
54. R.A. Millikan, H.V. Neher, Phys. Rev. **47**, 205 (1935)
55. I.S. Bowen, R.A. Millikan, H.V. Neher, Phys. Rev. **52**, 80 (1937)
56. I.S. Bowen, R.A. Millikan, H.V. Neher, Phys. Rev. **53**, 855 (1938)
57. R.A. Millikan, H.V. Neher, W.H. Pickering, Phys. Rev. **63**, 234 (1943)
58. S.E. Forbush, Phys. Rev. **51**, 1108 (1937)
59. S. Kusawa, Phys. Rev. **67**, 50 (1945)
60. B. Rossi, Phys. Rev. **36**, 606 (1930)
61. B. Rossi, *Cosmic Rays* (McGraw Hill, New York, 1964)
62. B. Rossi, Rend. Lincei **13**, 47 (1931)
63. B. Rossi, Nuovo Cimento **8**, 85 (1931)

64. E. Fermi, B. Rossi, Rend. Lincei **17**, 346 (1933)
65. B. Rossi, Phys. Rev. **45**, 212 (1934)
66. B. Rossi, Ric. Sci. **5**, 569 (1934)
67. T.H. Johnson, Phys. Rev. **43**, 834 (1933)
68. L. Alvarez, A.H. Compton, Phys. Rev. **43**, 835 (1933)
69. T.H. Johnson, J.C. Street, Phys. Rev. **43**, 381 (1933)
70. T.H. Johnson, J. Franklin Inst. **214**, 689 (1932)
71. T.H. Johnson, Phys. Rev. **48**, 290 (1935)
72. R.A. Millikan, H.V. Neher, Proc. Nat. Acad. Sci. **21**, 313 (1935)
73. R.A. Millikan, H.V. Neher, Phys. Rev. **21**, 313 (1935)
74. J. Clay, P.M. van Alphen, C.G. 'T. Hooft, Physica **1**, 829 (1934)
75. J. Clay, E.M. Bruns, J.T.J. Wiersma, Physica **3**, 746 (1936)
76. V.F. Hess, J. Eugster, *Cosmic Radiation and its Biological Effects* (Fordham University Press, New York, 1949)
77. S.E. Forbush, Phys. Rev. **51**, 1108 (1937)
78. S.E. Forbush, Phys. Rev. **54**, 975 (1938)
79. S.E. Forbush, Terr. Magn. **43**, 203 (1938)
80. S.E. Forbush, Rev. Mod. Phys. **11**, 168 (1939)
81. V.F. Hess, A. Demmelmair, Nature **140**, 316 (1937)
82. V.F. Hess, R. Steinmaurer, A. Demmelmair, Nature **141**, 686 (1938)
83. I. Lange, S.E. Forbush, Terr. Magn. **47**, 331 (1942)
84. S.E. Forbush, Phys. Rev. **70**, 771 (1946)
85. A. Ehmert, Z. Naturforsch. **3a**, 264 (1948)
86. S.E. Forbush, J. Geophys. Res. **63**, 651 (1958)
87. S.E. Forbush, J. Geophys. Res. **59**, 525 (1954)
88. J.A. Simpson, K.B. Frenton, J. Katzman, D.C. Rose, Phys. Rev. **102**, 1648 (1956)
89. K. Rypdal, T.B. Brundtland, J. Phys. **III** (Suppl. 7), C4-113 (1997)

Chapter 6
The Positive Electron (Positron)

6.1 1932, Annus Mirabilis for Nuclear Physics: A Digression Useful for a Better Understanding of What Follows

The year 1932 is usually considered the *annus mirabilis* of nuclear physics. In that year, in Cambridge, the *neutron* was discovered by J. Chadwick [1] by bombarding beryllium and boron with the alpha particles from polonium; in the USA, the *positive electron* was discovered by Anderson in cosmic rays; and the heavy hydrogen (the *deuterium* or hydrogen of atomic weight 2: whose nucleus is called *deuteron*) was found by H. Urey. Also in the USA, J. Cockcroft and E. Walton completed their *particle accelerator*; E.O. Lawrence and S.M. Livingstone built the *cyclotron*. In Italy, E. Fermi started to elaborate his theory on beta rays that was published the following year.

The discovery of the *neutron*[1] is a long story of which we give only a brief summary. Rutherford had always believed that a neutral particle could exist consisting of an electron fallen on an hydrogen nucleus, annihilating the electric charge and creating a particle of a mass about the one of the proton, but deprived of a charge. In his work, Chadwick [1] attributed the fatherhood of the name to Rutherford, quoting the Bakerian lecture [6] held at the Royal Society in 1920 on the nuclear constitution of atoms. In it Rutherford suggested that the nucleus of atoms was composed of positive charges and negative electrons. He said:

---

[1]The name neutron was introduced long before its effective discovery. Walter Nernst in the fourth edition of his book Theoretische Chemie, Stuttgart, 1898, introduced it to name electrically neutral molecules in electrolytic conductivity in liquids (see N. Feather [2] and B. Kroger, [3]).The name was then used by W. Sutherland [4] and Stark in Jahrbuch der Radioaktivitaet und Elektronik, 1904. In 1921, W.D. Harkins [5], an American Professor of physical chemistry, named the neutron to represent one proton plus one electron.

M. Bertolotti, *Celestial Messengers: Cosmic Rays*, Astronomers' Universe, DOI 10.1007/978-3-642-28371-0_6, © Springer-Verlag Berlin Heidelberg 2013

"In considering the possible constitution of the elements, it is natural to suppose that they are built up ultimately of hydrogen nuclei and electrons" and farther *"... it seems very likely that one electron can also bind two H nuclei and possibly also one H. In the one case, this entails the possible existence of an atom of mass nearly 2 carrying one charge, which is to be regarded as an isotope of hydrogen. In the other case, it involves the idea of the possible existence of an atom of mass 1 which has zero nucleus charge .... Under some conditions, however, it may be possible for an electron to combine much more closely with the H nucleus, forming a kind of neutral doublet. Such an atom would have very novel properties. Its external field would be practically zero, except very close to the nucleus, and in consequence it should be able to move freely through matter..."*. However he did never use the word neutron.

His proposal aroused a race among his collaborators for the discovery of the hypothetic particle. In 1921, J.L. Glasson [7] published a paper describing the unsuccessful results. He wrote:

"Such a particle, to which the name neutron has been given by Prof. Rutherford would have novel and important properties".

In none of Rutherford's papers from 1920 to 1921 the word neutron appears. The biography of Rutherford written by Eve [8] contains a long description of the search for the neutron and describes Rutherford's idea that the particle should exist. There are speculations [9] that Rutherford had proposed the name in the laboratory and that the world has been liked, but the first appearance of the name in the literature is in 1921. Meanwhile, bombarding nitrogen with alpha particles from polonium, Rutherford found that hydrogen nuclei were generated; the discovery of the phenomenon of artificial transmutations [10] which started to be studied in many laboratories.

Later, in 1928, W. Bothe and his student, H. Becker, were studying, in Berlin-Charlottenburg, the gamma rays obtained by bombarding several nuclei of light elements (Li, Be, B, F, etc.) with the alpha rays from polonium. After a series of experiments they found that the beryllium nuclei emitted an extremely penetrating radiation, more energetic than the alpha particles used as bullets, that they interpreted [11] as gamma rays, interpreting their results as a nuclear disintegration in agreement with the disintegrations found by Rutherford.

The results were discussed at the International Conference on Nuclear Physics in Rome, on October 1931, and stimulated I. Curie and F. Joliot to continue the study of this radiation in their laboratory in Paris. First Irene Curie [12] alone and then with her husband Frederic Joliot [13] repeated the experiment, and on 18 January 1932 reported a surprising observation: the produced radiation was able to eject protons of high energy (about 5 MeV) from a paraffin wax layer on which it was made to fall. On 22 February, they published this result.

Also H.C. Webster [14] found that the radiation emitted by boron bombarded by alpha particles from polonium consisted in part of a penetrating radiation that he too interpreted as gamma ray.

However the interpretation of these results in terms of hard gamma rays encountered a number of difficulties as no process was known able to transfer such a high energy from gamma photons to heavy particles. These difficulties became

Fig. 6.1 J. Chadwick

insurmountable when Chadwick[2] (Fig. 6.1) at Cavendish Laboratory demonstrated experimentally that the penetrating radiation was able to transfer energies of the same order of magnitude also to nitrogen nuclei that have a mass 14 times the mass of the proton. Chadwick immediately understood that these results could be explained by assuming that the penetrating radiation was made, at least in part, by a new kind of particle electrically neutral, with a mass close to the one of the proton. He published the result in a letter to Nature [15] on 27 February 1932 under the title "*On the possible existence of the neutron*" followed by the complete work [1]. Three years later James Chadwick, meantime knighted, was awarded the Nobel Prize for physics and in his acceptance speech declared to have been influenced by the Rutherford prediction that such a neutral particle could exists.

The discovery of the neutron had vast and deep consequences for nuclear physics. Rutherford and all physicists of that time believed that the nucleus was composed of electrons and protons. This hypothesis was a natural one because electrons were found to come out from the nucleus in the beta decay. The new particle was requiring a new formulation of the nucleus constitution which should account for the presence in its interior of protons and neutrons.

---

[2]James Chadwick (1891–1974) was an English physicist, a pupil of E. Rutherford. After graduating, in 1913, he went to Berlin to work with Hans Geiger and there he published the first accurate observations of the continuous spectrum of beta radiation emitted in radioactive decays. At the outbreak of war, in August 1914, still in Berlin he was interned in a camp and returned to England only in 1918, joining Rutherford laboratory. In 1935, he was appointed a Professor at Liverpool and since 1948 at Cambridge, where he was the director of the nuclear physics laboratory. He participated as the head of the English mission to the Manhattan project for the realization of the atomic bomb.

On 18 February 1932, Harold Clayton Urey[3] (1893–1981), F.G. Brickwedde (1903–1989) and G.M. Murphy [16] in a paper on Physical Review announced the discovery of the hydrogen isotope of mass 2: deuterium. The mass number of its nucleus, the deuteron, being two and its charge one, it was natural to think of it as composed of a proton and a neutron. This was confirmed experimentally in 1934, when Chadwick and M. Goldhaber (1911–2011) [17] bombarded the deuteron with gamma rays and dissociated it in a proton and a neutron. This experiment allowed to determine the mass of the neutron.

In April, J.D. Cockcroft (1897–1967) and E.T.S. Walton (1903–1995) completed the construction of the second version of their particle accelerator [18] and described its application to the disintegration of some elements by using the protons accelerated to 150,000 V [19]. The same year, bombarding Li with protons they produced two atoms of helium, obtaining the first artificial nuclear disintegration, being awarded the Nobel Prize in physics in 1951. Later that summer E.O. Lawrence (1901–1958) and M.S. Livingstone [20] (1905–1986) described a new type of particle accelerator: the *cyclotron*. It employed a magnetic field to make particles turn on a circumference, whose radius gradually increases at each turn, while being accelerated by a radiofrequency electric field. Lawrence was awarded the Nobel Prize in physics for this in 1939.

After the discovery of the neutron, the idea that the atomic nucleus could be composed by protons and neutrons was considered by many people. On 28 April 1932, Dmitrij Dmitrievic Iwanenko [21] (1904–1994) submitted a short note to Nature, just after Chadwick's note on the neutron [22]. The fundamental contribution was however given by Werner Heisenberg (1901–1976), who offered a clear explanation of the nuclear properties in three papers published in 1932. His first contribution to the theory of the nuclear structure was a fundamental paper submitted to Zeitschrift fur Physik in June 1932 [23]. In this paper he not only supports the basic idea of the nucleus composed of protons and neutrons, but also suggests that the force acting between protons and neutrons is an *exchange force*. Heisenberg wrote a second work in July [24] and a third one in December [25]. He believed that inside the nucleus a proton and a neutron alternatively loose or gain their charge so that a neutron turns into a proton and contemporarily a proton changes into a neutron, thus interchanging their role (because of this the term *exchange force*). Since the three particles, proton, neutron, and electron all have[4] spin $1/2$ the change cannot be realized through the exchange of a normal electron, as this would be inconsistent with the necessity to conserve the spin. Therefore, Heisenberg banned the electron from the scene and, instead of considering proton and neutron as different particles, considered them as different states of the same elementary particle, that later was called *nucleon*. To describe these processes, he introduced the concept of what was later called *isospin*.

---

[3] American chemist, he was awarded in 1934 the Nobel Prize for his discovery.

[4] The spin characterizes the intrinsic angular momentum of a particle. It was originally introduced with regard to the electron.

The previous idea that the nucleus was made by protons and electrons was overthrew by Heisenberg, who expelled the electrons from the nucleus, living open the problem of the *beta disintegration*, but introducing a new nuclear force with short range. This force explained through the charge exchange the interaction between protons and neutrons in the nucleus and gave a reason of its stability. That winter, a new visitor came to Lipsia, Ettore Majorana[5] who remained with Heisenberg until the beginning of summer and published a paper [26] in which he simplified the Heisenberg model in a very elegant way.

The fraction of the irradiation by a radioactive source that was strongly deviated by a magnetic field [27] placed perpendicularly to its travelling direction had been called *beta rays*. They were found to be formed by particles of a negative charge, emitted by the atomic nuclei of the source, soon identified with electrons, similar to the electrons of cathode rays but much more energetic.

Beta rays have a strange characteristic, found by the young Chadwick, then in Germany, that is a continuous distribution of their energy [28]. It was established that for each kind of nucleus undergoing beta decay, the electrons are emitted with an energy that spans from zero and a precise upper limit [29], while the energy of the nucleus decreases always of a quantity equal to this upper value. Initially this discovery did not attract much attention: one believed that electrons coming from different distances from the surface of the sample could have been attenuated differently [30, 31]. It was only 13 after Chadwick's results that the British physicists C.D. Ellis and W.A. Wooster [32, 33] were able to measure the heat liberated by the radioactive decay both inside and outside the parent body. The result was that the average heat energy per particle coincided exactly with the average energy of electrons as estimated from direct observations. This could only means that the electrons emitted were not loosing any appreciable energy within the parent body, but were actually leaving the nuclei with widely varying velocities.

This could not be understood. If a parent nucleus transforms into a new nucleus, emitting in the process a beta ray, due to the energy conservation principle the difference between their rest energy mc^2 must be equal to the energy of the beta ray, which should have accordingly a well defined value. Actually this is not the case and Bohr [34] to explain this—as well as Compton scattering, as reported in Chap. 4—assumed that energy conservation does not hold on such a small scale as the one of the nucleus.

---

[5]Ettore Majorana was born on 5 August 1906 in Catania, Italy, a nephew of the physicist Quirino Majorana (1871–1957). He studied in Roma first enrolling to engineering and turning to physics in 1928 to study under E. Fermi. In 1932 he was appointed an assistant at the University of Roma. Edoardo Amaldi (in Varenna Intern. School, 1966) writes that Majorana, seeing the works of Joliot and Curie immediately realized that the results had to be interpreted as the "recoil of protons produced by a heavy neutral particle" and that independently from Iwanenko and Heisenberg he had the idea of the proton–neutron composition of the nuclei. In November 1937 he accepted the chair of theoretical physics in Naples. He disappeared in mysterious circumstances in 1938 while travelling on a ferry from Naples to Palermo, Sicily.

Fig. 6.2 Pauli

Fig. 6.2 Pauli

W. Pauli[6] (Fig. 6.2) had a different idea. In 1930, before the discovery of the neutron, he believed that in the nucleus particles existed he named "neutrons" which were electrically neutral and had a practically null mass, and were therefore very difficult to be detected, and that when a nucleus suffers a beta disintegration, an electron is expelled together with this "neutron". In this case the sum of the energy of the two particles is established by the energy conservation law, but the energy of the electron or that one of the neutral particle may be distributed with continuity between zero and a maximum.

After discussing with Bohr, who on the contrary was for a non-conservation of energy in nuclear processes [35], Pauli ([36], see also [37]), on 4 December 1930 wrote from Zurich a letter to Hans Geiger and Lise Meitner (1878–1968) who were attending a Physics congress in Tubingen:

"Dear Radioactive Ladies and Gentlemen...
I have, in connection with... the continuous beta spectrum, hit upon a desperate remedy... the possibility that there might exist in the nuclei electrically neutral particles, which I shall call neutrons, which have spin 1/2, obey the exclusion principle and moreover

---

[6]Wolfang Pauli was born on 25 April 1900 at Vienna. He moved to the University of Munich to study theoretical physics under the guidance of Arnold Sommerfeld (1868–1951) and after his degree, for 1 year he was an assistant of Born in Gottingen (1921–1922). Having met Bohr, on his invitation, he went to Copenhagen (1922–1923) and then to Hamburg, where he remained until 1928 when he accepted a chair at the Polytechnic in Zurich. Except for the war years, which he passed in the United States at the Institute for Advanced Studies in Princeton, he remained in Zurich until his death in 1958. He began his scientific carrier by writing, at 21, a book on Einstein's relativity theory that still to day is one of the best on the subject. He made very important contribution to quantum theory. He is considered one of the most important theoretical physicist of the twentieth century. He was awarded the Nobel Prize for physics in 1945.

differ from the light quanta in not travelling with the velocity of light. The mass of the neutron would have to be of the same order of the electronic mass. The continuous beta-spectrum would then be understandable on the assumption that in beta decay, along with the electron a neutron is emitted as well, in such a way that the sum of the energies of neutron and electron is constant. Unfortunately I cannot appear personally in Tubingen, since on account of a dance which takes place in Zurich on the night of 6 to 7 December I cannot get away from here".

Indeed there was a Ball of the Studenti Italiani, the Italian students in Zurich, which took place on the night of Saturday 6/7 December at Baur au Lac, the most distinguished hotel in downtown Zurich, facing the lake.

Pauli discussed his hypothesis also at several congresses among which was the Physics Conference held in Roma in October 1931, and Enrico Fermi (Fig. 6.3) was strongly interested to the question. In 1932 at the Fifth International Conference on Electricity in Paris, Fermi[7] ([38], see also [39]) mentioned the Pauli "neutrons" explaining the difference with the neutron discovered by Chadwick. He also suggested that the new particle involved in the beta disintegration was a different neutral particle and, to distinguish it from the new particle discovered by Chadwick, proposed the name *neutrino*, that was universally accepted already at the Solvay Conference in October 1933. Immediately after this conference he developed a complete theory of the beta decay in which the neutrino [40,41] plays a fundamental role, which was favourably welcome by Pauli and Heisenberg.

Fermi [42] formulated for the first time a coherent theory of the neutrino and of the other particles known at the time (electron, proton, and neutron). He was espousing the view point that all nuclei consist only of massive particles, protons and neutrons: they are the two states of the same particle. In his theory, the beta particles (electrons) are created at the time of their emission in the same way as a light quantum is created in an atomic transition. Accordingly, the beta radiation becomes analogous to the electromagnetic radiation, with some differences. In particular, while in the atomic transition the corresponding energy is emitted in a single light quantum, in the beta disintegration the available energy is divided in all the possible ways between the electron and a neutrino. The Fermi theory postulates that proton and neutron interact with the combined field of electrons and neutrinos in such a

---

[7]Enrico Fermi was born in Roma on 29 September 1901. He studied physics at the University of Pisa (1918–1922) obtaining his doctorate in 1922. After a year's work as a private tutor in Roma, he received a travelling scholarship from the Italian Government, with the help of which he spent periods of stay at the University of Gottingen (Born 1923–1924) and at the University of Leiden (Ehrenfest 1924). Then he became a lecturer in physics at the University of Florence and, in 1926, Professor of theoretical physics at the University of Roma. He left Italy for the United States in 1938. He first accepted a professorship at Columbia University (1939–1942), then he moved to University of Chicago (1942–1945). He worked on the atomic bomb project. In 1946–1954 he was professor at the Institute of Nuclear Studies. He died in 1954 in Chicago. Fermi did outstanding theoretical and experimental research in many fields of physics: spectroscopy, statistical mechanics, quantum mechanics and quantum field theory, neutron physics, nuclear fission, and cosmic rays. He discovered radioactivity induced by neutrons. He received the Nobel Prize in Physics in 1938.

Fig. 6.3 Enrico Fermi

way that an electron and a neutrino are created and irradiated when a neutron turns to a proton. Conversely a positive electron (a positron) and an antineutrino[8] are radiated when a proton changes in a neutron. Today, in the modern terminology, the two transitions are written as

$$p \rightarrow n + e^+ + \underline{\nu} \quad \text{and} \quad n \rightarrow p + e^- + \nu,$$

where p and n stay for proton and neutron, e^- is the ordinary negative electron, e^+ is a positron (that is a particle with the same mass of the electron and a charge equal in absolute value to the electron charge, but positive, called positive electron or *positron*), and $\underline{\nu}$ and ν represent an antineutrino and a neutrino, respectively.[9]

The theory of *weak interactions* was born, in which the neutrino plays a crucial role.

The free proton seems to be stable and up to now no disintegration of it has been found. On the contrary the free neutron disintegrates with a lifetime τ that was measured in 1948 [43–46]. Later the τ value has been more precisely assessed and found to be 15 min [47].

6.2   The Discovery of the Positive Electron

All the previously discussed advances in experimental nuclear physics were occurring as the result of researches made in the laboratory. The discovery of the *positive electron* was, on the contrary, one of the most exciting results of the cosmic rays

---

[8]Positrons and antiparticles will be introduced later in this chapter; be patient.

[9]We explain later in this chapter this terminology.

research, increasing the number of "elementary" particles then known, giving a proof of the existence of *antiparticles*, and showing how fruitful could be the study of cosmic rays both for quantum mechanics and for the understanding of the interaction between matter and photons.

The credit of the discovery goes to a young PhD student of Millikan, Carl Anderson.

Carl David Anderson was born in New York of Swedish parents on 3 September 1905. He graduated with a BSc in physics and engineering from the California Institute of Technology in 1927, and took his PhD in the same Institute in 1930. For the period 1930–1933, he was research fellow there, subsequently Assistant Professor (1933) and eventually physics Professor (1939). During the war years (1941–1945), he worked for the American defence.

Principally at the request of his son and daughter-in-law, he wrote an autobiography that is an important information source [48]. During 1927–1930, he worked at his PhD thesis [49], studying by means of a Wilson chamber the space distribution of photoelectrons emitted by several gases irradiated with X rays, under the supervision of R.A. Millikan who, as he wrote in his autobiography [49] *"not once during the three years of my graduate thesis work did he visit my laboratory or discuss the work with me"*.

He retired in 1978 and died on 11 January 1991.

While Anderson was working on his thesis, Chung-Yao Chao, in a nearby room, was using an electroscope to measure the absorption and scattering of gamma rays from ThC''. At that time the absorption of high energy gamma rays (ThC'' emits gamma rays of 2.6 MeV) was believed to occur principally through Compton scattering, according to calculations made by two theoreticians Oscar Klein and Yoshio Nishina. Chao results, however, showed clearly that the absorption and scattering were larger than calculated [50].

Anderson remembers [51] that when he was finishing his PhD, on June 1930, he went to Millikan to ask if it was possible to stay one more year at Caltech as a postdoc to continue his research, suggesting to study the interaction of gamma rays from ThC'' with matter, by using a cloud chamber with a magnet to investigate the secondary electrons produced in a thin lead plate inserted in the cloud chamber, continuing in this way the Chao measurements. However Millikan decided that Anderson should continue his research somewhere else and Anderson applied to A.H. Compton at the University of Chicago. A few months later, Millikan changed his mind and called Anderson back to spend one more year at Caltech in order to built a Wilson chamber and measure the energies of the electrons in cosmic rays.

The Skobelzyn results of 1927, that demonstrated the utility of the cloud chamber, had eventually convinced Millikan that the Wilson chamber could be a powerful means of investigation. Millikan hoped that by studying the energy of the corpuscular radiation emitted by the photons of cosmic rays (that he thought were the primary rays) by using the cloud chamber technique, with which Anderson was familiar, he could obtain better data on the photon energies with respect to the values he obtained with his absorption experiments described in Chap. 3. He had in fact realized from the information given by Oppenheimer that the relation between incident photon energy and absorption was still an open question. Possibly Millikan expected that the secondary electrons that Anderson could observe in the experiments with the cloud chamber would exhibit the assumed band structure of the primary cosmic ray photons [52].

Anderson started to build the chamber that Millikan had proposed and in the spring of 1930 they had completed a cloud chamber apparatus, suitably designed for the study of cosmic rays. The chamber of dimensions $17 \times 17 \times 3 \, \mathrm{cm}^3$ was arranged with its long dimension vertical, and incorporated into a powerful electromagnet capable of maintaining a uniform magnetic field of 24,000 gauss strength, able to produce measurable curvatures of electrons with energies of a few billion eV (one billion eV is 1 GeV). The apparatus was mounted on a wheel carriage so to be transportable.

The aim of the experiment was to directly measure the energy spectrum of the secondary electrons produced in the atmosphere and in other materials by the incoming cosmic radiation [52].

In the summer of 1931 the first results were obtained [53]. The direct measurement of the energies of atomic particles was extended from about 15 MeV, the highest energy measured before that time, to 5 GeV. Out of about 1,000 exposures, 34 showed measurable cosmic ray tracks. The great number of useless photographs (966) was the consequence of the fact that at that time the photos were taken at random and only casually a cosmic ray was detected. The notable fact was that of the 34 tracks, 11 were single positive tracks.

In November 1931 Millikan at the Institut Henri Poincarè in Paris and at the Cavendish Laboratory, Cambridge [54], presented and discussed these first 11 photos. He maintained that the positive particles were protons ejected simultaneously by the nucleus together with an electron. His conclusion that all tracks could be interpreted with the assumption that the cosmic rays were photons was not well received by the European experts [55].

In 1931 he had concluded his series of measurements with balloons bearing self-registering electroscopes. The flights had attained an altitude of 16 km. These measurements, Millikan maintained in Paris, showed that the rate of ionization reached a maximum between 9 and 16 km, which *"is precisely what we would expect if the rays penetrating in the atmosphere were gamma rays which necessarily penetrated to a certain depth in the atmosphere before coming to equilibrium with their secondaries"* [54].

There is a maximum of ionization in the upper atmosphere and part of Millikan's conclusions was ineluctable. If particles reach a maximum inside the atmosphere, the primary rays must be producing secondary ones. However, as it was realized only several years later, the primaries were protons and not photons.

Since the only positive particle known at the time was the proton, Millikan interpreted the positive particle tracks detected in their Wilson chamber as protons. Of course the first critics came from Skobeltzyn [56] and also Anderson had several discussions with Millikan about the nature of the positive tracks and the value of their energy.

The paper on these experiments with the cloud chamber, received by Physical Review on 12 April 1932, was a conjunct effort by Millikan and Anderson [53], in which they were discussing the positive particles found in the photos, interpreting the results as a proof of nuclear disintegrations and identifying the positive particles as protons. In this way they introduced a new process by which the radiation could be absorbed, as added to Compton scattering and photoelectric emission. However, the discussion published by the two authors on the nucleus structure was wrong and in disagreement with quantum mechanics. They concluded the paper writing: *"In a word, then, on the assumption that the tracks are due in all cases either to protons or to electrons, nine-tenths of all the observed encounters yield energies which lie within the range computed from Einstein equation and the atom-building hypothesis"* (the *birth-cry* theory). Millikan was so convinced that his theory was correct that he speculated that even the one-tenth of the secondary protons and electron above 216 MeV might well turn out to be only apparently so energetic, having been straightened by the turbulence that is produced by the cloud chamber expansion which does not allow precise measurements of the curvature of tracks in the magnetic field. In his recollections, Anderson wrote that he had argued in favour of the existence of much more energetic particles, but with no success.[10]

The first result of these experiments was to show that the Compton process did not play an important role in the absorption of cosmic radiation, but that some new process that they believed to be of a nuclear type was present. This was the consequence of the fact that about half of the observed cosmic particles were charged positively and therefore could not be Compton electrons. Millikan wanted them to be protons, perhaps resulting from photonuclear disintegrations. Obviously it was important to give a clear identification of these unexpected positive particles and this could be obtained by the knowledge of their mass, inasmuch as all photos clearly showed that in all cases these particles carried a unitary charge. The experimental conditions, however, did not allow to have information on their mass, excepted in the cases in which the velocity of the particle was appreciably less than that of light, and this occurred only for a small fraction of the events. In a second paper, received on 28 June and carrying only the name of Anderson [57], the positive particles were still interpreted as protons.

Some of the particles with low velocity were clearly identifiable and identified as protons. However, as more data were accumulated, the evidence that some particles had too small a mass to be protons was growing.

Anderson soon succeeded in obtaining clearer photographs by minimizing turbulence and improving the illumination in the chamber. As a consequence he obtained better measurements of curvature and ionization density. In this way, he calculated the mass of the positive particle to be the same as that of the electron.

Two different interpretations could be given: either the particles were electrons of negative charge moving upward or some unknown light-weight particle of positive charge was moving downward. The more conservative interpretation was, of course,

---

[10]Quoted in Gallison [55].

that the particles moved upward. This lead to frequent and at time somewhat heated discussions with Millikan in which he repeatedly pointed out that everyone knew that cosmic ray particles travel downward, and not upward, except in extremely rare instances, and that therefore, these particles must be downward moving protons [52]. This point of view was however at odd with Anderson's results.

Eventually, with the help of his first graduated student, Seth H. Neddermeyer, Anderson convinced Millikan of the existence of energies that in some cases could be of the order of 10^9 eV, while Millikan had always insisted they were not higher than 400–500 MeV. Moreover Anderson gradually disagreed with Millikan about the proton nature of the positive particles.

In measurements done together with Seth Neddermeyer, Anderson found tracks of particles of both positive and negative sign nearly with the same abundance and in many cases saw several positive and negative particles ejected simultaneously by a single centre. Measurements of the specific ionization of the particles both positive and negative, by counting the number of drops per unit length along the track, showed that the great majority of particles both positive and negative had a unitary electric charge (that is the charge of the electron both positive and negative). The particles with negative charge were rapidly identified as electrons, while the positive charged ones were tentatively first identified as protons, then the sole known particles of unitary positive charge. But this hypothesis had to be discarded on the basis of their ionization.

To differentiate with certainty between the particles of positive and negative charge, it was necessary only to determine without ambiguity their direction of motion. To accomplish this purpose a plate of lead was inserted across a horizontal diameter of the chamber. After traversing the plate, the particle loses energy and therefore its trajectory, if a magnetic field is applied, has a greater curvature.

It was not long after the insertion of the plate that a fine example was obtained in which a low-energy light-weight particle of positive charge was observed to traverse the plate, entering the chamber from below and moving upward through the lead plate. Ionization and curvature measurements clearly showed this particle to posses a mass much smaller than that of a proton and, indeed, a mass entirely consistent with an electron mass. Curiously enough, despite the strong admonitions of Millikan that upward-moving cosmic ray particles were rare, this indeed was an example of one of these very rare upward-moving cosmic ray particles [58].

The first results were published by Anderson in a paper signed 1st September on Science [59] on September 1932 under the sole name of Anderson. Later, he discussed the discovery in a paper on Physical Review by the title "*The positive electron*" reporting the measurements made in August 1932 [58]. A successive paper [60] gave further details.

The photo of Fig. 6.4 is the one published in the paper [58] and shows a positively charged particle that traverses a lead plate of 6 mm thickness. If one assigns an electronic mass to this particle, its energy before traversing the plate is 63 MeV which decreases to 23 MeV on the outer side. The possibility that this particle of positive charge could be a proton is eliminated on the basis of its range and of the curvature. The energy of a proton of that curvature comes out to be 300,000 eV, but a

Fig. 6.4 The positron photo by Anderson [58] (Reprinted with permission from C.D. Anderson, Phys. Rev. **43**, 491 (1933). Copyright 1933 by the American Physical Society)

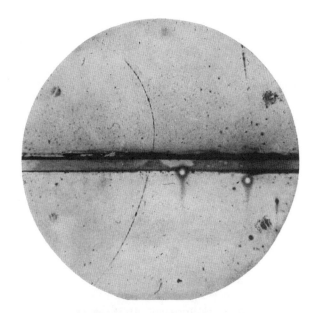

Fig. 6.5 From C.D. Anderson and S.H. Neddermeyer [61] (Reprinted with permission from C.D.Anderson and S.H.Neddermeyer, Phys. Rev. **50**, 263 (1936). Copyright 1936 by the American Physical Society)

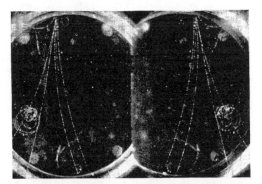

Fig. 1. Pike's Peak, 7900 gauss. An electron shower of three negatrons and three positrons of energies, respectively from left to right, 3.5, 55, 190, 78, 70, 90 MEV. The low energy electrons coincident in time with the shower represent the absorption of low energy photons accompanying the shower electrons. In all illustrations *the direct image is at the left.* The magnetic field is directed into the paper.

proton of that energy has a total range of 5 mm in air, while that portion of the range actually visible in this case exceeds 5 cm. The only escape from this conclusion was a positively charged electron.

Further examples similar to that one and others, in which two or more particles seemed produced from a single centre, gave further evidence of the existence of particles of positive charge and mass small compared to that of the proton. We present here a very beautiful photo obtained some years later (see Fig. 6.5).

Anderson was awarded the Nobel Prize in Physics in 1936, for *his discovery of the positron.*

No one had before imagined the existence of the positive twin of the electron. And indeed the proton, with a mass 2,000 times larger than the negative electron mass, had been considered until that time the sole existing positive electricity unity. Anderson's discovery implied a radical change in the conception of the final constitution of matter. The Wilson chamber experiments demonstrated that tracks of positive electrons appear frequently, as a result of cosmic rays encounters with atomic nuclei.

6.3   A Theoretical Digression: Dirac and the Negative Energy States

Six months after Anderson's paper on Science, the generation of electron pairs (one positive and one negative), produced when a photon interact with sufficient energy with an atomic nucleus, was confirmed by Blackett and Occhialini.

Before discussing this result, we ought to make a digression in the field of theoretical physics. In the 1920s, Erwin Schrödinger [62, 63] found an equation which was able to explain the behaviour of atomic systems in terms of quantum mechanics. His equation explained marvellously well, for example, the hydrogen atom. However it referred only to particles moving with a speed much smaller than the light velocity, and therefore, did not satisfy Einstein special relativity theory.

Many attempts were done to find the suitable equation for the relativistic case, but no one worked. Eventually a young English theoretician, Paul Adrien Maurice Dirac, found the right equation.

Paul Adrien Maurice Dirac[11] was born in Bristol on 1902 from a Swiss father, who emigrated into England where he was teaching French, and an English mother. He studied electro-technical engineering at Bristol turning later to pure mathematics, first in Bristol and then at St. John's College in Cambridge. In 1925, Heisenberg, after a visit to Cambridge, sent him the proofs of his first paper on *Matrizenmechanik.* The young Dirac, inspired by the uncertainty principle, gave a very beautiful and general formulation of quantum mechanics.

In 1932, he was appointed "Lucasian Professor" of mathematics at Cambridge, covering therefore, at only 30, the same chair of Newton, and remained there until he retired.

A man of few words, he had a very important influence through his writings. He was very concise and difficult to understand. Once, at the end of a seminar, he asked, as usual, if there were questions. One person from the audience said: "*I have not understood how one goes from A to B*" and pointed at two equations. Dirac replied sharply "*This is not a question; it is an ascertainment. The next question, please*".

---

[11] A brief biography on this complex man is G. Farmelo [64]. G. Farmelo is also the author of the book: The Strangest Man: The HiddenLife of Paul Dirac, Quantum Genius [65].

In 1928, Dirac [66, 67] found a way to write a relativistic equation for the electron. In the limit of small velocity it reduces to the Schroedinger equation, but it is more general. Dirac's equation gives directly the spin and the magnetic moment of the electron, two quantities that in the non-relativistic theories had to be postulated and added separately. However, the new theory encountered a big difficulty: its solutions admit particles with positive energy (normal particles) and particles with negative energy. These last particles may be derived from the former ones simply by changing the sign of their mass, from positive to negative. Particles with an inertial negative mass do not correspond, however, to anything observed in nature: in fact assigning a negative sign to their mass, they would be accelerated in an opposite direction to the applied force. To stop a particle of this kind, it would be necessary to push it in the direction of motion and not against it. If we consider two charged particles, for example, two normal electrons, that initially travel near each other, due to the Coulomb force they are repelled one from the other by electrostatic forces with the same absolute strength but acting on opposite side, and would escape one from the other with increasing velocities. If instead one of the particles has a negative mass, it would be accelerated in the same direction of the other particle and they would escape together remaining at a constant distance from each other and increasing indefinitely their velocity (that remains always less than that of the light, of course).

The negative solutions could not be eliminated, as Oscar Klein and Yoshio Nishina demonstrated in a study on Compton effect by using the Dirac equation and finding a strange paradox: when electrons with negative mass were reflected from a potential wall, a greater intensity was returned than was going in [68] (*Klein paradox*).

In the later months of 1928 and in early 1929, Dirac was busy to write a fundamental book on quantum mechanics [69] (that is still a textbook on the subject) and at the end of March he left Cambridge for a tour around the world. Coming back, in order to solve the problem of the negative energy solutions, Dirac considered the trajectory of negative energy states in the classical theory and found that the motion of an electron with negative energy is identical to the one of an electron with positive mass and a charge +e instead of −e, a result that could be transferred also to quantum mechanics, and concluded that an electron with negative energy behaves somehow as a proton, but cannot be a proton since certainly the proton has no negative energy. However, he suggested a link between electrons with negative energy and protons by proposing that nearly all of the states with negative energy in the universe are filled: the empty places are the protons [70].

This point of view was making use of the Pauli *exclusion principle* (two identical particles cannot stay in the same quantum state). The situation introduced the idea of a "filled" vacuum, that later would be termed "Dirac sea", and Dirac described the vacant places in this sea as "holes". These ideas were published under the title "*A theory of electrons and protons*" on the Proceedings of the Royal Society [71].

Dirac inquired if this distribution of negatively charged electrons with negative mass was observable. The answer was not. However if a place in the sea of negative electrons and negative mass is empty, the free level behaves as if it is a particle of positive mass and positive charge.

Dirac threw himself somehow too far in his speculation and identified this empty place with the proton, calculating the annihilation process [72] of electrons and protons on the basis of the hole theory. This process took place via the emission of two photons (this is because of the conservation of energy and momentum). The work roused an uproar.

The Russian physicist Igor Tamm (1895–1971), with whom Dirac had correspondence, observed that something was not going in the right direction because an atom in which there is an electron and a proton should annihilate in an extremely short time [73]. The American physicist J. Robert Oppenheimer performed some calculation and found that the recombination time of matter should be of the order of 10^{-10} s [74]. Other critics came from Hermann Weyl [75] (1885–1955) and many others.

Eventually in 1931, Dirac [76] retired the proton hypothesis and wrote:

"...we must abandon the identification of the holes with protons and must find some other interpretation for them.. A hole, if there were one, would be a new kind of particle, unknown to experimental physics, having the same mass and opposite charge to an electron. We may call such a particle an anti-electron".

He then explained that the particle had not yet been observed experimentally because it should immediately recombine with electrons. However, if it had been produced in vacuum it would have been possible to examine it and an encounter between two hard gamma rays (of energy at least half MeV each, because the mass of an electron corresponds to the materialization of this energy) could lead to the simultaneous creation of an electron and an antielectron. Negative energy states should exist also for protons which could be interpreted as antiprotons. Thus Dirac expounded the concept of antimatter for particles obeying his relativistic equation. In his work Dirac introduced also the notion of a magnetic monopole, a still unfound particle, which may not exist at all, but that sometime pops out because it is liked by some theoreticians.

People joked a lot about Dirac theory, but when Anderson discovered the positive electron, first called positon and then *positron*, this particle behaved exactly as predicted by Dirac's theory.

6.4 The Confirmation and the Official Birth of Antiparticles

In the fall of the same year in which Anderson published the work in which the positive particle observed in cosmic rays was recognized as a positive electron (1932), in Cambridge, P.M.S. Blackettt and G. Occhialini confirmed this result, and interpreted it on the basis of Dirac theory.

Patrick Maynard Blackett was born in London in 1897, where he died in 1974.

He enrolled at the Royal Naval College in Dartmouth and served with the Navy during the First World War. Then he came to Cambridge and in the 1920s worked with Rutherford, as an expert of the cloud chamber.

In 1933 he was appointed physics Professor at the London University. Later, in 1937, he moved to Manchester University, coming back in 1953 to the Imperial College in London where he remained until retiring in 1963. Since 1933 he was a member of the Royal Society.

During the Second World War, he worked in a commission directed by Professor H.T. Tizard (1885–1959) which started the research on radar in Great Britain.

In July 1931, a new collaborator arrived in Cambridge from Arcetri: Giuseppe Occhialini.

Giuseppe Occhialini (1907–1993) (Fig. 6.6) was born in Fossombrone (Pesaro, Italy) on 5 December 1907, the son of the physicist Augusto (1878–1951), and studied in Florence, earning his degree in physics under Bruno Rossi, in 1929, with a thesis on cosmic rays. Occhialini remained at the Istituto di Fisica in Arcetri, first as a voluntary assistant (it means no money) and then as Assistant Professor (1932–1937), in a group of enthusiastic young people who, under the guidance of Bruno Rossi, moved from the traditional spectroscopy to the more modern activities of nuclear physics and cosmic rays. As an assistant under Bruno Rossi, Occhialini started to make research using the Geiger–Müller counters. Rossi, in the summer of 1930 met in Bothe's laboratory Patrick Blackett, and being interested in the cloud chamber, that Skobeltzyn had shown to be a good research instrument for cosmic rays, decided to send one of the Arcetri researchers in the Blackett's laboratory. The choice had fallen on Gilberto Bernardini,[12] another of the young physicists of the group, but he could not go because he had to serve in the Army, and so Occhialini took his place. In 1931, he arrived at Cavendish Laboratory in Cambridge, where he obtained together with Blackett results that have a place in the history of physics.

In 1934, Occhialini came back to Florence and after a while he accepted an invitation by Gleb Wataghin (1899–1986)[13] to join a research group on cosmic rays at the University of Sao Paulo in Brazil.

After the war, in 1944, he went to Bristol and became a master in the new technique of nuclear emulsions with which he did fundamental discoveries.

He also was for a brief period at the Free University in Bruxelles, where he established one of the more active groups in nuclear emulsions; he then came back definitely in Italy, first in Genova and then in Milan where he created and directed a group for the study of elementary particles by using nuclear emulsions realizing original research solutions and participating to the G-stack Collaboration, an international collaboration for the cosmic ray study with nuclear emulsions sent by balloons. In 1961, after a year at MIT with Bruno Rossi, he established in Milan

---

[12]Gilberto Bernardini (1906–1995) was Professor of physics (fisica superiore) in Bologna (1937), then in Roma and eventually the director of the Scuola Normale di Pisa (1964). He was also the director of the Testa Grigia Laboratory at Cervinia for the study of elementary particles (cosmic rays), president of the Istituto Nazionale di Fisica Nucleare, research director at CERN, national member of the Accademia dei Lincei (1949). He made studies on cosmic rays.

[13]Gleb Watagin was born at Birsula, Ucraina, on 1899 and died in Torino on 1986. He was Professor of Physics since 1939. He performed research on cosmic rays.

Fig. 6.6 G.P.S. Occhialini

a new group devoted to research on cosmic radiation in the frame of the European Space Agency (ESA) working in gamma ray astronomy using satellites.

Once he was asked of the meaning of his initials *"No meaning*—he answered— *Blackett signed P.M.S. Blackett, and so I, to be of no less importance, started to sign G.P.S"*.

In Cambridge, in 1932, Occhialini was responsible for a great improvement in the use of the Wilson chamber, by introducing a new method of control of the expansion that used Geiger–Müller counters to detect the arrival of a cosmic ray particle and the coincidence technique learned from Rossi.

In the summer of that year, Blackett and Occhialini [77] built a cloud chamber with two Geiger–Müller counters placed in coincidence, one above and the other below the chamber, so that when a cosmic particle passed through, the coincidence was used to operate the expansion. A magnet to measure the energy of the charged particles was also provided. On 21 August, they sent to Nature a preliminary paper with some results.

As cosmic rays arrive randomly, the current practice was to take a photo every 15 s, hoping to catch by chance some cosmic ray that was traversing the chamber, with a great waste of photographic material. With the Geiger–Müller counters placed in coincidence one above and the other below the chamber, this was activated only when the ray passed along the line through the two counters and therefore the track could be photographed at the right time. Nearly each photo showed an event: now 76% of the images displayed tracks of particles and a photo every 2 min could be taken. For this work, as Blackett himself remembered, the electronic technique due to the Occhialini formation in Arcetri was of fundamental importance.

For comparison Skobelzyn had one cosmic ray track every ten expansions and Anderson had about one measurable track in 50 photographs. A correlation between the occurrence of tracks in a Wilson chamber and the discharge of a counter had been found by Mott Smith and Locher [78, 79] and Johnson et al. [80] had used the coincidence of the discharge of two counters to operate the flash which illuminated a continuously working cloud chamber.

In the fall, Blacket and Occhialini started to take photos and, as Occhialini wrote later, some tracks were deflected by the applied magnetic field in a direction corresponding to a negative particle while other tracks were deflected as positive particles. By considering both the penetration and the ionization of the positive tracks, one could deduce they could not be produced by protons [81]. On 7 February 1933 a report was sent to the Proceedings of the Royal Society on "*Some photographs of the tracks of penetrating radiation*" communicated by Rutherford [82] in which the two authors, referring to Anderson's paper, agreed with him and declared that they found 14 tracks of showers that should be attributed to positive electrons, plus others more doubtful.

From an analysis of their results, the two authors concluded that one could think that negative and positive electrons were created in pairs during collisions with nuclei. They answered also the obvious question why the positive electrons had not been observed before, saying that the free particles should have a very limited life, probably due to annihilation with negative electrons, converting the energy of their mass in radiation quanta. They suggested that the negative–positive electron pair could be created by the annihilation of a photon nearby a nucleus. For this process, the photon should have an energy $>2m_0c^2(=1\,\text{MeV})$. They also referred to the work of their Cambridge colleague, Dirac, and clearly invoked the phenomenon of particle–antiparticle pair formation.

The Blackett and Occhialini work was presented by Rutherford at the Royal Society meeting on 16 February 1933, arousing great sensation. The results and conclusions turned the whole theoretical discussion in favour of the brilliant Dirac theory of antiparticles. Watson Davis, the director of the American Science Service, informed Anderson about these results and suggested the name "*positron*" for the new particle that Anderson accepted and hastily finished the paper for Physical Review [58], where however he was still uncertain whether to accept Dirac theory. Also in a paper in which he studied positrons produced by gamma rays from radiothorium he did not mention Dirac theory [83]. Only later [60] he accepted the English hypothesis of the antielectron, still doubtful [84] because of Millikan continued scepticism.

In 1933, Millikan maintained that both positive and negative tracks were born together by the disintegration of the nucleus of an atom [85], without any reference to Dirac. On the contrary he quoted the Anderson energy measurements as the "*most complete demonstration*" that primary cosmic rays were photons (as Anderson had shown the vast majority of measured charged particles to be below 600 MeV, still compatible with the "*birth cry*" of created atoms).

Millikan point of view was that positrons and electrons were already in the nucleus and could be ejected by the collision with a gamma ray. This point of view was inconsistent with relativistic quantum mechanics because he affirmed having found that more negative than positive particles were emitted. He concluded maintaining that this "*seems difficult to reconcile with the Dirac theory, as interpreted by*

Blackett and Occhialini of the creation of electron pairs out of the incident photons, and point strongly to the existence of nuclear reactions of a type in which the nucleus plays a more active role than merely that of a catalyst" [86].

The Blackett and Occhialini suggestion was later confirmed by using gamma rays emitted by radioactive substances [83, 87–89].

Dirac was convinced that his theory was right and welcomed the discovery with great joy.

In America, a definitive support came from J. Robert Oppenheimer and M.S. Plesset [90] who explicitly calculated the pair creation by gamma rays in the electrostatic field of nuclei by using a simplified model and finding rather good agreement with the Anderson and Neddermeyer results obtained with gamma rays of ThC″ [83].

Soon it was found that it was possible to produce positrons in the laboratory simply shutting gamma rays against metallic plates. Entering in collision with an atomic nucleus, a gamma ray quantum disappears and its whole energy is transformed into two (one negative and one positive) electrons. As the mass of an electron expressed in energy units is 0.5 MeV, the process takes place only if the gamma photon energy $h\nu$ is larger than 1 MeV. The excess energy $h\nu - 2m_0c^2$ is given as kinetic energy to the electron pair created in the collision. The fate of these two electrons is very different. The (ordinary) negative electron is gradually slowed down by collisions with the other negative electrons that form the matter. The positive electron does not live for long, but is destroyed (*annihilated*) by a collision with one negative electron, emitting two gamma quanta. J. Thibaud [91] demonstrated that the positron has exactly the same mass of the negative electron.

The positron is the first example of an antimatter particle. Today we know that all particles have their corresponding antiparticle and the antiproton [92, 93] and antineutron [94] have been found. Several events were recorded also in cosmic ray investigations which might be due to antiprotons [95–97].

In retrospect we may note that, if people had taken for seriously the Dirac theory, the positron could have been discovered in one afternoon. However, real life is not so simple! It is also worthwhile to note that at that time, positive electrons were necessary to explain nothing.

References

1. J. Chadwick, Proc. R. Soc. Lond. A **136**, 692 (1932)
2. N. Feather, Contemp. Phys. **1**, 191, 257 (1960)
3. B. Kroger, Physis **22**, 175 (1980)
4. W. Sutherland, Phil. Mag. **47**(5), 269 (1899)
5. W.D. Harkins, Phil. Mag. **42**(6), 305 (1921)
6. E. Rutherford, Proc. R. Soc. Lond. A **97**, 374 (1920)
7. J.L. Glasson, Phil. Mag. **42**(6), 596 (1921)
8. A.S. Eve, *Rutherford* (Cambridge University Press, Cambridge, 1939)
9. C.T. Walker, G.A. Slack, Am. J. Phys. **38**, 1380 (1970)

10. E. Rutherford, Phil. Mag. **37**(6), 537 (1919)
11. W. Bothe, H. Becker, Z. Phys. **66**, 289 (1930)
12. I. Curie, CR **193**, 1412 (1931)
13. I. Curie, F. Joliot, CR **194**, 273 (1932)
14. H.C. Webster, Proc. R. Soc. Lond. A **136**, 428 (1932)
15. J. Chadwick, Nature **129**, 312 (1932)
16. H. Urey, F. Brickwedde, G. Murphy, Phys. Rev. **39**, 164 (1932)
17. J. Chadwick, M. Goldhaber, Proc. R. Soc. Lond. **151**, 479 (1935)
18. J.D. Cockcroft, E.T.S. Walton, Proc. R. Soc. Lond. A **136**, 619 (1932)
19. J.D. Cockcroft, E.T.S. Walton, Proc. R. Soc. Lond. A **137**, 229 (1932)
20. E.O. Lawrence, M.S. Livingstone, Phys. Rev. **40**, 19 (1932)
21. D. Iwanenko, Nature **129**, 798 (1932)
22. J. Chadwick, Nature **129**, 312 (1932)
23. W. Heisenberg, Z. Phys. **77**, 1 (1932)
24. W. Heisenberg, Z. Phys. **78**, 156 (1932)
25. W. Heisenberg, Z. Phys. **80**, 587 (1933)
26. E. Majorana, Z. Physik. **82**, 137 (1933)
27. H. Becquerel, CR **130**, 206 (1900)
28. J. Chadwick, Verh. Deutsch. Phys. Ges. **16**, 383 (1914)
29. L. Meitner, W. Orthmann, Z. Phys. **60**, 143 (1930)
30. S.A. Watkins, Lise Meitner and the beta-ray controversy: An historical perspective. Am. J. Phys. **51**, 551 (1983)
31. J. Carsten, F. Auserud, *Controversy and Consensus: Nuclear Beta Decay 1911–1914* (Birkhauser Verlag, Basel, 2000)
32. C.D. Ellis, W.A. Wooster, Proc. R. Soc. Lond. **117A**, 109 (1927)
33. C.D. Ellis, W.A. Wooster, Nature **119**, 563 (1927)
34. N. Bohr, H.A. Kramers, J.C. Slater, Z. Phys. **24**, 69 (1924)
35. C.P. Enz, *No Time to be Brief, A Scientific Biography of Wolfang Pauli* (Oxford University Press, Oxford, 2002), p. 212 (et seg)
36. W. Pauli, in *Address to a group of Radioactivity (Tubingen, 4 December 1930)*, ed. by C.P. Enz, K. von Meyenn. Wolfang Pauli writings on Physics and Philosophy (Springer, Heidelberg, 1994), p. 193–217
37. W. Pauli, *Rappts Septieme Conseil Phys. Solvay, Bruxelles 1933* (Gauthier-Villars, Paris, 1934)
38. E. Fermi, La Ricerca Scientifica **3**, 161 (1932)
39. E. Fermi, *Collected Works*, vol. I (Accademia Nazionale dei Lincei, Rome, 1962), p. 498
40. E. Fermi, La Ricerca Scientifica **4**, 491 (1933)
41. E. Fermi, Z. Phys. **88**, 161 (1934)
42. E. Fermi, Nuovo Cimento **11**, 1 (1934)
43. A.H. Snell, L.C. Miller, Phys. Rev. **74**, 1217 (1948)
44. A.H. Snell, F. Pleasonton, R.V. McCord, Phys. Rev. **78**, 310 (1950)
45. J.M. Robson, Phys. Rev. **78**, 311 (1950)
46. J.M. Robson, Phys. Rev. **83**, 349 (1951)
47. L. Montanet et al. Phys. Rev. D **50**, 1173 (1994)
48. R.J. Weiss (ed.), *The Discovery of Anti-matter: The Autobiography of Carl David Anderson, the Youngest Man to Win the Nobel Prize* (World Scientific, River Edge, NJ, 1999)
49. C.D. Anderson, Phys. Rev. **35**, 1139 (1930)
50. Chung-Yao, Chao, Phys. Rev. **36**, 1519 (1930)
51. C.D. Anderson, H.L. Anderson, in *The Birth of Particle Physics*, ed. by L.M. Brown, L. Hoddeson (Cambridge Press, Cambridge, 1983), p. 131
52. C.D. Anderson, Am. J. Phys. **29**, 825 (1961)
53. R.A. Millikan, C.D. Anderson, Phys. Rev. **40**, 325 (1932)
54. R.A. Millikan, Annales de l'Institut H. Poicarè **3**, 447 (1932)
55. P. Galison, Centaurus **26**, 262 (1983)
56. M. De Maria, A. Russo, Rivista di storia della scienza **2**, 237 (1985)

57. C.D. Anderson, Phys. Rev. **41**, 405 (1932)
58. C.D. Anderson, Phys. Rev. **43**, 491 (1933)
59. C.D. Anderson, Science **76**, 238 (1932)
60. C.D. Anderson, Phys. Rev. **44**, 406 (1933)
61. C.D. Anderson, S.H. Neddermeyer, Phys. Rev. **50**, 263 (1936)
62. E. Schroedinger, Ann. Phys. **79**, 361 and 482 (1926)
63. E. Schroedinger, Ann. Phys. **81**, 109 (1926)
64. G. Farmelo, Paul Dirac a man apart. Phys. Today **62**(11), 46 (2009)
65. G. Farmelo, *The Strangest Man: The HiddenLife of Paul Dirac, Quantum Genius* (Faber & Faber, UK, 2009)
66. P.A.M. Dirac, Proc. R. Soc. Lond. A **117**, 610 (1928)
67. P.A.M. Dirac, Proc. R. Soc. Lond. A **118**, 351 (1928)
68. O. Klein, Z. Phys. **53**, 157 (1929)
69. P.A.M. Dirac, *The Principles of Quantum Mechanics* (Oxford University Press, Oxford, 1930)
70. P.A.M. Dirac, Annales de l'Institut H. Poicarè **1**, 357 (1931)
71. P.A.M. Dirac, Proc. R. Soc. Lond. A **126**, 360 (1930)
72. P.A.M. Dirac, Proc. Camb. Phil. Soc. **26**, 361 (1930)
73. J. Mehra, H. Rechenberg, *The Historical Development of Quantum Theory*, vol. 6 (Springer, Heidelberg, 2001), p. 788
74. J.R. Oppenheimer, Phys. Rev. **35**, 562, 939 (1930)
75. H. Weyl, *Gruppentheorie und Quantummechanik* (Hirzel, Leipzig, 1931) (English edition: *The Theory of Groups and Quantum Mechanics*) (Dover, New York), p. 263
76. P.A.M. Dirac, Proc. R. Soc. Lond. A **133**, 60 (1931)
77. P.M.S. Blacket, G.P.S. Occhialini, Nature **130**, 363 (1932)
78. L.M. Mott Smith, G.L. Locher, Phys. Rev. **38**, 1399 (1931)
79. L.M. Mott Smith, G.L. Locher, Phys. Rev. **39**, 883 (1932)
80. T.H. Johnson, W. Fleicher, J.C. Street, Phys. Rev. **40**, 1048 (1932)
81. G.P.S. Occhialini, La Ricerca Scientifica, 373 (1933)
82. P.M.S. Blacket, G.P.S. Occhialini, Proc. R. Soc. Lond. A **139**, 699 (1933)
83. C.D. Anderson, S.H. Neddermeyer, Phys. Rev. **43**, 1034 (1933)
84. C.D. Anderson, Nature **133**, 313 (1934)
85. R.A. Millikan, Phys. Rev. **43**, 661 (1933)
86. C.D. Anderson, R.A. Millikan, S. Neddermeyer, W. Pickering, Phys. Rev. **45**, 352 (1934)
87. J. Chadwick, P.M.S. Blacket, G.P.S. Occhialini, Nature **131**, 473 (1933)
88. L. Meitner, K. Philipp, Naturwissenschaften **21**, 286 (1933)
89. I. Curie, F. Joliot, CR **196**, 1581 (1933)
90. J.R. Oppeneimer, M.S. Plesset, Phys. Rev. **44**, 53 (1933)
91. J. Thibaud, CR **197**, 237 (1933)
92. J.M. Brabant et al., Phys. Rev. **101**, 498 (1956)
93. O. Chamberlain, E. Segre, C. Wiegand, T. Ypsilantis, Phys. Rev. **100**, 947 (1955)
94. B. Cork, G.R. Lambertson, O. Piccioni, W.A. Wensel, Phys. Rev. **104**, 1193 (1956)
95. E. Hayward, Phys. Rev. **72**, 937 (1947)
96. E. Amaldi, C. Castagnoli, G. Cortini, C. Franzinetti, A. Manfredini, Nuovo Cimento **1**, 492 (1955)
97. H.S. Bridge, H. Courant, H.C. DeStaebler, B. Rossi, Phys. Rev. **95**, 1101 (1954)

Chapter 7
Electromagnetic Showers

7.1 Introduction

The years after 1928 witnessed a great development of the cloud chamber and of the coincidence technique applied to cosmic ray research. The use of these detecting methods brought to new important discoveries. One was the individuation, since 1929, of the presence of tracks that apparently came from a common point and were said to belong to a *shower*.

A shower may be defined as that ensemble of radiations which comes out from a non-radioactive material piece exposed to cosmic rays that may be seen with a cloud chamber or that discharges simultaneously three non-aligned Geiger–Müller counters. In a cloud chamber in which an absorbing layer is inserted, a shower may manifest itself as two or more tracks which seem to come from the same point in the material. As we will discuss later, a shower may be produced by a photon or a ionizing particle and may be formed by photons and ionizing particles.

One of the first indications of an interaction between cosmic rays and matter, besides ordinary ionization and absorption, was evidenced in the work by Millikan and Otis [1]. They, using lead screens a few centimetres thick around an electroscope, detected an absorption coefficient surprisingly high for the rays in the lead. Although this result was misinterpreted at the time, it may easily be explained if the air–lead *transition effect* is taken into account. This effect was found and investigated, in 1927, by G. Hoffman [2]. Cosmic rays passing through matter by interacting with it produce a cascade of secondaries travelling with them. Different materials are not equally efficient in producing these cascades and when the radiation passes from one material to another one, the relation between incident and secondary rays is usually changed, with the result that the total intensity changes. To this change of intensity of the beam the name *transition effect* was given.

Still in 1927, Hoffman [3, 4] registered also the first ionization burst in a ionization chamber, communicating later this result at a conference on cosmic rays in London in 1934. Many other researchers, since 1930 [5–9], studied this phenomenon. A burst is a sudden and transient increase of the current in a ionization

Fig. 7.1 The first cloud chamber photographs were of poor quality. To show a good example of a shower we have therefore chosen to present this later photo from Anderson and Neddermeyer [18] (Reprinted with permission from C.D. Anderson and S.H. Neddermeyer, Phys. Rev. **50**, 263 (1936). Copyright 1936 by the American Physical Society)

FIG. 3. Pike's Peak, 7900 gauss. A shower in which eight electrons (+ and −) strike the upper surface of the 0.35 cm lead plate. More than twenty-four electrons emerge from the lower face of the lead plate. This photograph is an example of the multiplication of shower tracks in a thin piece of absorbing material. Observation of many cases of this type shows that most of the additional tracks arise from the absorption of photons which accompany the electrons, but occasionally an electron itself may produce several positron-negatron secondaries.

chamber, several times larger than the statistical fluctuations. Bursts were observed in concomitance with large cascades.

In our exposition that tries to trace the story of the evolution of the study of cosmic rays, we should consider two kinds of showers. The most common [10] and the first discovered one, dealt of in this chapter, consists of photons, electrons, and positrons, and the high-energy particles are well collimated and have a momentum in the horizontal plane that corresponds to energies of only a few MeV. These showers increase passing through matter.

The other kind of shower, less frequent, has particles with momentum in the horizontal plane that corresponds to energies of the order of 100 MeV, the particles are not collimated, there are heavy recoil particles and the spatial density is small. However because the shower arrives on a surface on earth that may be very extended, the total number of particles may be very large. These are named *extended showers* and we discuss them later.

Actually there is only one type of shower that is the extended shower, and the first kind of showers is simply a local manifestation of a portion of the extended one, but we will clarify this point later in Chaps. 10 and 12.

Skobelzyn [11, 12], in his experiments with the cloud chamber, found, in 1929, that cosmic rays often arrive in groups and confirmed these results in research made in collaboration with Pierre Auger [13]. However, because people believed these phenomena were the product of Compton electrons, the cloud chambers were not used for their study for several years.

Eventually, in 1932, in their cloud chambers, Anderson [14], G.L. Locher [15], P. Kunze [16], and Skobelzyn [12, 17] evidenced cases of two or three particles that were ionizing as beta rays and came from the same point (later examples are Fig. 6.6 and Fig. 7.1). As it is extremely improbable that two or more Compton fast recoil electrons are generated very near to each other, eventually people recognized that the chamber was registering showers.

Fig. 7.2 A shower (is from
P.M.A. Blackett and
G. Occhialini, Proc. Roy. Soc.
London A **139**, 699 (1933))

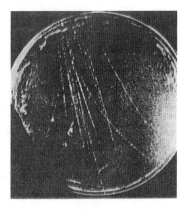

In 1931, L.M. Mott-Smith and G.L. Locher [19] used a cloud chamber and demonstrated that the simultaneous discharge of two counters always occurred in coincidence with the passage of an incident ray and so, first Rossi ([20]; see also [21]) showed, in 1932, that cascades of particles could be detected by using three Geiger–Müller counters in a triangle disposition, and then also J.C. Street and T.H. Johnson [22] showed that showers may be detected with counters.

The cloud chamber built by Blackett and Occhialini which made photos only when at least one particle arrived, showed soon tracks of many particles that clearly resulted from the interaction of a single high-energy cosmic ray somewhere nearby the chamber (see Fig. 7.2). The authors gave to these multiple happenings the name *showers*.

In 1933 the presence of showers of cosmic ray particles was so independently established from the results of both cloud chamber—which showed unequivocally that the showers consist of particles ejected from an apparently common origin—and counters. A huge number of papers appeared in a few years discussing a number of different situations. We discuss here only some of the most important results[1].

7.2   The Rossi Curve

Because Skobelzyn had found multiple traces, Rossi [20, 21] placed three counters in a triangle disposition (see Fig. 7.3). In this way the three counters could not be discharged by a single particle travelling along a straight line; however, when the system was surrounded with lead it registered a great number of coincidences (around 35 in an hour). Taking out the upper part of the screening the coincidence number dropped nearly to zero.

---

[1]Research up to 1938 is well documented in D.K. Froman and J.C. Stearns [9]

Fig. 7.3 Triangle
disposition. At least two
particles are needed to
discharge the three counters
in coincidence. One of these
particles may be a primary
particle but the other should
have been produced in the
lead, in fact if the lead is
removed the rate of
coincidences falls to
nearly zero

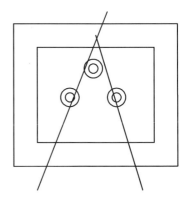

The coincidences could be only the result of two or more ionizing particles emerging simultaneously from the lead, a direct proof that cosmic radiation gave place to ionizing secondary rays.

Taking out the upper lead screen, about two coincidences per hour remained, a circumstance that Rossi was unable to explain at that time and that today we know is due to the production of secondary rays in the atmosphere.

Also the production of secondary particles in the lead could not be explained at that time, because the only known effects were ionization (but the electrons emitted by the incoming particle in traversing the lead had not sufficient energy to escape outside) and the Compton effect which could explain the emission of an electron in the knock-on of a high-energy photon in the lead but could not explain the multiple emission of particles. The possibility to have two Compton collisions one after the other is a so rare phenomenon that may safely be neglected. Therefore there was no explanation for the abundant production of secondary particles as detected by the experiment. The results seemed so incredible that the editor of the journal Naturwissenschaft to whom Rossi submitted the paper refused it. The work was published on another journal [20].

Rossi investigated the production of showers by cosmic radiation, that is the number of showers detected by a system of three or more non-aligned counters in coincidence, as a function of the thickness of the absorber placed over the counter disposition (the *Rossi curve* as it is known today).

This curve is shown in Fig. 7.4, that is an arrangement of that published in the paper by Rossi [21] made by Rossi himself in his book [24]. The material over the counters was lead. We may see that by increasing the lead thickness the number of coincidences due to the showers first increases, reaches a maximum between 1 and 2 cm of lead, and then decreases rapidly.

The common opinion at that time was that the penetrating particles found in the atmosphere were the primary cosmic rays. These particles were believed to start the shower by interacting with the atomic nuclei of the atmospheric gases. The shape of the curve demonstrated that the hypothesis was wrong. The penetrating particles in

Fig. 7.4 Shower curve. The number of coincidences per hour is plotted as a function of the thickness of lead above the counters. The experimental arrangement is shown schematically in the insert. The circles are experimental points. This drawn is based on the original Rossi's paper as arranged by himself in his book ref. [24] (is fig. 7-1 from B. Rossi *Cosmic Rays* (McGraw Hill, New York, 1964))

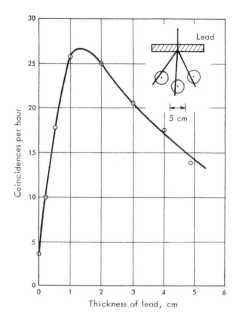

the cosmic rays had in lead a range of the order of a few metres, their number and energy should not appreciably change penetrating a few metal centimetres. However the number of showers that came from the lead layer decreased to about one half when the lead thickness was increased from 2 to 5 cm. Clearly the radiation responsible for the showers was much easily absorbed in lead than the penetrating corpuscular radiation.

A second interesting result was that the number of showers depends on the nature of the metal. Placing over the counter different metals (lead, iron, or aluminium), chosen of a thickness such as to have the same mass per unit area (several g cm²), Rossi found a ratio about 4:2:1 for the three metals. He concluded showers were produced more abundantly in a given mass of a heavy element (for ex. lead) than in the same mass of a light element (for ex. aluminium). Other experiments showed that the heavier elements were also more effective absorbers of the radiation responsible for the showers. This different behaviour between heavy and light elements was a novelty. The ionization losses for charged particles, as well as the Compton collisions, are approximately the same in different materials when the thickness is measured in g/cm² instead than cm.

It was soon clear that the matter–radiation interaction gave place to more complex and unusual effects than those which were expected, and the counter-controlled cloud chamber was really pivotal for this.

7.3   The Interpretation of Showers and the Development
of Quantum Electrodynamics

The abundant production of secondary rays meant that a significant fraction of the particles observed at sea level were produced in the atmosphere by the high-energy primary rays which had a very low probability to pass through it without suffering interaction and produce the soft radiation. Therefore the penetrating component present at sea level should not be there. The problem was starting to be very complex.

In earlier 1930s, quantum electrodynamics, then in full development, found itself to front a series of problems and the description of the phenomena observed in the high-energy radiation of the cosmic rays was an excellent proof bench to verify the theories then developed.

It was known that fast electrons when abruptly stopped in the matter produce photons. This is in fact one of the two modes in which X-rays are generated. However it was generally believed that the energy lost by electrons through the photon emission was but a small fraction of the energy lost by ionization. The remaining part of the interaction was through the Compton effect.

Dirac [25] and W. Gordon [26] had considered how the energy exchange between photons and electrons took place through the Compton effect by using the new Dirac relativistic equation for the electron. Later, in Copenhagen, Oscar Klein[2] and Yoshio Nishina[3] studied the intensity of the scattered radiation from electrons, obtaining an expression for the probability that a photon with a given energy suffers a Compton collision in some material thickness, that corrected Dirac formula, and sent a letter to Nature [27]. Finding some agreement with the experimental results, they submitted

---

[2]Oscar Benjamin Klein was born in Stockholm on September 15, 1894. He studied at the University of that town, graduating in 1918 and earning his Ph.D. in 1922. He started teaching in Stockholm. In 1922 he went to the University of Michigan at Ann Arbor. From 1926 to 1933 he was lecturer at the University of Copenhagen and was then professor and director of the Institute of Mechanics of the University of Stockholm. He retired in 1968 and died on February 5, 1977 in Stockholm. He worked on problems of quantum theory, non-relativistic and relativistic quantum mechanics, meson theory and general relativity.

[3]The Japan physicist Yoshio Nishina was born on December 6, 1890 in the Okayama Prefecture. After graduating from Tokyo University in 1918, he joined the Institute for Chemical and Physical Research (RIKEN) in Tokyo. In 1921 he went to Europe, first going to Ernest Rutherford at Cavendish Laboratory, where he studied the electrons produced by X-ray scattering on atoms with the help of a Geiger counter, then to James Frank in Gottingen and finally from 1923 to Niels Bohr in Copenhagen. After many years, he returned to Tokyo in December 1928, working again at the RIKEN. In May 1931 he lectured on quantum mechanics at Kyoto University, and in July of the same year, he started a laboratory of nuclear physics at RIKEN, created a theoretical group and helped to establish a group of nuclear and cosmic ray physics at the University of Osaka which was directed by Seichi Kikuchi (who in 1933 hired Hideri Yukawa as a theoretician. Nishina has been recognized as the father of Japanese nuclear physics. After the Second World War he started reconstruction of science in Japan but died in Tokyo on January 10, 1951, before he could achieve his goals.

an extended version of their calculations [28]. The Kleein and Nishina work showed the necessity to take into account the negative Dirac energy states in the calculations, even if the problem was not yet fully solved. It was in any case an improvement and in fact, in a work on the absorption of cosmic rays, Louis Harold Gray [29] from Trinity College, Cambridge, comparing the experimental data with either the Dirac or Klein–Nishina theory found that the last theory fitted better the results.

Chung-Yao Chao—as Anderson remembered (see Chap. 6)—had studied the absorption and scattering of ThC" gamma rays (2.6 MeV) which were essentially governed by Compton scattering, and compared them with the Klein and Nishina formula, finding that the absorption and scattering were larger than predicted by the theory. However, Chao was late to publish his results [30, 31] and was preceded by Lise Meitner and H.H. Hupfeld [32] who independently found similar results. The observed deviation of gamma-ray scattering received so the name of *Meitner–Hupfeld effect*.

In February 1932, Werner Heisenberg submitted under the title *"Theoretical considerations on cosmic radiation"* [33] the first of a series of important papers on cosmic rays. There he discussed in detail the most important experiments on cosmic radiation from the point of view of the existing theories to verify at which point experiments agreed with theory. He then discussed, in particular, the deceleration of electrons in their passage through matter, and some typical cosmic ray phenomena, as the ones observed in the absorption curves and underlined the discrepancies with theory (especially the Klein–Nishina theory) and the problems connected with the use of the Dirac theory.

According to classical electrodynamics, a charged particle emits a radiation when it is deflected in the nuclear electric field, the intensity of radiation being proportional to the square of acceleration. This process is known as *bremsstrahlung* (braking radiation).

At about the same time, Walter Heitler[4] (1904–1981) in Gottingen, after a series of studies in quantum chemistry, moved into the field of quantum electrodynamics. He started to work out the problem of bremsstrahlung and the related absorption processes, first in Gottingen and then, after Hitler's persecution, in England, where he took refuge. Meantime the Dirac theory on holes and the discovery of the positive

---

[4]Walter Heitler was born in Karlsruhe in 1904. A pupil of A. Sommerfeld, he lectured at the University of Gottingen (1929), then in Bristol, Dublin, and since 1946 Zurich, where he died at Zollikon in 1981. He has developed the quantum theory of chemical valence and the theory of the energy loss of charged particles in radiative collisions together with H. Bethe.

electron occurred and he realized that the processes were intimately connected. Therefore he included the electron pair creation and annihilation in his calculations.

Heitler submitted his first study *"On the radiation emitted by extremely fast collisions"* on June 1933 [34]. In it he calculated the bremsstrahlung for collisions of electrons of energy much greater than mc^2 (m is the electron mass) finding that the cross-section increases indefinitely with energy. Independently also Fritz Sauter [35] in Berlin extended the theory to relativistic electrons, subsequently underlying, in a work made in collaboration with Heitler [36], the importance of bremsstrahlung for cosmic rays. The young Hans Bethe[5], a German refugee, joined in these studies.

The passage of electrons through matter dominated Bethe's theoretical work since the start of his career. When, in 1926, he asked Sommerfeld a subject for study, Sommerfeld assigned him the task to explain some anomalies of electron diffraction in crystals, suggesting to make analogies with X-ray diffraction ([37], see also [38]). After this exercise, Bethe in his PhD thesis, came back on the problem from the quantum mechanics point of view. He then came to Frankfurt and then Stuttgart. Here he continued to work to what later he considered his better work *"Towards a theory of the passage of fast corpuscular radiation through matter"* [39].

In the fall of 1930, he arrived at Cambridge where Blackett pushed him to calculate the energy-range relations for electrons. The following year he was in Roma, a guest of Enrico Fermi, where he extended the calculations to the relativistic case [40]. He came back to Germany but, after the racial laws of April 1933, he emigrated to England at University of Manchester, having also frequent contacts with Cambridge. Here he met Heitler, who was working on the absorption of fast particles in the matter by using the Dirac cross-section for the production of pairs, and associated to the work.

It is so that Heitler and Hans Bethe, in 1934, published the results of more refined calculations [41]. By defining a radiation length as that length (in g/cm^2) in which a fast electron loses in the mean 63% of its energy due to bremsstrahlung, they calculated the probability that an electron of a given total energy, traversing a given thickness of a material, emits a photon of some given energy.

To calculate the effect precisely, Bethe and Heitler needed to use quantum mechanics—because they had to do with individual atomic particles—and the relativistic equations, because the particles moved at nearly the light speed.

Their main results showed the radiation losses are enormously greater for light particles (such as electrons) than for heavy particles (as protons). This result is easy to understand because the particle irradiates as a result of acceleration, and for a given force, the acceleration is inversely proportional to the mass. The larger is the

---

[5]Hans Albrecht Bethe was born at Strasburg in 1906 and was professor at the Munich University in 1930. In 1931 he collaborated with E. Fermi in Roma; he emigrated to the USA and was offered a chair at Cornel University in Ithaca in 1937. He worked on several problems of nuclear physics. His studies on nuclear reactions and scattering of elementary particles are fundamental. In 1938 he formulated a theory on the origin of stellar energy based on a cycle of thermonuclear reactions. From 1943 to 1946 he was chair of the theoretical physics division in the Manhattan project at Los Alamos. In 1967 he was awarded the Nobel Prize for Physics. He died on 2005.

mass, the smaller is the acceleration. They also found that for a given mass, losses per unit area are larger in high atomic number than in low atomic number elements. This too may be understood by considering that acceleration depends on the nucleus charge and the atomic number tells which is the nuclear charge.

Radiation losses grow rapidly increasing energy. By increasing energy the losses due to bremsstrahlung become much more important than the ones due to ionization. For electrons, bremsstrahlung losses are larger than the ones of ionization at about 10 MeV in lead and at nearly 100 MeV in air.

Bethe and Heitler considered also what was occurring for the pair production and found that:

> For a given photon energy, the probability for pair production, in thicknesses of different elements for the same mass per unit area, increases with the atomic number Z of the element.

From an energy of 1 MeV, at which pair production begins to be possible, the pair production in a given thickness of material, first increases rapidly by increasing the photon energy and then levels at a nearly constant value.

Therefore, because the probability for Compton collisions decreases constantly by increasing energy, at low energies, photons are absorbed principally through Compton effect and at high energies, through pair production. The energy value at which the pair production overcomes Compton effect is nearly 5 MeV in lead, and about 20 MeV in air.

On October 1934, in London, an International Conference on Cosmic Rays was held, in which Millikan, Bowen, Neher, Anderson, Neddermeyer, Bethe, Rossi, and others participated. Anderson was continuing to measure the absorption coefficients and the energies of cosmic rays, now, however, not anymore to find proves of the Millikan "*birth cry*", but to compare the obtained results with the new quantum calculation results. The results did not agree with theory. In the cosmic ray section, C. Anderson and S. Neddermeyer presented results that seemed at odd with theory [42]. At that time people believed that all charged particles in the cosmic radiation were negative or positive electrons. The circumstance that many of these particles were able to traverse a metre of lead seemed to contradict the Bethe and Heitler theory. If the theory was correct only electrons of an absurdly great energy could be able to traverse those thicknesses. Moreover, Anderson in his cloud chamber observed that charged particles of an energy about 300 MeV did not lose so much energy as predicted by the theory. Bethe and Heitler at first concluded that the quantum theory for electrons at those energies was not suitable. Bethe was very dissatisfied with this admission, because he was convinced his theory was correct.

In the spring of 1933, Bohr had given a series of conferences at Caltech where a young researcher, Oppenheimer, was very much interested to the problems linked to the pair production and the two men discussed the problem.

J. Robert Oppenheimer (1904–1967) took his physics degree with laude at Harvard in 1925 and spent the following year in Cambridge at Cavendish Lab, working in experimental physics without any result, maturing there the decision to interest himself to theoretical physics. Therefore, in 1927, he accepted the offer of Max Born to come at the University of Gottingen, where he earned his Ph.D.

Here together with Born he wrote a work on the quantum theory of molecules in which the Born–Oppenheimer approximation used still today was introduced [43]. In 1928 he went to Harvard and in 1929 obtained a teaching position at the California University at Berkeley and simultaneously at Caltech, Pasadena. In 1930 he showed that the positive Dirac particles should have the same mass of electrons and could not be protons [44] as Dirac suggested, concluding that the proton should be an independent elementary particle with its own antiparticle. He made so the first prediction of the antiproton. He then considered the problem of the S.H. Chao anomalous absorption of ThC'' gamma rays, working at Caltech and together with Harvey Hall [45] calculated the relativistic photo-effect. An error brought them to the conclusion that the Dirac theory should be wrong for energies greater than mc^2. In 1932, together with J. Franklin Carlson [46, 47], in an effort to try to understand the great penetration of cosmic rays, he studied the ionization losses of relativistic electrons and of Pauli "neutrons". The conclusion was that neither the electrons nor the "neutrons" had the properties of the penetrating radiation. After the discovery of the positron and the discussions with Bohr, he gave the first correct description of the pair production by gamma rays [48] and showed that the theory explained quantitatively the absorption excess of ThC'' gamma rays in heavy elements. However he still believed the theory could be wrong at high energies. In 1934, he published a paper asking himself if the formulae for the absorption of high-energy radiation were valid [49]. His reply was in negative. He observed the problem was that the Bethe–Heitler theory was predicting an increase of ionization with energy in the cloud chamber that was not observed neither by Anderson and Nedderemeyer nor by Paul Kunze [50] of the University of Rostock. In addition, also the specific energy losses, as measured by Anderson and Neddermeyer, seemed too low to be compatible with the Bethe–Heitler theory. As a consequence, Oppenheimer observed: "*It is therefore possible to do justice to the great penetration of the cosmic rays only by admitting that the formulae are wrong or by postulating some other and less absorbable component of the rays to account for their penetration*" [51].

Successively he wrote a short theoretical note on the production of pairs of high-energy charged particles [52]. We will see in the next chapter that the solution to the many contradictions, was the existence of a new particle, the muon, which was definitely found only in 1938. However a decisive advance was made in 1936, when Oppenheimer and J. Franklin Carlson [10] considered what happens in showers made exclusively by photons and electron pairs.

We will see in Chapter 8 that the key to interpret the apparent contradictions between theory and experiments lies just in separating the *soft* from the *hard* component. Simple inspection of the Rossi curve, shows that the maximum occurs exactly at the lead thickness at which the soft component is absorbed. The conclusion should then be that the soft component produces many showers in thin layers of lead, while the hard component produces less of them in greater thicknesses. Therefore the soft component is the one responsible of showers. If calculations are applied only to this component, a theory of showers may be developed as shown in Fig. 7.5. We may, for example, assume a photon of an energy of several GeV enters a lead block. After

Fig. 7.5 Sketch of the development of an electron–photon shower. The *broken lines* represent photons and the *full lines* electrons

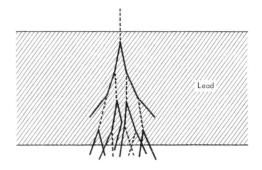

travelling a short distance (about 7 mm according to the theory) it disappears giving place to two electrons, one positive and one negative, that share the photon energy among them. The two electrons do not go very far as each one of them irradiates by bremsstrahlung a photon, so losing a large fraction of their energy.

The new photons soon materialize in electron pairs and the process repeats itself. At each interaction two particles spring from one. Two electrons come out from one photon, an electron and a photon come out from a single electron. Correspondingly the energy of an individual particle in the mean divides by half, and at the beginning the particles increase in number while their energy decreases.

At the end, when the original energy is divided among an increasing number of new created particles, most of the electrons have a too low energy to efficiently irradiate and are rapidly stopped by ionization. Similarly, also photons lose their energy and at some point are no more able to produce pairs and are absorbed by Compton collisions. Therefore the shower grows and then dies as clearly shown in Fig. 7.6.

This kind of showers takes the name of *electromagnetic* or *electrophotonic shower*.

The interpretation of showers along these lines was developed almost simultaneously by Homi J. Bhabha (1909–1966)[6] and Walter Heitler [53] in England, and J. F. Carson and Oppenheimer [10] in the United States and the results agreed with the experiment.

---

[6]Homi Jehangir Bhabha was born in Bombay and enrolled in Cambridge in 1927 where he was expected to study mechanical engineering. However, his great talent for pure science led him to a BA in theoretical physics in 1930. He received his Ph.D. in 1935 for a research at Cavendish Laboratory on cosmic ray produced electron showers. He paid extended visits to meet Fermi in Roma, Pauli in Zurich, and Bohr in Copenhagen. He was elected a fellow of the Royal Society in 1940. At the outbreak of World War II in 1939 he was in vacation in India where he remained. In 1942 he was appointed professor of cosmic ray physics at the Indian Institute of Science in Bangalore. In 1945 under his proposal the Tata Institute for Fundamental Research, for which he was the first director, was founded in Bangalore. In 1948 he was appointed chairman of the Indian Atomic Energy Commission. In 1954 the commission was reorganized as the Department of Atomic Energy with Bhabha as its secretary. The same year he was appointed director of the Atomic Energy Research Center. He died in January 1966 in an airplane crash.

Fig. 7.6 A shower developing through a number of brass plates 1.25 cm thick placed across a cloud chamber. The shower was initiated in the top plate by an incident high-energy electron or photon. The photograph was taken by the MIT cosmic ray group. (is fig. 7-5 from B. Rossi *Cosmic Rays* (McGraw Hill, New York, 1964))

Oppenheimer and Carlson in their work concluded that the success of the shower theory proved the validity of the Dirac electron theory and required the existence of a new type of particle in cosmic rays.

Almost simultaneously, Nevill F. Mott communicated the paper *"The passage of fast electrons and the theory of cosmic showers"* by Homi Bhabha and Walter Heitler to the London Royal Society [53]. In their work the authors affirmed:

"More recent experiments of Anderson *and Neddermeyer [Phys. Rev. 50 (1936) 263] have… led them to revise their former conclusion, and their new and more accurate experiments show that up to energies of 300 MeV (the highest energies measured in their experiments) and probably higher, the experimentally measured*

Fig. 7.7 Development of showers produced by electrons of 1.1 and 3 GeV in lead, according to the theory of Snyder. The number of shower electrons is plotted against the lead thickness. (is fig. 7-6 from B. Rossi *Cosmic Rays* (McGraw Hill, New York, 1964))

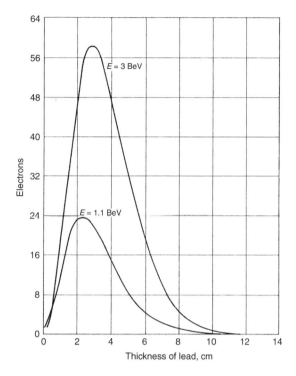

energy loss of fast electrons is in agreement with that predicted theoretically. In fact, one may say that at the moment there are no direct measurements of energy loss by fast electrons which conclusively prove a breakdown of theory... Under these circumstances, and in view of the experimental evidence mentioned above, it is reasonable.. to assume the theoretical formulae for energy loss and pair creation to be valid for all energies, however high, and workout the consequences which result from them".

The detailed theory of the electromagnetic shower process through the gradual growing by pair production and radiative collisions presented a mathematically difficult problem. Many people contributed to its solution: Lev D. Landau [54], Igor E. Tamm and S. Z. Belenky [55] in Russia, Hartland S. Snyder [56], Robert Serber [57] and Wendell H. Furry [58] in America.

The problem is of a statistical nature. The exact point where a specific photon materializes or a specific electron irradiates is at random. How the photon or the electron energy is shared between the two particles produced in a single event is also at random. Therefore not all showers started by photons (or electrons) of a given energy look similar. However the mean behaviour of showers may be inquired. For example Fig. 7.7 shows the mean number of electrons in a shower that may be found under different lead thicknesses when electrons of 1.1 BeV and 3 BeV have started the shower. These curves show a strong similarity with the one of Fig. 7.2.

The Bhabha and Heitler, and Carlson and Oppenheimer theories are so proved and a study of showers allowed further to conclude that:

The local cosmic radiation contains electrons and photons with energies of billions electron volt

The observed showers result from processes started by these electrons and photons

The individual interactions responsible for showers are the radiative collisions of electrons and the pair production by photons. These processes occur nearby atomic nuclei; however, they do not involve any change in the nucleus structure, contrary to the first hypothesis that showers were the result of disintegrations. Each interaction gives place to only two particles

The ionizing particles that constitute the soft component of the local cosmic radiation are the electrons of the showers originated in the atmosphere or in the roof of the buildings where the experiments are made

The theory of the electromagnetic (that is via bremsstrhalung and pair production) produced showers in air and other materials as described by standard quantum electrodynamics was so a great triumph for the theory.

When G. Pfotzer [59] found that the vertical intensity of cosmic rays at high altitudes reaches a maximum at about one-tenth of atmosphere and then decreases, the result, at least qualitatively, was well explained by the multiplicative theory of showers and was considered a great success.

So at the end, the quantitative explanation of showers as cascade processes of elementary events, was generally accepted with its quantitative treatment well given in the two independently developed theories of Carlson and Oppenheimer [10] and Bhabha and Heitler [53].

7.4 Heisenberg's Explosion Theory

Before the satisfactory shower theory we have described in the previous paragraph was developed, some physicists believed a shower could be produced in a single interaction of very high energy. In this way an explanation was given of the bursts (already found in 1928, by Gerhard Hoffmann at high altitudes [60]), that the shower theory was unable to explain and continued to intrigue the observers until 1953, when it was found they are not initiated by photons or electrons but by very energetic nuclear fragments.

This alternative explanation of the origin of showers, then rapidly eliminated, was proposed by Heisenberg [61] in a paper "*On the theory of showers in cosmic radiation* (Hoehnstrahlung)" in which he suggested a kind of strong nuclear interactions in which the simultaneous production of multiple pairs of neutrinos and electrons arose giving rise to an "*explosive shower*", as he called it. Hans Euler, who was the theoretical expert on bursts in Lipsia, analyzed the situation in his thesis [62], in a strict cooperation with the experimentalist Gerhard Hoffmann who meanwhile had become the successor of Peter Debye on the experimental chair in Lipsia. Euler admitted that a fraction of bursts could be attributed to electromagnetic showers, but a substantial fraction of them was clearly created in an explosive way. And indeed the hard, very penetrating component appeared linked to the explosive creation of many particles. However, the understanding of the hard component of

Fig. 7.8 The growth of showers. (Reprinted with permission from D.K. Froman and J.C. Stearns, Rev. Mod. Phys. **10**, 133 (1937). Copyright 1937 by the American Physical Society)

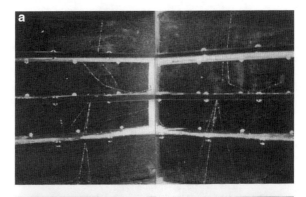

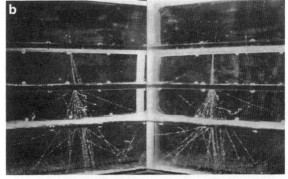

cosmic rays changed rapidly with the discovery of muons and in 1939, Heisenberg developed a different approach of explosive showers through meson theory.

Anyhow, the theory of showers generated in an explosive way received a hard stroke when L. Fussell Jr. [63] at MIT, by using a Wilson chamber into which three separated plates had been placed, showed that the most of showers develops and multiplies in successive steps through the plates (see Fig. 7.8 and also Fig. 7.6). The results showed definitely that showers do not develop in a single point, except very rare cases, and therefore the divergence of the showers from a point inside the layer is often misleading.

Out of 900 photographs, only three showed showers developing from a single point.

References

1. R.M. Otis, R.A. Millikan, Phys. Rev. **23**, 778 (1924)
2. G. Hoffmann, Ann. Phys. **82**, 413 (1927)
3. G. Hoffmann, Int. Conf. Nuclear Phys. (Lond.) **1**, 226 (1934)
4. G. Hoffmann, Phys. Z. **17**, 633 (1932)
5. E. Steinker, Phys. Z. **31**, 1019 (1930)
6. H. Schindler, Z. Phys. **72**, 685 (1931)

7. E. Steinker, H. Schindler, Z. Phys. **75**, 115 (1932)
8. E. Steinker, H. Schindler, Naturwissenschaften **20**, 491 (1932)
9. D.K. Froman, J.C. Stearns, Rev. Mod. Phys. **10**, 133 (1938)
10. J.F. Carlson, R.J. Oppenheimer, Phys. Rev. **51**, 220 (1937)
11. D. Skobelzyn, Z. Phys. **54**, 686 (1929)
12. D. Skobelzyn, C. R. Acad. Sci. **195**, 315 (1932)
13. P. Auger, D. Skobelzyn, C. R. Acad. Sci. **189**, 55 (1929)
14. C.D. Anderson, Phys. Rev. **41**, 405 (1932)
15. G.L. Locher, Phys. Rev. **39**, 883 (1932)
16. P. Kunze, Z. Phys. **79**, 203 (1932)
17. D. Skobelzyn, C. R. Acad. Sci. **194**, 118 (1932)
18. C.D. Anderson, S. Neddermeyer, Phys. Rev. **50**, 263 (1936)
19. L.M. Mott-Smith, G.L. Locher, Phys. Rev. **38**, 1399 (1931)
20. B. Rossi, Phys. Z. **33**, 304, 405 (1932)
21. B. Rossi, Z. Phys. **82**, 151 (1933)
22. J.C. Street, T.H. Johnson, Phys. Rev. **40**, 1048 (1932)
23. P.M.A. Blackett, G. Occhialini, Proc. Roy. Soc. (Lond.) **A139**, 699 (1933)
24. B. Rossi, *Cosmic Rays* (McGraw Hill, New York, 1964)
25. P.A.M. Dirac, Proc. Roy. Soc. (Lond.) **A111**, 405 (1926)
26. W. Gordon, Z. Phys. **40**, 117 (1926)
27. O. Klein, Y. Nishina, Nature **122**, 398 (1928)
28. O. Klein, Y. Nishina, Z. Phys. **52**, 853 (1929)
29. L.H. Gray, Proc. Roy. Soc. (Lond.) **A122**, 646 (1929)
30. C. Chao, Proc. Natl. Acad. Sci. U.S.A. **16**, 431 (1930)
31. C. Chao, Phys. Rev. **36**, 1519 (1930)
32. L. Meitner, H.H. Hupfeld, Naturwissenschaften **18**, 534 (1930)
33. W. Heisenberg, Ann. Phys. **13** (5), 430 (1932)
34. W. Heitler, Z. Phys. **84**, 145 (1933)
35. F. Sauter, Ann. Phys. **18**(5), 486 (1933)
36. W. Heitler, F. Sauter, Nature **132**, 892 (1933)
37. J. Bernstein, *H. Bethe, Prophet of Energy* (Basic Books, New York, 1980)
38. P. Gallison, Centaurus **26**, 262 (1983)
39. H. Bethe, Ann. Phys. **5**, 325 (1930)
40. H. Bethe, Z. Phys. **76**, 293 (1932)
41. H.A. Bethe, W. Heitler, Proc. Roy. Soc. (Lond.) **A146**, 83 (1934)
42. C.D. Anderson, S. Neddermeyer, International Conference on Physics, vol. I (Cambridge University Press, London, 1934) p. 171
43. M. Born, J.R. Oppenheimer, Ann. Phys. **84**, 457 (1927)
44. J.R. Oppenheimer, Phys. Rev. **35**, 562 (1930)
45. H. Hall, J.R. Oppenheimer, Phys. Rev. **38**, 57 (1931)
46. J.F. Carlson, J.R. Oppenheimer, Phys. Rev. **38**, 1787 (1931)
47. J.F. Carlson, J.R. Oppenheimer, Phys. Rev. **41**, 763 (1932)
48. J.R. Oppenheimer, M.S. Plesset, Phys. Rev. **44**, 53 (1933)
49. J.R. Oppenheimer, Phys. Rev. **47**, 44 (1935)
50. P. Kunze, Z. Phys. **83**, 1 (1933)
51. J.R. Oppenheimer, Phys. Rev. **47**, 44 (1935)
52. J.R. Oppenheimer, Phys. Rev. **47**, 146 (1935)
53. H.J. Bhabha, W. Heitler, Proc. Roy. Soc. (Lond.) **A159**, 432 (1937)
54. L. Landau, G. Rumer, Proc. Roy. Soc. (Lond.) **166**, 213 (1938)
55. I. Tamm, S. Belenky, J. Phys. (USSR) **1**, 177 (1939)
56. H. Snyder, Phys. Rev. **53**, 960 (1938)
57. R. Serber, Phys. Rev. **54**, 317 (1938)
58. W.H. Furry, Phys. Rev. **52**, 569 (1937)
59. G. Pfotzer, Z. Phys. **102**, 23 (1936)

60. G. Hoffmann, F. Lindholm, Gerlands Beitrage zur Geophysik **20**, 12 (1928)
61. W. Heisenberg, Z. Phys. **101**, 533 (1936)
62. H. Euler, Z. Phys. **110**, 450, 692 (1938)
63. L. Fussell Jr., American Physical Society meeting in Washington, April 29–30, 1937. Phys. Rev. **51**, 1005 (1937)

Chapter 8
The Muon

8.1 Introduction

The *muon* has been the first discovered particle with a mass between an electron and a proton. Originally named *mesotron* [1][1] to indicate this characteristic, was later called *meson* (actually, after π *meson* was discovered, at first the muon was called *meson* μ).

Today, the name *meson* is reserved to particles with a strong interaction with atomic nuclei, are unstable and have integer spin (therefore obeying the Bose–Einstein statistics). According to the present views, they are formed by a combination of *quark* and *antiquark*[2] and therefore the denomination "meson" for the μ meson has become improper. Thus the decision to call it *muon* to clarify that it is not a meson, but a member of the *leptons* family, to which also belong the electron and the photon, all particles presumably without an elementary structure, not subjected to the strong interaction, that is to the forces present inside the nuclei. The muon interacts only through electromagnetic interactions.

As time went by, with the use of particle accelerators, hundreds of particles have been found with intermediate masses between electron and proton or even with masses larger than the proton.

---

[1] A name proposed by C.D. Anderson.

[2] According to now-a-day most accepted theory, called *standard theory*, protons, neutrons and mesons are actually particles composed by smaller entities, called *quark*. There exist three families of quark, each with two quarks. The first family contains two quarks called *up* and *down* together with their antiparticles. The second is made up by two quarks, *top* and *bottom*, together with their antiparticles and the third one has two quarks, named *charm* and *strange* with their antiparticles. The up, charm and top quarks have an electric charge 2/3 of the electron charge, and the other three quarks have a charge $-1/3$. The nuclear or *hadronic* forces exert essentially among quarks. A proton, is made by two up and one down quark and a neutron has a up and two down quark. In the beta disintegration the neutron emits an electron and turns into one proton, that is a down quark changes into an up. This process lies therefore the two leptons electron, neutrino to the quark up and down. The π meson is formed by an up quark and one down antiquark.

M. Bertolotti, *Celestial Messengers: Cosmic Rays*, Astronomers' Universe, DOI 10.1007/978-3-642-28371-0_8, © Springer-Verlag Berlin Heidelberg 2013

The discovery of the muon, a particle with a charge equal to the electronic charge, positive or negative, and of an intermediate mass between a proton and an electron, was another example of the great fertility of the cosmic ray field. At variance with the positron discovery, that occurred essentially through a single photograph, the discovery of the muon was the final step of a number of systematic and accurate investigations lasted over 2 years and devoted to solve some apparent paradoxes present in the study of comic rays. The credit of his discovery is usually given to S.H. Neddermeyer and C. D. Anderson [2], and to J. C. Street and E. C. Stevenson [3–5].

Let us see how it happened.

8.2   Something Does Not Fit

Following the positron discovery, there was the tendency to believe that the charged particles in the cosmic radiation were only electrons and positrons plus some protons.

However, the sole particles surely identified as electrons and positrons, by studying the deflection produced by a magnetic field in a Wilson chamber, were those with relatively low energy, up to about 10 MeV. High-energy particles, that form the majority of the flux of cosmic rays, suffered extremely small deflections, difficult to measure with enough accuracy, so that from their deflection it was difficult to extract the information on their energy.

Another way to gain information on their energy was offered by the study of their ability to traverse the matter, and was therefore on the study of this characteristic that researchers devoted much attention.

We already discussed the results obtained by Rossi, and his absorption curve as a function of the thickness of absorbers, which shows that in the charged particles flux at sea level, two components may be individuated: a *soft* component that is absorbed in the first 1 or 2 cm of lead, and a *hard* component, much more penetrating, able to traverse even 1 m of lead. These results were confirmed by a number of repeated measurements. Here, shown in Fig. 8.1, is a typical absorption curve from Street et al. [6] The curve was obtained interposing a lead thickness between a counter telescope made with four Geiger–Müller counters and registering the number of quadruple counts as a function of the thickness of the absorber expressed in electrons cm^{-2}, that is a quantity proportional to gr cm^{-2}. Because absorption is roughly proportional to density, the results for absorbers of different atomic weight, fall on a single curve if they are reduced to this common unit. From the curve, after some calculation, one finds that about 15% of particles are stopped during the first 1.5 cm of lead, but a further thickness of 1.5 cm of lead does not stop a tenth of the remaining particles.

It was then natural to describe observations in terms of a *soft component*, easily absorbed, and a *hard component*, made by the penetrating particles. This distinction was pointed out by P. Auger who suggested the two components could be two different species of particles [7].

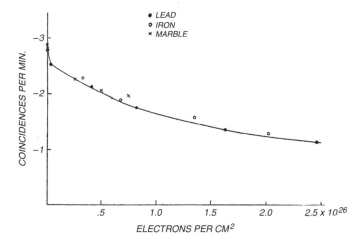

Fig. 8.1 The absorption of cosmic rays in lead, iron and marble. (Reprinted with permission from J.C. Street, R.H. Woodward, E.C. Stevenson, Phys. Rev. **47**, 891 (1935). Copyright 1935 by the American Physical Society)

The two components may be identified as separate ones because the soft component that produces the showers is characterized by the following properties:

1. It is absorbed in a few centimetres of lead
2. It produces showers in a few centimetres of lead
3. Its intensity grows rapidly with altitude in the atmosphere and decreases quickly underground
4. The atomic absorption is proportional to the atomic number squared (Z^2)

On the contrary the hard component is able to penetrate down to more than 100 cm of lead, produces very few showers, has a slower variation with altitude in the atmosphere and with depth underground and its absorption per atom is proportional to Z.

8.3 More on the Quantum Theory and the Energy Loss by Fast Particles

Since 1934, experiments near sea level were showing that the greater part of both positive and negative cosmic rays, with momentum of the order of 300 MeV/c, suffered an energy loss in solid absorbers that could be explained on the basis of the sole ionization, without recurring to bremsstrahlung radiation. However, the just-developed quantum electrodynamics was predicting that for electrons the losses due to bremsstrahlung at those values of momenta should be very important, while for protons this loss should be negligible, as heavy particles have a low probability to emit braking radiation.

It was, however, difficult to admit that the greater part of particles had a proton mass, as the number of observed slow protons, could hardly be reconciled with the great number of fast protons needed to explain the abundant hard component; and, what is more, the existence of negative protons, that at the time were completely outside any theory, should have been postulated. These difficulties induced many people to believe that the quantum theory of radiation was not valid at high electronic energies greater that about 300 MeV, and that the penetrating particles were actually positrons and negative electrons. However, there was a number of contradictions which were solved convincingly only in 1937.

The most dramatic phenomenon observed in the cloud chamber and with counters were the showers. Now the question was whether the shower-forming particles were ordinary electrons, another kind of electrons (Anderson [8] remembers that speaking with Neddermeyer they indicated this new electron as "*red*" to distinguish it from the ordinary one nicknamed "*green*") or even new particles. Both Rossi and Curry Street later remembered that in those days, in 1935, they believed the particles in the showers being a new kind of particles and the penetrating ones ordinary high-energy electrons which simply did not obey the Bethe–Heitler theory. This was a natural conclusion due to the strange behaviour of showers where often many particles were ejected from a single point [9].

The current theory did not offer any explanation of simultaneously multiple emission of particles. Blackett, Heisenberg, Bohr and Pauli all believed that a new theory was needed.

In the United States, Curry Street (1906–1989) had, in order to study showers, built, during his last year at Bartol Institute (1931–1932), logic circuits based on Rossi publications. Placing a suitable number of counters with coincidence or anticoincidence circuits, it was possible to study the flux of particles with great precision and, relaying on these techniques, he measured together with Johnson the east-west effect in late 1933 (thus defeating Rossi).

When Street arrived at Harvard in the fall of 1933, as an instructor, together with two students, E.C. Stevenson and L. Fussell, he showed that coincidences occurred even between counters separated by a great thickness of lead (tens of centimetres). So, when the paper by Anderson, Millikan, Neddermeyer, and Pickering et al. [10] appeared in 1934, criticizing the work of Rossi, maintaining that counters in coincidence did not detect a single particle that had traversed a notable thickness of lead, but that coincidences were due to showers; Street also was implicated. Millikan was convinced that nothing could pass through such a great thickness provided by a meter of lead, because this would have been against his birth cry theory.

Street did not agree and began to build a Wilson chamber controlled with two counters in coincidence. With this configuration he showed that individual charged particles crossed at least 45 cm of lead [11, 12], confuting Millikan's observation.

Figure 8.2 shows the used configuration [6]. Three counters were placed one above the other in a vertical plane with their axes horizontal. The absorbing material was introduced between the central and the upper counter and a vertical cloud chamber between the central and lower counter. Figure 8.2 shows that actually the upper counter consisted of three counters in parallel to increase the acceptance of

Fig. 8.2 Geometrical arrangement of counters (C_1,C_2,C_3), absorbing material (Pb), and cloud chamber (Ch). On the left the side-view, while on the right is the prospect. (Reprinted with permission from J.C. Street, R.H. Woodward, E.C. Stevenson, Phys. Rev. **47**, 891 (1935). Copyright 1935 by the American Physical Society)

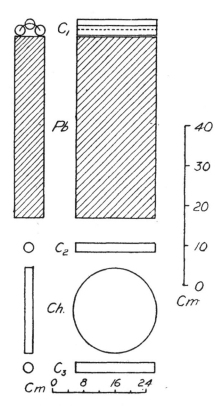

the system. The whole system was placed far from materials that if struck could produce lateral showers.

More than 200 photos, obtained in the chamber when the three upper/central/lower counters were in coincidence and by using interposed lead thicknesses of 45 or 15 cm, showed that in nearly 90% of cases the chamber showed a single track that crossed the counters (see Fig. 8.3). Street concluded: "*These direct observations with thick absorbers show that the first objection of Anderson, Millikan et al. is not valid. Our results are not contradictory to their actual observation that the relative importance of shower produced coincidences is increased by placing one cm of lead between a pair of counters. However, their assumption that showers become increasingly important with thicker absorbers is definitely in error. Auger and Ehrenfest [13] have recently reached the same conclusion from a similar experiment*".

Street put also in evidence that with the single tracks obtained after 45 cm of lead it is sporadic to find associated tracks, and assembled another counter system with lead thicknesses in between to investigate the abundance of rays associated to penetrating particles.

With an ingenious system able to detect—via counters—the presence of straight tracks associated with lateral tracks, utilizing a disposition of the kind of the one realized by Johnson and and called by them *hodoscope* [14], Street showed also

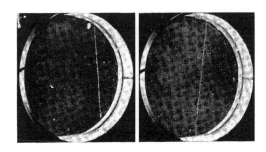

that lateral tracks were present only in a small percentage and could not explain the contemporary discharge of the four counters placed all along the same line. A conclusion that once more confirmed that the argument of Anderson, Millikan et al. was ungrounded.

Some authors interpreted the increase of the shower production up to a maximum after 1.5 cm of lead as an indication that the particles forming the showers had a penetration length of 1.5 cm of lead. Street and his students [15, 16] showed, with their apparatus, that it was not the case and that the particles forming the shower were absorbed with an exponential law. Doing this, for the first time, the authors put attention on the shower particle properties as distinct from the shower properties. In the short summary of the work presented at the American Physical Society meeting in Washington, April 25–27, 1935, it was written:

"A group of Geiger–Müller counters has been arranged to investigate the production of showers. A slight modification of this arrangement has been used to measure directly the absorption of the shower particles. The shower production curve for lead obtained with this arrangement is in agreement with similar data reported by numerous observers but the directly observed absorption of the shower particles is quite different from the usual assumption that these particles have a penetrating power of about 1.5 cm of lead (thickness for maximum observed number of showers). The shower particles are absorbed approximately exponentially and fall to one-third value in three mm of lead".

Meanwhile people found that the most of particles existing in showers or which produced showers, lose their energy in agreement with the quantum theory of radiation, while most of the individual particles with the same momentum lose their energy at a rate which could very well be explained by the sole ionization. Therefore, the particles of the showers are electrons, while the single particles have some property different from electrons. Most of these particles could not be protons, as in the considered momentum range they ionized much less than a proton of the same momentum. Neither they could have a multiple electronic charge because when they were fast they displayed a ionization very similar to the one of an electron. As a working hypothesis, the existence of positive and negative singly charged particles of a mass intermediate between that of a proton and of an electron was considered.

The success of the application of quantum electrodynamics to the shower theory and to the transition effects (Bethe–Heitler and Openheimer–Carlson theories) and the behaviour of the curve altitude-ionization (Pfoster) gave a strong confirmation of the validity of quantum theory of bremsstrahlung and eventually favoured an interpretation of the results on the basis of an intermediate mass particle.

In a further study, Street, with an ingenious configuration with two Wilson chambers [4, 17], demonstrated that the fraction of particles not producing showers (of the order of ten thousands or more) which penetrated 6 cm of lead was much greater than that allowed by the Bethe–Heitler theory for electrons of the same momentum and concluded they were not electrons and did not ionize enough to be protons.

At last, in October 1937, Street and Stevenson [5] in a short letter to Physical Review published some photos taken with the Wilson chamber, showing a track of a particle with the charge of a negative electron and a mass intermediate between the electron and proton mass.

8.4 The Street and Stevenson Experiment

To measure the mass of the new particle it was necessary that it possessed such a low energy to have an appreciable curvature in a magnetic field. Because the low-energy particles were rare, J.C. Street and E.C. Stevenson [5] at Harvard University, made an arrangement that allowed to take pictures only when a particle stopped in the chamber. To this aim, they arranged their Wilson chamber with a counter disposition able to tell when a particle had entered the chamber but has not come out from the walls. With this procedure, in 1937, they obtained the desired picture. The drops concentration along the track indicated an ionization density about six times as larger as normal thin tracks and the track curvature indicated a magnetic rigidity of 9.6×10^4 gauss cm. The rest mass was found approximately 130 times the rest mass of the electron.

The used set-up is shown in Fig. 8.4. Three counters 1, 2, and 3 are aligned one above the other in a vertical plane with their axes horizontal. Between counter 2 and 3 a lead filter L is inserted for removing shower particles. The particles which pass the three counters in coincidence pass also the Wilson chamber C where a magnetic field of 3,500 gauss is present. In order to select only particles at the end of their range, the three counters 4 below the Wilson chamber were in anticoincidence.

Among 1,000 photos, one track was obtained which could be assigned to a negative particle of a mass of approximately 130 electronic masses, within a 25% approximation due to the uncertainty in the ion count (see Fig. 8.5). It was interpreted as a particle of negative charge coming from above. An up-travelling proton would have an energy of 440,000 eV and a range of approximately one cm in the chamber while the observed track extended for 7 cm.

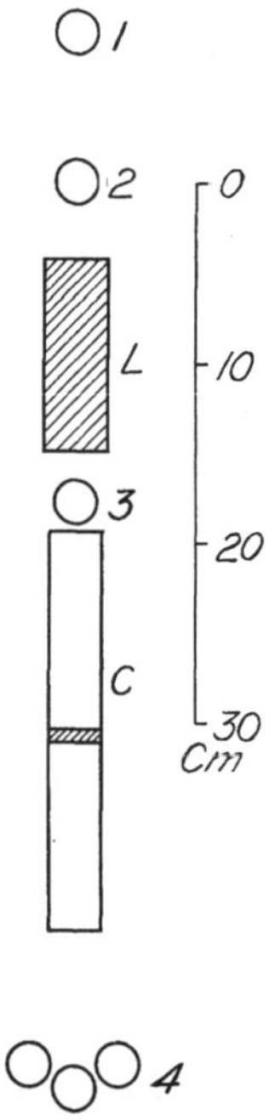

Fig. 8.4 The Street set-up
(Reprinted with permission
from J.C. Street and
E.C. Stevenson, Phys. Rev.
52, 1003 (1937). Copyright
1937 by the American
Physical Society)

Fig. 8.5 The photo shows
only the right quadrant
(Reprinted with permission
from J.C. Street and
E.C. Stevenson, Phys. Rev.
52, 1003 (1937). Copyright
1937 by the American
Physical Society)

8.5   Neddermeyer and Anderson Measurements

To study high-energy cosmic rays and to infer the nature of their particles, Anderson
and Neddermeyer, after the discovery of the positron, moved their cloud chamber
with magnetic field on the summit of Pikes Peak, in Colorado, where high-energy

cosmic rays were more abundant than at sea level. Analyzing the cloud chamber photos after exposition on the summit lasting the whole Summer, they found tracks of both positive and negative particles which looked different from the ones of electrons and protons and seemed to have an intermediate mass. While they were still studying these tracks, Millikan ordered the cloud chamber with its equipment to be brought to Coco Solo, in the Panama Canal zone, in order to investigate the dependence of cosmic radiation from latitude at sea level, and so the research suffered an interruption.

When back home, in June 1936, the two researchers sent an extended study to Physical Review [18] discussing nearly 10,000 photos and making a comparison between the measurements done on Pike Peak at an elevation of 4,300 m above sea level, and in Pasadena. They showed that the ratio between the number of showers produced per unit time on Pikes Pike and those in Pasadena (near sea level) was larger than the corresponding ratio between the number of single particles in the two locations. This was in agreement with other measurements performed with counters and ionization chambers [19–26]. They performed also a direct measurement of the losses in lead of those particles which were found exclusively in the showers. These measurements, extended up to 400 MeV, offered the surprising result that if the measurements were limited to the particles in the shower, the experimental loss values were in agreement with the Bethe–Heitler theory. On the contrary, the behaviour of the single particles at sea level that made up the hard component, was in disagreement. For larger energies, the authors had not reliable data, but something did not seem right. By quoting measurements by Rossi [27] and Street et al. [6], who found the presence of particles too penetrating to permit their identification with electrons obeying the Bethe–Heitler formula, they concluded [18]: "*It is obvious that either the theory of absorption breaks down for energies greater than about 1000 MeV, or else that these high energy particles are not electrons*".

Incorrectly they added that in the photos they found there was "*evidence of disintegration of a heavy element, vis., lead, by photons. The probability of this type of nuclear photo-absorption, although apparently much smaller than the probability of producing positive-negative electron pairs and showers, seems to be large enough to account for many of the strongly ionizing particles observed*" and concluded "*the majority of the heavy tracks seem to represent protons*".

In their research, Anderson and Neddermeyer placed a lead plate of 3.5 mm thickness across the Wilson chamber to study the energy lost by cosmic rays in matter. Many photos showed particles with magnetic rigidities between 10^5 and 10^6 gauss cm which seemed to ionize not so heavily as protons of the same rigidity, but that were losing, crossing the layer, less energy than the bremsstrahlung theory predicted for electrons.

However, the energy lost through radiation is a stochastic effect. While the theory predicts, on the average, a large energy loss in 3.5 mm of lead, it would always be possible for some electron to travel through the absorber without passing so near to an atomic nucleus to undergo appreciable losses of energy. Therefore, the Anderson and Neddermeyer experiments, while very interesting, did not allow to derive any sure conclusion, and in their first paper they reached the wrong conclusion. To obtain

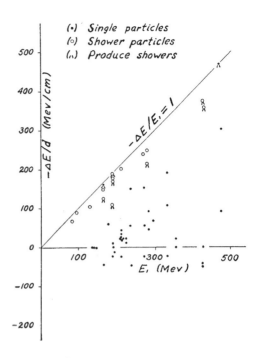

Fig. 8.6 Energy loss in 1 cm of platinum. (Reprinted with permission from S.H. Neddermeyer and C.D. Anderson, Phys. Rev. **51**, 884 (1937). Copyright 1937 by the American Physical Society)

more definite proofs, they replaced the lead plate with a platinum plate of 1 cm thickness, which was equivalent to about 2 cm of lead, as far as radiation losses are concerned [2]. The probability that an electron passes through such a plate without suffering a radiative collision is negligible. They made 6,000 photos.

The two researchers plotted the momentum lost by particles traversing the platinum plate measured separately for single particles and for the particles belonging to showers, or able to produce shower (called *shower-particle*). The graph in Fig. 8.6 shows these measurements as a function of the initial energy, which is proportional to the momentum. Simple inspection shows that the particles can be divided into two rather well-defined groups. The energy (momentum) loss is sensibly constant for single particles, while it is nearly proportional to energy for shower particles. Moreover, single particles never produce showers when crossing the metallic plate, as, on the contrary, often do the shower particles. Finally, going from the sea level to 4,500 m, the shower particle number becomes three times larger than that of single particles.

This deep difference in behaviour was correctly explained assuming that the shower particles were electrons or positrons (the *soft component* of cosmic rays) and that the single particles (the *hard component*) had a mass between 100 and 300 electronic masses. These experiments, which offered the first experimental evidence of the muon were published in the Spring of 1937 and established that two well-defined groups of particles exist. The first one consists of penetrating particles that lose a small fraction of their energy in traversing the metal plate. The second group consists of particles easily absorbed, losing a great fraction of their energy while crossing the metal plate. The energy loss of these particles is in agreement with the values calculated for electrons that irradiate and ionize according to the theory. Therefore these particles are likely to be electrons. The energy loss of penetrating particles, on the contrary, is entirely justified by the sole ionization and therefore these particles do not irradiate appreciably.

The easily absorbed particles, because often present in groups, are presumably secondary particles of a shower originated above the chamber. They often produce showers, confirming the hypothesis that they are electrons. On the contrary the penetrating particles usually appear as single tracks in the cloud-chamber photos.

Among particles exhibiting the same curvature in the cloud chamber, some, traversing the plate, lose great quantities of energy and others a small quantity. Electrons cannot behave as easily absorbed particles below some critical energy and as penetrating ones above it, because if they did so, the losses of energy would be always the same for particles of the same rigidity (that is of the same curvature). The penetrating particles are therefore not electrons.

Many of the penetrating particles had a magnetic rigidity lower than 1.5×10^6 gauss cm. Below this value protons ionize at least three times more than fast electrons. The ionization density along the tracks of the penetrating particles was, however, nearly the same of that of electrons. Because the penetrating particles ionized less than protons their mass should be smaller than the one of the proton. On the other hand because they did not irradiate as much as electrons, their mass should be greater than the one of electrons.

From these observations, Neddermeyer and Anderson concluded that very likely particles with unitary charge but with mass greater than that of a normal electron and smaller than a proton should exist. They [2] wrote: *"The present data appear to constitute the first experimental evidence for the existence of particles of both penetrating and non-penetrating character in the energy range extending below 500 MeV. Moreover, the penetrating particles in this range do not ionize perceptibly more than the non-penetrating ones, and cannot therefore be assumed to be of protonic mass"*. The non-penetrating particles are positive and negative electrons.

Then they observed that the penetrating particles presented themselves with both the negative and positive sign and at the end of the paper, in a note added in proof, they mentioned the measurements presented by Street at the 40th Meeting of the American Physical Society April 29, 1937. Anderson and Neddermeyer, however, were not able to measure the mass with a reasonable accuracy.

The discovery of the muon is usually associated to a couple of photographs. One is the one published by Street and Stevenson [5], shown here as Fig. 8.5, showing a dense track with an ionization and curvature indicating a mass of approximately 130 electronic masses. The other photograph was obtained by Neddermeyer and Anderson [28] (see Fig. 8.7). To obtain this photo, Anderson and Neddermeyer placed a Geiger–Müller counter inside the chamber coupled in coincidence to a second counter placed above the chamber. The photo shows a positively charged particle which after traversing the counter emerges with an energy low enough for it to be brought to rest in the chamber. In the upper part, the track is fine and discontinue showing a weak ionization which is consistent with a great energy.

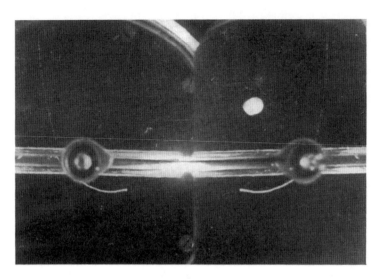

Fig. 8.7 A stopping meson (Reprinted with permission from S.H. Neddermeyer and C.D. Anderson, Phys. Rev. **54**, 88 (1938). Copyright 1938 by the American Physical Society)

Below the counter, the track is thick and cannot be interpreted as an electron track but must belong to a particle of greater mass. From the characteristics of the track and the knowledge of the applied magnetic field, the mass was estimated approximately 240 electronic masses.

The comment to the photo is the following: "*A positively charged particle of about 240 electron-masses and 10 MeV energy passes through the glass walls and copper cylinder of a tube-counter and emerges with an energy of about 0.21 MeV. The magnetic field is 7900 gauss. The residual range of the particle after it emerges from the counter is 2.9 cm in the chamber (equivalent to a range of 1.5 cm in standard air). It comes to rest in the gas and may disintegrate by the emission of a positive electron not clearly shown in the photograph. It is clear from the following considerations that the track cannot possibly be due to a particle of either electronic or protonic mass. Above the counter the specific ionization of the particle is too great to permit ascribing it to an electron of the curvature shown.*

The curvature of the particle above the counter would correspond to that of a proton of 1.4 MeV and specific ionization about 7000 ion-pairs/cm, which is at last 30 times greater than the specific ionization exhibited in the photograph. The curvature (radius nearly 3 cm) of the portion of the track below the counter would correspond to an energy of 7 MeV if the track were due to an electron. An electron of this energy would have a specific ionization imperceptibly different from that of a usual high energy particle which produces a thin track, and in addition would have a range of at least 3000 cm in standard air instead of the 1.5 cm actually observed. Moreover if the particle had electronic mass and emerged from the counter with a velocity such that its specific ionization were great enough to correspond to that exhibited on the photograph, its residual range (in standard air) would be less than

Fig. 8.8 A muon track obtained (Reprinted with permission from Y. Nishina, M. Takeuchi, T. Ichimiya, Phys. Rev. **55**, 585 (1939). Copyright 1939 by the American Physical Society)

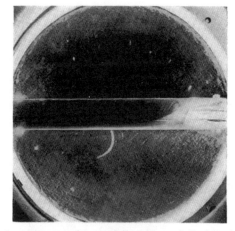

FIG. 1. Wilson track of a mesotron. H = 12,600 oersteds. $H\rho$ = 3.88 × 10⁴ oersted·cm. Observed range = 6.15 cm.

Fig. 8.9 This is may be the first photo of a muon (Kunze [30]) (Reprinted with permission from S.H. Neddermeyer and C.D. Anderson, Rev. Mod. Phys. **11**, 191 (1939). Copyright 1939 by the American Physical Society)

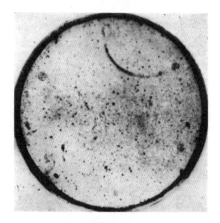

FIG. 18. 15,000 gauss. An early photograph (1931) of a particle with an $H\rho$ of 6×10⁴ gauss cm, whose ionization quite certainly exceeds that of a fast electron. A value 10 times the minimum should correspond to a mass of 150, and 15.5 times the minimum to a mass 200. Either of these ionization values could be consistent with the photograph. Range and $H\rho$ give a rough upper limit of 300. (See also Fig. 5 by Kunze, Zeits. f. Physik **83**, 10 (1933).)

0.05 cm instead of the 1.5 cm observed. A proton of the curvature of the track below the counter would have an energy of only 25,000 eV and a range in standard air of less than 0.02 cm".

Another beautiful photo, obtained by Nishina et al. [29], is shown in Fig. 8.8.

Another photo is also often shown (Fig. 8.9), which may be considered the ancestor of the previous ones, in which a track is present that a posteriori may be interpreted as a muon.

This photo is also reported by Anderson in a paper on Reviews of Modern Physics [31].

The new particle received many names: mesotron, barion, yukon, and meson. This last name was accepted for a while. Later, when the existence of other mesons was found, to the one discovered by Neddermeyer and Anderson and by Street and Stevenson was given the name meson μ. Today it is called simply muon, as explained at the beginning of this chapter.

Precise measurements of the muon mass were done some years later by Robert B. Brode [32] and collaborators at the University of California, Berkeley, in 1949, by using an ensemble of cloud chambers with a magnetic field and lead plates, consistent with a value of 215 ± 4 electronic masses. Since 1950, muons were produced also in the lab by using particle accelerators. Today, there exist production factories of intense muon beams obtained by collisions of protons accelerated in the large accelerators.

8.6   The Theoretical Prediction of the Existence of Mesons

Dirac, with his relativistic theory, was able to predict the existence of the positive electron before its actual experimental discovery. Analogously, the Japan physicist Hideri Yukawa had already introduced a particle of mass intermediate between those of the electron and proton to explain the forces exerted in a nucleus between protons and neutrons.

The west physics entered Japan at the time of the Meji revolution (1868). The first physics scholars were exclusively samurai. In particular 27 of them attended three schools in Tokyo: one French, one German and one English, in which physics was taught in the respective languages. Some samurai went in Europe for studying. One of them, Yamagawa, was sent by his feudal master in 1871 to study in Europe and, after coming back, wrote an insulting autobiography against foreign people.

The physicist Hantaro Nagaoka (1865–1950) travelled in Germany and England around 1900. He was the author of a planetary theory of the atom similar to the one of Rutherford, which was, however, unsuccessful. Coming back in Japan he had great influence and settled in Tokyo as a professor of theoretical physics in 1906, becoming many years later the dean of Osaka University. After the First World War, Y. Nishina, coming back from Europe, established the Scientific Research Institute (RIKEN) in Tokyo giving a great impulse to research, with the first important Japanese contributions to Physics.

Hideki Yukawa is one of the first Japan physicists to receive his entire education in Japan. He was born in Tokyo, January 23, 1907, the son of a Geology professor at the University of Tokyo, and studied in Kyoto, where he graduated in 1929. Together with his classmate Sin-Itiro Tomonaga (1906–1979), who later began famous for his studies on quantum electrodynamics, he remained at University as unpaid assistant until 1932. Tomonaga went then to Tokyo under Nishina at RIKEN, and Yukawa was appointed lecturer at the Imperial University of Kyoto between 1932 and 1939 and also lecturer at Osaka University. He earned his Ph.D. in 1938 and since 1939 was appointed professor of theoretical physics at the Kyoto University. While at Osaka University he published the paper "*On the interaction of elementary particles*" [33] in which he proposed a new theory for the nuclear forces and predicted the existence of the meson.

He was awarded the Physics Nobel Prize for his work in 1949 (he was the first Japanese to receive this prize). From 1948 to 1953 he was in the United States, first as a visiting professor at the Institute for Advanced Studies, Princeton, and then (since 1949) as professor of Physics at Columbia University. In 1953 he came back in Japan at the Kyoto University as the head of the Institute for Fundamental Research. He died on September 8, 1981 in Kyoto.

When Hideki Yukawa in 1935 published his paper, people already knew of the existence inside the atomic nucleus of protons and neutrons. Protons repel each other because of the Coulomb force between electrical charges of the same sign, while neutrons do not interact electrically among themselves or protons because they lack of an electric charge. However the atomic nuclei are notably stable structures. This circumstance made it necessary to assume the existence of attractive forces of a non-electrical nature between protons and neutrons. At the short nuclear distances (the dimension of a nucleus is of the order of 10^{-15} m) these nuclear forces are much stronger than the electric forces, while they should be much weaker at larger distances. In other words, the nuclear forces decrease with distance much more rapidly than the electrical ones, inversely proportional to the distance between particles squared.

In the electromagnetic case, considering two charged particles A and B placed at some distance between them, the electric mutual force is carried by a field, the electric field to which are associated the photons, the quanta of electromagnetic energy. In the case of two nucleons, Yukawa assumed that a nuclear field existed to which a suitable quantum was associated, the *meson*. However, at variance with the electrical forces that extend over very large distances, the nuclear forces extend only on distances of the order of the nuclear dimension and go to zero very rapidly for larger distances. This has the consequence that their quantum should have a mass. Let us consider a proton and a neutron at some distance. Each one is surrounded by its own field which interacts with the other particle. To be surrounded by its own field means to emit quanta of this field, and therefore the particles should emit mesons. However, emitting a meson costs energy and, if it has a mass m, this energy is mc^2. In quantum mechanics energy may be available for a short time, thank to the Heisenberg uncertainty principle. The new quanta should also be instable, that is had to decay as radioactive substances do. In fact, to explain the beta decay, in the Yukawa theory it was necessary to postulate a radioactive decay of the new particle, with a lifetime of the order of a microsecond to agree with the Fermi theory.

H. Yukawa [33–35] built his theory, based on the exchange force scheme already considered by Fermi [36], but replacing the Fermi electron and neutrino pair with a new particle. The distance over which the nuclear forces were found experimentally to operate required a mass of this hypothetical particle about 200 times the electron mass.

This estimate may be done very simply, albeit not very correctly. Nuclear forces exert over a distance of the order of the nuclear radius, that is 1.3×10^{-13} cm. The force between two nucleons may be interpreted as due to the emission of one quantum by one of them and its absorption by the other nucleon. The quantum to travel from one nucleon to the other employs a time larger than r/c where r is the distance between the two nucleons. During this time, the energy conservation is violated because a quantum of energy mc^2, if m is its mass, has been emitted. However, according to the Heisenberg uncertainty principle, it is impossible to determine the energy with a precision better than

$$\Delta E \cdot \Delta t \approx h/2\pi,$$

If we put $\Delta E = mc^2$ and $\Delta t = r/c$ we find

$$m = h/2\pi rc = 6 \times 62 \times 10^{-34} \times 10^{15}/3 \times 10^8 = 240 \text{ electronic masses}$$

(the electron mass is 9×10^{-31} kg). So for the short time Δt, no conservation principle is violated.

These quanta must be neutral or positively or negatively electrically charged to account for the forces between proton–proton, neutron–neutron, and proton–neutron, nearly equal as experimentally found.

Anderson was not aware of the Yukawa work, and the war isolated Japan. When Yukawa knew the results of Anderson, he submitted a letter to the English journal Nature [37]. In it he suggested that the particle observed by Anderson could be his quantum. However the editor of Nature rejected the letter that was later published by a Japanese journal [38]. Yukawa research was first quoted in the west by Oppenheimer and Serber [39].

The discovery of the muon, in 1937, closed the second period in the cosmic ray history which opened with the Bothe and Kolhoerster experiments. The nature of the local cosmic radiation observed in the atmosphere was now clear: the penetrating particles are muons, the easily absorbed particles are electrons. The non-ionizing particles are photons. Muons, electrons and photons accounted for practically all radiation at sea level.

8.7   The Experimental Discovery of the Muon Decay

Between 1936 and 1937 in England, France, Germany and Italy, researchers made precise measurements of cosmic ray particles at different altitudes in the atmosphere. The results were very strange. At variance with Millikan's first measurements, it seemed that, when layers of the same mass per unit area were compared, air absorbed cosmic rays more effectively than solid or liquid matter.

Moreover the less dense air at very high altitudes seemed to be a better absorber of the denser air in the low atmosphere.

According to Yukawa his particles were unstable. If mu mesons were the Yukawa particles, they should decay spontaneously. In 1938, the German physicist H. Kuhlenkampff [40] pointed out that this property offered a natural explanation for the anomalous absorption of the cosmic ray particles in air.

The argument is very simple. Let us consider, for example, a water layer of 10 cm thickness. At sea level an air layer of the same mass per unit area is thick 8,000 cm, because air is 800 times less dense than water. At an altitude of 5,100 m, where the air density is half than the sea level, the air equivalent thickness is 16,000 cm.

A particle travelling at a speed close to that of light crosses 10 cm in 3.3×10^{-10} s or 3.3×10^{-4} μs. The same particle needs 0.265 μs to cross 8,000 cm and 0.53 μs to travel for a distance of 16,000 cm.

Let now assume that mesons have a lifetime of 1 μs. Practically, no one of them will decay crossing 10 cm of water. Only mesons with a range less than 10 cm of water would be apparently stopped by this absorber. On the contrary, many mesons will decay spontaneously while they traverse 8,000 cm of air and even more traversing 16,000 cm of air. Therefore to the normal absorption of mesons, the effect of spontaneous decay adds and as a result air appears a more efficient absorber than condensed matter.

Kuhlenkampff's idea was considered by many investigators, including Werner Heisenberg and H. Euler [41] in Germany and Blackett in England. Heisenberg calculated a lifetime of 2.7 μs. The atmospheric results were suggestive but not a direct proof, and the experiments aimed at finding a decay event in the Wilson chamber had failed. In 1938–1939 the question of the muon decay became the object of a strong debate.

H. Euler [42] indirectly demonstrated muon decay from the number of secondary electrons in the cosmic radiation. He observed that the distribution of electrons in the lower atmosphere could be explained assuming that some were struck by muons, some were produced in the muon decay and the remaining ones were due to the transmission of the soft component in the atmosphere. By taking into account the ratio of the number of electrons to the number of muons below an absorbing material one finds that it is smaller than in the atmosphere. The reason, he argued, is that the contribution due to the muon decay is present in the atmosphere but practically absent below an absorber, because muons pass through the absorber in a time too short to produce electrons by their decay.

The muon decay provided also a natural explanation of the anomalous variation of the intensity with the zenith angle as observed by Auger and collaborators [43].

The large anomalous absorption of mesons in a long column of atmosphere compared with that of a shorter column of a denser absorber with the same stopping power had been observed by A. Ehmert [44, 45].

Another proof was connected to the observation that the intensity of cosmic rays at sea level decreases with increasing temperature. In fact, an increase of temperature means an expansion of the atmosphere and therefore muons are

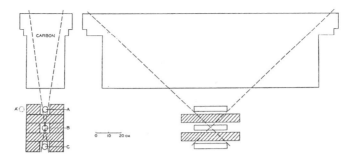

Fig. 8.10 The Rossi's set-up for measuring the muon lifetime. (Reprinted with permission from B. Rossi, N. Hilbert, J. Barton Hoag, Phys. Rev. **57**, 461 (1940). Copyright 1940 by the American Physical Society)

produced at a higher altitude and have more time to decay before reaching the sea level. Accordingly, the intensity decreases [46].

A tentative to measure directly the muon decay with arrays of counters was performed by C.G. Montgomery et al. [47] with no success.

In 1938, Bruno Rossi and his wife left Italy taking first shelter in Copenhagen with Bohr, who received them cordially. There, Rossi met Blackett who proposed him to visit Manchester. The following year Arthur Compton invited him to participate in a summer conference on cosmic rays at the University of Chicago. Rossi and his wife arrived at America in June 1939. After the conference, they visited the Compton's in their country house in Michigan and Rossi, discussing with Compton, proposed an experiment to measure the muon decay. Compton pushed him to do it and Rossi, in little more than a month, assembled the necessary counters and breadboard circuits and set out for Mt. Evans in Colorado in a borrowed bus loaded with equipment and a ton of lead and graphite, travelling over a thousand miles of rough roads across the Great Plains. He wanted to complete the experiment before the snow blocked the access to the mountain. Shifting the equipment up and down the mountain up to an altitude of 4,300 m, he obtained results showing the intensity of muons to be attenuated more rapidly in air than in an equivalent quantity of graphite [48].

His ingenious experiment [49] allowed him to measure the lifetime of the meson by using the absorption of the atmosphere, finding it to be 2 μs. It consisted of three Geiger–Müller counters in coincidence placed one above the other, with some lead in between, and a gross lead shield all around to cut all the electrons (see Fig. 8.10). With this apparatus [50] he performed measurements in Chicago (180 m above sea level), Denver, Colorado (1,600 m), Echo lake (3,240 m), and Mount Evans (4,300 m). In every place the meson number was counted with and without a graphite layer (about 87 g/cm^2) above the counters.

Figure 8.11 shows the logarithm of the observed intensities N against the total mass per cm^2, h, of air and carbon above the Geiger–Müller tubes. The circles refer to measurements taken without graphite. Hence, the solid curve connecting the

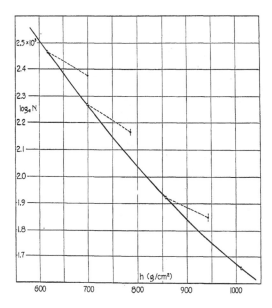

Fig. 8.11 Results of meson-absorption measurements in air and graphite (Reprinted with permission from B. Rossi, N. Hilbert, J. Barton Hoag, Phys. Rev. **57**, 461 (1940). Copyright 1940 by the American Physical Society)

circles represent, on a logarithmic scale, the variation of the vertical meson intensity as a function of the depth below the top of the atmosphere. The solid dots refer to measurements taken under the graphite absorber. The dotted lines, connecting the points taken at the same altitude with and without graphite, give, therefore, the initial slopes of the logarithmic absorption curves in carbon of the meson beam under 616, 699, and 856 g/cm^2 of air. These slopes are much smaller than the corresponding slopes of the air absorption curve, showing that the meson intensity is reduced much more by a given mass of air than by the same mass of carbon.

The difference in the behaviour was due to the meson decay. To estimate the lifetime, the researchers took into account that, according to relativity theory, the determination of a time interval depends on the reference frame in which the measurement is made. For example, if a clock moves with a velocity v with respect to an observer, it looks to count at a rate slower to that observed by another observer moving with the clock. The time interval between two ticks, as measured by the first and second observer are in the ratio $1/(1 - v^2/c^2)^{1/2}$ where c is the light velocity.

It is easily seen that the difference is appreciable only when the clock velocity in very near the light velocity. Now, the unstable mesons of cosmic rays are subatomic clocks which move at a very high speed. The lifetime τ' of these mesons is therefore notably different from the lifetime τ of a meson at rest. For example, a muon with an energy of 500 MeV has $\tau'/\tau = 4.7$. In other words the measured mean life of this muon is 4.7 times longer than the one measured if the muon was at rest.

Therefore, to deduce the lifetime of a meson at rest from measurements on moving mesons it is necessary to know their energy. By taking this into account,

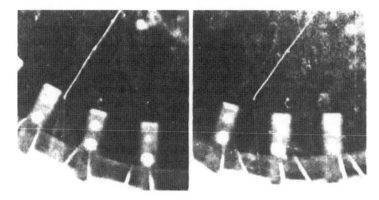

Fig. 8.12 The first photo in a cloud chamber of the decay of a muon. From Williams and Roberts [53]. The figure is a larger reproduction of a more extended photo. The thick track on the *upper right* which stops after about 5 cm before reaching the end of the chamber is a slow meson. From its end there emerges a fast electron track. The *rectangular* electrodes visible on the side are supports to hold wires to establish an electric field to maintain clean the chamber. The estimated mass was about 250 electronic masses. Both particles are positive. (Reprinted with permission from E.J. Williams and G.E. Roberts, Nature **145**, 102 (1940) Copyright 1940 by Nature Publishing Group)

Rossi measured a lifetime of 2 μs. It was the first proof of the instability of a fundamental particle.

More accurate measurements were done [51] to verify the relativistic transformation formula. It was calculated a value for the meanlife of 2.4 μs.

The anomalous absorption of muons in air was also confirmed by comparing the vertical intensity measured at two different altitudes (Chatillon at 500 m and Pian Rosà, at 3,460 m) by M. Ageno collaborators [52].[3]

Meanwhile E.J. Williams and G.E. Roberts [53] in England, succeeded in obtaining the first photo in a cloud chamber of the decay of a muon (Fig. 8.12). It shows the track of a positive muon that ends in the gas of the Wilson chamber; from its end the track of a fast positive electron emerges with a kinetic energy much greater of the muon kinetic energy, but comparable with its mass energy. This famous photo, besides giving the definitive proof of the muon decay, also showed that at its death an electron is generated.

Three main problems arose when trying to make direct measurements of muon lifetime. The first one is that the available number of muons for the experiment is always small. The second is that the involved times are very small, a few microseconds, so that their measurement was difficult as the necessary electronics has not yet been well developed. The third problem was that direct measurements were difficult to make on the mesons with energies typical of cosmic rays, because they may travel thousands of metres before disintegrating. Direct measurements can be done only on very slow mesons that stop in the detecting apparatus.

---

[3]Mario Ageno (1915–1992) was a pupil of Fermi.

A quantitative series of measurements on the mean life was made by employing a measurement method which used counters with delayed coincidences. A muon that stops in matter does not decay immediately, but waits a time in the mean equal to its decay time. If stopped in an absorber, the exit electron should appear with a delay of about 1 μs.

On March 15, 1941, Franco Rasetti [54][4] then at Quebec, Canada, performed a measurement in which both the muon decay and the emission of the decay electron was studied, finding 3.1 ± 1.5 μs. Later he published a complete account [55] in which the mean lifetime value resulted to be 1.5 μs with more precision.

He measured the coincidences between the muons that stopped in an absorber and their decay electrons, by using coincidence circuits with different resolution times. The experimental disposition is shown in Fig. 8.13. The coincidence system ABCD select a muon beam that passes through 15 cm of lead.

The counters G, in anticoincidence reduce the countings due to showers and the counters F in anticoincidence select the muons that stop in the absorber or have been scattered out from the beam. The G counters are not very clearly visible in the figure: they are the two counters at right and left of the counter C that is placed where the two dotted lines that define the acceptance angle encounter themselves. Unit 1 selects all coincidences ABCDE with a resolution time of 15 μs. To be registered in the Recorder 1, it is required that they are in anticoincidence with F and G. Such an event is presumably a muon that decays in the absorber or is scattered by it. To be registered in Recorder 2, the more strict request of a coincidence ADE in unit 2 it must be also satisfied, with a resolution time of only 1.95 μs. Similarly, Recorder 3 requires a coincidence DE in unit 2 with a resolution time of 0.95 μs.

These results give directly the lifetime because $N(15) - N(1.95)$ is a measure of the number of muons which decay after 1.95 μs and $N(15) - N(0.95)$ gives the number after 0.95 μs. However, since virtually all mesons decay before 15 μs, these numbers may be used as the number of muons present after 1.95 and 0.95 μs, that is $N(0) \exp(-1.95/\tau)$ and $N(0) \exp(-0.95/\tau)$.

Accordingly, τ is obtained from

$$(N(15) - N(1.95)) / (N(15) - N(0.95))$$

$$= N(0) \exp(-1.95/\tau)/N(0) \exp(-0.95/\tau) = \exp(-1/\tau).$$

In this way Rasetti found 1.5 ± 0.3 μs.

More precise measurements were done by Rossi and Nerenson [56–58] who wanted to determine the disintegration curve of muons at rest. In the fall 1940, Rossi with the help of Hans Bethe, was given a position of associate professor at Cornell

---

[4]Franco Rasetti (1901–2001) Professor of spectroscopy at Roma (1934–1938) then of physics at Laval University in Quebec (1939–1947) and since 1947 at Johns Hopkins University of Baltimora. After studies on spectroscopy, he collaborated with E. Fermi on researches on neutrons. Later he interested himself to cosmic rays.

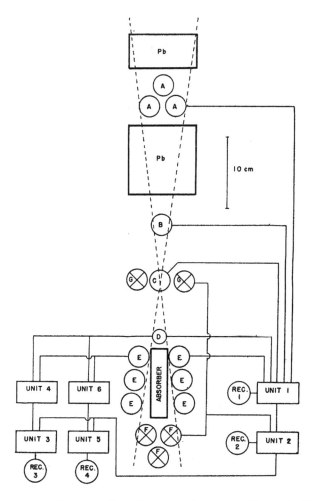

Fig. 8.13 Arrangement of counters, illustrating connections to Amplifier units. (Reprinted with permission from F. Rasetti, Phys. Rev. **60**, 198 (1941). Copyright 1941 by the American Physical Society)

University. There he made the first precise measurement of the muon lifetime with the disposition of Fig. 8.14. For this experiment he invented another fundamental device, the time-amplitude converter, or TAC.

The previous experiments, like the one of Rasetti, had been devised with special coincidence circuits designed in such a way as to register only those electrons that were emitted from the absorber at preselected times after the arrival of the muons. Rossi together with his student Norris Nerenson succeeded in increasing the selectivity and the accuracy of the method by registering all the decayed electrons and measuring the time interval between the muon arrival and the emission of the corresponding electron.

Fig. 8.14 Experimental
arrangement for the
determination of the
Disintegration curve of
muons. (Reprinted with
permission from B. Rossi,
N. Nerenson, Phys. Rev. **64**,
199 (1943). Copyright 1943
by the American Physical
Society)

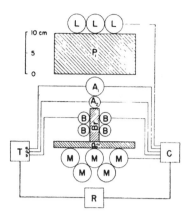

In Fig. 8.14 the coincidence circuit required a coincidence $LA_1 A_2 B$ with anti-coincidence with M. A special circuit gives an impulse whose height measures the time interval between $A_1 A_2$ and the pulse from B. The results obtained with lead, brass and aluminium are combined in a single curve in Fig. 8.15. In this figure, in the inferior curve the number of muons with lifetime between τ and $\tau + \Delta \tau$ is shown on a logarithmic scale as a function of $\tau + \Delta \tau / 2$, where $\Delta \tau$ is taken as $0.4\,\mu$s. The points all lay on a straight line, which gives the lifetime of the muon:

$$\tau = 2.15 \pm 0.07\mu s$$

In this way, it was definitely proved that muons decay with an exponential law. The circuit of TAC was employed in the secret researches performed at Los Alamos and classified until the end of the war.

8.8   Which Are the Products of Muon Decay?

At the beginning, when people believed muons to be the meson predicted by Yukawa, it was natural to expect, to explain the beta decay, that its disintegration products were an electron and a neutrino. This point of view was easily accepted as the decay particle must be lighter than the muon and the electron is the only particle with this characteristics.

From momentum conservation, another particle should escape when the electron is emitted by a meson at rest. Because the cloud chamber shows that only one charged particle is produced, the other should be a neutral particle. Two possibilities existed: either it was a neutrino or it was a photon or, may be more than one of these particles was involved.

Experiments were performed to check if photons were emitted [59–61], by using various ways to detect the electrons they should subsequently produce. The results

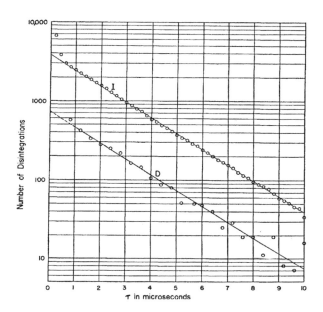

Fig. 8.15 Differential (*D*) and integral (*I*) disintegration curves of mesotrons obtained from the combine date of the lead, brass, and aluminum absorbers. The straight line through the experimental points for the differential curve is draw according to the method of least squares. (Reprinted with permission from B. Rossi, N. Nerenson, Phys. Rev. **64**, 199 (1943). Copyright 1943 by the American Physical Society)

allowed to rule out the emission of a photon. The hypothesis of a decay into an electron and a neutrino was therefore reasonable. However, there was still a check to be done, since, if only two particles were emitted, the energy of the electron should be fixed: 50 MeV with the other particle having the same energy. If a different energy were to be found, the neutral particle should have a different mass. A distribution of energies of the electron, would mean that more than one neutral particle was emitted.

Two principal methods were used. The more direct one, involved the measurement of the curvature of the electron track in a cloud chamber with magnetic field. The other method involved a measure, which used counters, of the range of the decaying electron. Using the cloud chamber, the difficulty is that photos showing the decay are extremely rare. With counters it is difficult to define exactly the range. By using a delayed coincidence scheme, Thompson [62] succeeded in obtaining several photos of decay and concluded that his data were consistent with a single decay energy.

A number of further measurements were done by many researchers, with conflicting results. Eventually Leighton et al. [63] solved the question, obtaining 75 photos of decays over a total of 15,000, obtaining, albeit with large errors, an energy distribution as shown in Fig. 8.16.

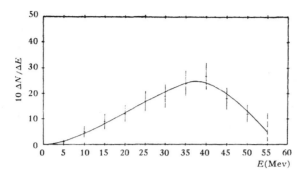

Fig. 8.16 Energy spectrum of mesotron decay electons. Each point represents the number of electrons per energy interval of 10 Mev. The vertical lines indicate the expected statistical spread. A smooth curve is draw through the observed points. (Reprinted with permission from R.B. Leighton, C.D. Anderson, A.J. Seriff, Phys. Rev. **75**, 1432 (1949). Copyright 1949 by the American Physical Society)

Further measurements of the electron range were done by Steinberger [64, 65] and Hincks and Pontecorvo [66–68], and also by others researchers, and all confirmed a distribution of energies. Sagane et al. [69] used a proton beam from the Berkeley cyclotron to produce muons and detected the decay electrons. The spectrum, similar to the one of Anderson, went smoothly to zero at a maximum energy of 53 ± 2 MeV.

Therefore the conclusion that the decay process involves two neutrinos was suggested [70] and that the process may be written as

$$\mu \rightarrow e + \nu + \underline{\nu},$$

where μ stands for the muon, e for the electron, ν for the neutrino and $\underline{\nu}$ for an antineutrino. As both the electron and the neutrino have spin $1/2$ and the antineutrino has a spin opposite to the one of the neutrino, the muon must have spin $1/2$.

8.9 The Leptons

The standard model expects six kinds of particles that are considered deprived of an internal structure and that may be viewed as *elementary particles*. They are the *electron*, the *muon* and the *tau particle*, plus their associated *neutrinos*. These particles, which do not interact strongly with the constituents of the atomic nucleus, interact via the weak and the electromagnetic force: an interaction unified in the electronweak interaction in 1967 by Glashow, Weinberg and Salam [71–73][5].

---

[5] All three were awarded the Nobel Prize for Physics in 1979.

The tau[6] (positive and negative) particle has a mass 3,477 times the electronic mass and was discovered at Stanford in 1975 studying the interactions produced at the electron–positron collision ring of SLAC (Stanford Linear Accelerator Center) by Martin Perl[7] with his collaborators [74]. To this new particle also is associated a neutrino.

Nearly 30 years after Pauli introduced the neutrino, Bruno Pontecorvo [75] suggested that the neutrino associated with the muon is different from the beta radioactivity neutrino. The two neutrinos are therefore indicated as *muonic* and *electronic neutrino*, respectively.

Bruno Pontecorvo (1913–1993) was a member, in the years 1933–1936, of the group of young researchers associated with Enrico Fermi, the so-called "ragazzi di via Panisperna", collaborating to the classic investigations of the properties of slow neutrons. He then moved to Paris working at the Radium Institute with F. Joliot-Curie, obtaining notable results in the nuclear physics field, and then (1940) to the USA where he developed a technology and an instrument for well logging, based on the properties of neutrons, still used in petroleum researches [76]. In 1943, in Canada, he participated to the realization of the nuclear reactor of Chalk River. In 1948, he assumed one of the technical directions of the British atomic laboratory at Harwell. Under mysterious circumstances, he moved to URSS in 1950 to lead one of the groups at the Nuclear Institute of Dubna (Moscow). A member of the Academy of Sciences of the URSS, he was awarded the Lenin prize in 1963. He was also foreign member of the Accademia dei Lincei since 1982. After his researches on the neutron physics, he got interested to several problems concerning elementary particles, in particular neutrinos, becoming a world expert in the field.

When, in 1950, he suggested the existence of two kinds of neutrinos (*electronic neutrino* and *muonic neutrino*), he proposed [77, 78] also a chemical method to experimentally detect them, today known as the *chlorine-argon method*, and became interested to the problem of the mass of the neutrino.

The experimental verification of the correctness of the two kinds of neutrinos came in 1962 by Jack Steinberger (1921–), Leon Lederman (1922–) and Melvin Schwartz (1932–2006) [79] with an experiment in which, by using the alternate gradient synchrotron of Brookhaven, a beam of high-energy protons was made to collide with a beryllium target [80][8]. The neutrinos so obtained, gave always place to muons and never to electrons, making it evident that neutrinos must be at least of two types. The three researchers were awarded in 1988 the Nobel Prize in Physics

---

[6]The name tau was suggested to Perl by his graduate student Petros Rapidis. In greek τρῐθόν stands for third: the third charged lepton.

[7]Martin Lewis Perl was born at New York in 1927. After obtaining his Ph.D. in physics at the Columbia University, New York, under Isidor Rabi, he was professor at the University of Michigan (1955–1962) and since 1963 at Stanford University, CA. In 1995 he was awarded together with F. Reines the Nobel Prize for physics.

[8]The history of this experiment is given by Schwartz [80].

for *the neutrino beam method and the demonstration of the double structure of the leptons through the discovery of the muon neutrino.*

When the tau particle was discovered it was natural to expect the existence of a third neutrino, the *tauonic neutrino*, that should be produced in couple with the tau particle. This neutrino was experimentally observed only 25 years later [81]. It completes the frame of the particles that operate in weak interactions and are generically called *leptons* from the Greek word for light: electron, muon and tau particle, each one with its different neutrino. The three doublets identify the three leptons generations in a similar way to what happens with quark, which are also distributed on three generations in a quark-leptons symmetry with important theoretical implications.

Accordingly, we have 12 particles which appear in the weak interactions considered by Fermi: the negative electron with its neutrino.

The antiparticle of the electron is the positron, with a mass identical but a positive charge. To it is associated an antineutrino. If an electron and a positron meet, they annihilate with the production of two gamma rays. If the encounter happens when the two particles are at rest, the two gamma photons are emitted in opposite directions. The inverse process consists in the interaction of high-energy radiation (approximately 1 MeV, that is gamma rays) with matter and the production of an electron–positron pair.

The muon is a lepton with both positive or negative charge with a mass 207 electronic masses, equivalent to an energy of 105.7 MeV, which decays to produce an electron or a positron (according to its charge) through the following scheme

$$\mu^+ \to e^+ + \nu_e + \underline{\nu}_\mu$$

and

$$\mu^- \to e^- + \underline{\nu}_e + \nu_\mu$$

where antiparticles are underlined. In the decay, one electron and one muonic neutrino are produced.

The mean lifetime of the muon is 2.20 μs; it is produced in the high atmosphere through the decay of pions produced by cosmic rays (see Chap. 9).

Muons constitute more than one half of cosmic radiation at sea level, the remaining being mostly electrons, positrons and photons produced in the showers. The mean flux at sea level is one muon per square centimetre per minute.

The tau is the heavier lepton. It exists with an electric charge both positive and negative and has a lifetime of 2.96×10^{-13} s.

It decays through several possible processes, as for example[9]

---

[9] π is the meson π introduced in the next chapter.

$$\tau^- \rightarrow \mu^- + \underline{\nu}_\mu + \nu_\tau$$
$$\tau^- \rightarrow e^- + \underline{\nu}_e + \nu_\tau$$
$$\tau^- \rightarrow \pi^- + \nu_\tau$$
$$\tau^+ \rightarrow \mu^+ + \nu_\mu + \underline{\nu}_\tau$$

etc.

The theory shows that the neutrino needed to explain the beta decay, the one of the muon and of the tau cannot be the same neutrino but three different kinds of particles, all possessing the property of having an extremely small mass (if any) and spin $1/2$. The present standard model assumes there are no other leptons besides the ones we mentioned. An experimental proof of this is given by the ratio of the abundance hydrogen/helium in the Universe. Considering the nucleo-synthesis process in the Big-Bang model, the number of the kinds of neutrinos influences the abundance of helium and the observed abundance is in agreement with three kinds of neutrinos.

An important rule in all the interactions is that the number of leptons and the number of heavy particles (barions) does not change after the interaction (conservation law of the lepton and barion numbers).

The fact that the decay of the muon occurs with three particles is an example of the conservation of the leptonic number. This number is 1 for a particle and -1 for an antiparticle. In the decay an electron neutrino and a muonic antineutrino appear or the other way around.

References

1. R.A. Millikan, Phys. Rev. **55**, 105 (1939)
2. S.H. Neddermeyer, C.D. Anderson, Phys. Rev. **51**, 884 (1937)
3. J.C. Street, E.C. Stevenson, *Communication at the American Physical Society meeting*, Washington, April 29–30 1937
4. J.C. Street, E.C. Stevenson, Phys. Rev. **51**, 1005 (1937)
5. J.C. Street, E.C. Stevenson, Phys. Rev. **52**, 1003 (1937)
6. J.C. Street, R.H. Woodward, E.C. Stevenson, Phys. Rev. **47**, 891 (1935)
7. P. Auger, C. R. Acad. Sci. **200**, 739 (1935)
8. C.D. Anderson, Am. J. Phys. **29**, 825 (1961)
9. P. Galison, Centaurus **26**, 262 (1983)
10. C. Anderson, R.A. Millikan, S. Neddermeyer, W. Pickering, Phys. Rev. **45**, 352 (1934)
11. C. Stevenson, J.C. Street, *Communication to the American Physical Society meeting*, New York, February 22–23 1935
12. C. Stevenson, J.C. Street, Phys. Rev. **47**, 643 (1935)
13. P. Auger, P. Ehrenfest, C. R. Acad. Sci. **199**, 1609 (1934)
14. T.H. Johnson, E.C. Stevenson, J. Franklin Inst. **216**, 329 (1933)
15. R.H. Woodward, J.C. Street, *Communication at the American Physical Society meeting*, Washington, April 25–27 1935
16. R.H. Woodward, J.C. Street, Phys. Rev. **47**, 800 (1935)
17. J.C. Street, E.C. Stevenson, Bull. Am. Phys. Soc. **12**, 13 (1937), abstract 40
18. C.D. Anderson, S.H. Neddermeyer, Phys. Rev. **50**, 263 (1936)

19. B. Rossi, Int. Conf. Phys. (Lond.) **1**, 233 (1934)
20. C. Montgomery, D. Montgomery, Phys. Rev. **47**, 429 (1935)
21. R.T. Young, Phys. Rev. **49**, 638 (1936)
22. T.H. Johnson, Phys. Rev. **47**, 318 (1935)
23. R.H. Woodward, Phys. Rev. **49**, 638 (1936)
24. R.D. Bennett, G.S. Brown, H.A. Rahmel, Phys. Rev. **47**, 437 (1935)
25. J.C. Street, R.T. Young, Phys. Rev. **46**, 823 (1934)
26. J.C. Street, R.T. Young, Phys. Rev. **47**, 572 (1935)
27. B. Rossi, Z. Phys. **82**, 151 (1933)
28. S.H. Neddermeyer, C.D. Anderson, Phys. Rev. **54**, 88 (1938)
29. Y. Nishina, M. Takeuchi, T. Ichimiya, Phys. Rev. **55**, 585 (1939)
30. P. Kunze, Z. Phys. **83**, 10 (1933)
31. S.H. Nedermeyer, C.D. Anderson, Rev. Mod. Phys. **11**, 191 (1939)
32. R.B. Brode, Rev. Mod. Phys. **21**, 37 (1949)
33. H. Yukawa, Proc. Phys. Math. Soc. Jpn **17**, 48 (1935)
34. H. Yukawa, Proc. Phys. Math. Soc. Jpn **19**, 712, 1084 (1937)
35. H. Yukawa, Proc. Phys. Math. Soc. Jpn **20**, 319, 720 (1938)
36. E. Fermi, Z. Phys. **88**, 161 (1934)
37. L.M. Brown, R. Kawabe, M. Konuma, Z. Maki (eds.) Elementary particle theory in Japan 1930–1960. Progr. Theor. Phys. (Suppl. 105), 182–185 (1991)
38. H. Yukawa, Proc. Phys. Math. Soc. Jpn **20**, 712 (1937)
39. J.R. Oppenheimer, R. Serber, Phys. Rev. **51**, 1113 (1937)
40. H. Kulenkampff, Verh. Dtsch. Phys. Ges. **19**, 92 (1938)
41. H. Euler, W. Heisenberg, Erg. exakt. Naturw. **17**, 1 (1938)
42. H. Euler, Naturwissenschaften **26**, 382 (1938)
43. P. Auger, P. Ehrenfest, A. Freon, A. Fourier, C. R. Acad. Sci. **204**, 257 (1937)
44. A. Ehmert, Z. Phys. **106**, 751 (1937)
45. A. Ehmert, Phys. Z. **38**, 975 (1937)
46. P.M.S. Blackett, Phys. Rev. **54**, 973 (1938)
47. C.G. Montgomery, W.E. Ramsey, D.B. Cowie, D.D. Montgomery, Phys. Rev. **56**, 635 (1939)
48. A. Pascolini (ed.), *The Scientific Legacy of Bruno Rossi* (Imprimenda, University of Padova, Italy, 2005)
49. B. Rossi, H. van Norman Hilberry, J. Barton Hoag, Phys. Rev. **56**, 837 (1939)
50. B. Rossi, N. Hilbert, J. Barton Hoag, Phys. Rev. **57**, 461 (1940)
51. B. Rossi, D.B. Hall, Phys. Rev. **59**, 223 (1941)
52. M. Ageno, G. Bernardini, N.B. Cacciapuoti, B. Ferretti, G.C. Wick, Phys. Rev. **57**, 945 (1940)
53. E.J. Williams, G.E. Roberts, Nature **145**, 102 (1940)
54. F. Rasetti, Phys. Rev. **59**, 613 (1941)
55. F. Rasetti, Phys. Rev. **60**, 198 (1941)
56. B. Rossi, N. Nerenson, Phys. Rev. **62**, 417 (1942)
57. B. Rossi, N. Nerenson, Phys. Rev. **64**, 199 (1943)
58. B. Rossi, N. Nerenson, Rev. Sci. Instrum. **17**, 65 (1946)
59. E.P. Hincks, B. Pontecorvo, Phys. Rev. **73**, 257 (1948)
60. R.D. Sard, E.J. Altheus, Phys. Rev. **74**, 1364 (1948)
61. O. Piccioni, Phys. Rev. **74**, 1754 (1948)
62. R.W. Thompson, Phys. Rev. **74**, 490 (1948)
63. R.B. Leighton, C.D. Anderson, A.J. Seriff, Phys. Rev. **75**, 1432 (1949)
64. J. Steinberger, Phys. Rev. **74**, 500 (1948)
65. J. Steinberger, Phys. Rev. **75**, 1136 (1949)
66. E.P. Hincks, B. Pontecorvo, Phys. Rev. **74**, 697 (1948)
67. E.P. Hincks, B. Pontecorvo, Phys. Rev. **75**, 698 (1949)
68. E.P. Hincks, B. Pontecorvo, Phys. Rev. **77**, 102 (1950)
69. R. Sagane, W.L. Gardner, H.W. Hubbard, Phys. Rev. **82**, 557 (1951)
70. O. Klein, Nature **161**, 897 (1948)

71. S. Weinberg, Phys. Rev. Lett. **19**, 1264 (1967)
72. A. Salam, Proceedings of the VIII Nobel symposium, Aspenagsarden (Almquist and Wiksell, Stockholm, 1968), p. 367
73. S.L. Glashow, Nucl. Phys. **22**, 579 (1961)
74. M.L. Perl, Phys. Rev. Lett. **35**, 1489 (1975)
75. B. Pontecorvo, Sov. Phys. JETP, **37**, 1236 (1960)
76. B. Pontecorvo, Oil Gas J., **40**, 32 (1940) reprinted in B. Pontecorvo, *Selected Scientific Works* (Società Italiana di Fisica, Bologna, 1997)
77. B. Pontecorvo, Chalk River Laboratory Report **PD-205** (1946) reprinted in ed. by J.N. Bahcall, *Neutrino Astrophysics* (Cambridge University Press, Cambridge, 1989)
78. B. Pontecorvo, *Selected Scientific Works*, ed. by S.M. Bilenky, T.D. Blokhintsava, I.G. Pokrovskaya, M.G. Sapozhnikov (Società Italiana di Fisica, Bologna, 1997)
79. G. Dunby, J.M. Gaillard, K. Goulianos, L. Lederman, N. Mistry, M. Schwartz, J. Steinberger, Phys. Rev. Lett. **9**, 36 (1962)
80. M. Schwartz, Rev. Mod. Phys. **61**, 527 (1989)
81. K. Kodama et al., Phys. Lett. **B504**, 218 (2001)

Chapter 9
The Discovery of the π Meson, the Nuclear Emulsions, and the First "Strange" Particles

9.1 Introduction. The mu Mesons Are Really Peculiar

In 1940, the Japan physicists Sin-Itiro Tomonaga[1] (1906–1979) and Gentaro Araki pointed out that the positive and negative muons, at the time still identified with the Yukawa mesons, should behave differently when they are stopped in matter [1].

They noted that a positive meson cannot come very near to an atomic nucleus because of the repulsive electric forces acting between charges of the same sign. Therefore, it should turn around until it decays spontaneously. On the contrary a negative meson is attracted by the positive charge of the nucleus and therefore has great probability to be captured before decaying. If the interaction with the nucleus is strong, practically all of these mesons are captured, while if it is weak a fraction of the mesons has time to decay spontaneously. Accordingly, the life expectancy of a negative meson has to be shorter than the one of a positive meson because while this last one dies of a natural death, the negative mesons have the additional risk of nuclear capture. Once captured by the nucleus, the negative meson should release its energy and cause an explosion of the nucleus.

Experimentally, assuming that an equal number of positive and negative mesons exist in the cosmic rays, one should expect that the number of decay electrons would come only from positive mesons and therefore should be one-half of the total meson number, irrespective of the material with which they have interacted.

Shortly after the publication of this paper, Rasetti performed the decay experiment we described in the previous chapter. By comparing the number of mesons stopped in an iron block with the number of decaying electrons coming out, Rasetti

---

[1] Sin-Itiro Tomonaga was born in Tokyo where he also died. After graduating in physics in Tokyo, he worked under H. Yukawa in Japan and then in Germany under W. Heisenberg. He was appointed physics professor at Tokyo in 1941. His interests were on quantum mechanics and quantum electrodynamics. He was awarded the physics Nobel Prize in 1965 together with R.P. Feynman and J. Schwinger all three *"for their fundamental work in quantum electrodynamics, with deep-ploughing consequences for the physics of elementary particles"*.

M. Bertolotti, *Celestial Messengers: Cosmic Rays*, Astronomers' Universe,
DOI 10.1007/978-3-642-28371-0_9, © Springer-Verlag Berlin Heidelberg 2013

found that only half mesons decayed into electrons. He therefore concluded that the decay electrons came from the positive mesons and the negative mesons disappeared due to nuclear capture, as predicted by Tomonaga and Araki.

Meanwhile in Roma, Marcello Conversi (1917–1988) and Oreste Piccioni (1915–2002), who were not aware of the Rossi and Nerenson lifetime measurements because the communications were interrupted due to the war, decided to improve the Rasetti lifetime measurement using a more sophisticated electronics.

They both graduated in physics in Roma; Conversi under Bruno Ferretti (1913–2010), in 1940, and Piccioni under Fermi, in 1938, and then they worked on cosmic rays in the Gilberto Bernardini group.

In 1943, in Roma, they assembled the set-up for the lifetime measurements with an apparatus placed inside the Institute of Physics in Via Panisperna. On July 19, 1943, Roma, for the first time, was bombed by the American air force. Nearly 80 bombs fell in the town perimeter and one of them exploded a few meters from the laboratory window a short while after Conversi had removed the instrumentation from it. The next day they decided to transfer the apparatus outside the University and found a place more secure near the Vatican City: a lecture hall in the basement of the Lyceum Virgilio ([2], see also [3]).

There Piccioni and Conversi installed their instrumentation and began a first series of measurements that, however, were soon interrupted due to the German occupation of Roma, immediately after the armistice on September 8, 1943. Piccioni was captured by Germans and fortunately released shortly after; Conversi joined the resistance but remained in Roma. The first results of their research, regarding the measurement of the muon lifetime, concluded in February 1943, were published on Nuovo Cimento [4], confirming that only one-half of the muons stopped in iron decayed as free. Resuming in an adventurous way the research, an extended version was then sent to Physical Review. The manuscript was prepared in 1944, but, on account of wartime conditions, it was received by Physical Review only on February 10, 1945 and published in the December 1946 issue [5]. They found an exponential decay that—not being aware of the previous publications—they believed to be the first ones to report and measured a lifetime $\tau = 2.33$ μs with an error of ±6.5%.

Successively, they associated Ettore Pancini (1915–1981) who had suggested to use magnetic lenses to select the sign of the muons. Ettore Pancini graduated in physics in Padova, in 1938, under Bruno Rossi and in 1940 was appointed assistant under Gilberto Bernardini in Roma. In 1950, he got the chair of professor of experimental physics in Genoa.

They were already familiar with the use of the special magnetic lenses having used in several experiments [6, 7]. So they designed a new experiment using a magnetic lens to divide the mesons of different charge one from the other. This so-called "magnetic lens" is simply a block of magnetized iron that, due to its magnetic field, deflects the charged particles that cross it. The story of these magnetic lenses went back to 1931 when the young Bruno Rossi ([8], see also

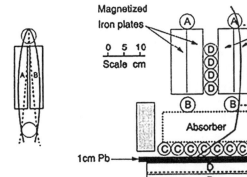

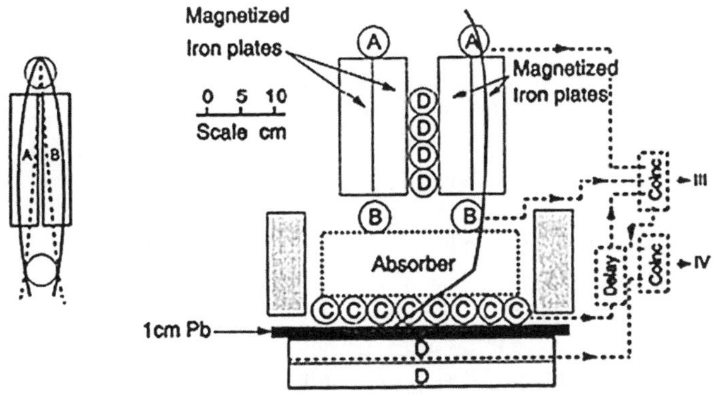

Fig. 1 Disposition of counters, absorber, and magnetised iron plates..
All counters "D" are connected in parallel.

Table 1. Results of measurements on β-decay rates
for positive and negative mesons.

Sign	Absorber	III	IV	Hours	M/100 hours
(a) +	5 cm Fe	213	106	155.00'	67± 6.5
(b) -	5 cm Fe	172	158	206.00'	3
(c) -	none	71	69	107.45	-1
(d) +	4 cm C	170	101	179.20'	36± 4.5
(e) -	4 cm C+5 cm Fe	218	146	243.00'	27± 3.5
(f) +	6.2 cm Fe	128	120	240.00'	0

Fig. 9.1 Up on the *left* two bars A and B with opposite magnetization deflect the muon paths in one way (*continuous line*) or the other (*dotted line*) according to the charge sign. The two counters *above* and *below* the two *bars* are shown. *Upside* on the *right*, the set-up used by Cocconi and collabs. Downward the table with results with Fe, paraffin or air are shown. (Elaboration of the Physics Dept. G. Marconi, Roma)

[9–11]) built them, following a suggestion of Luigi Puccianti (1875–1952), a professor in Pisa, with the hope to deflect cosmic rays. The Rossi experiment, aimed at seeing if the cosmic particles which were able to pass great thicknesses of material had positive or negative sign, failed. Concerning the use of such lenses a debate was made as to which of the two fields, H or B, was responsible for the deflection, which was solved by Weizsaecker [12] who demonstrated the deflection depends on B.

In their experiment, the three researchers mentioned above used a magnetized iron block that acted as a lens. Because the cosmic ray mesons are able to traverse large thicknesses of iron, the magnetic field present in it may deflect them and separate positive from negative mesons. The basic of the experiment is shown in Fig. 9.1. Two iron bars are placed side by side, one with its magnetization exiting from the paper and the other with its magnetization entering the paper.

Two Geiger–Müller counters, one above and the other below the bars, are connected in coincidence. The positive mesons travelling downward through the first counter are deflected towards the second counter by the magnetic bars, increasing the number of mesons crossing both counters. The negative mesons (with the exception of those that travel straight down between the bars), on the contrary, are deflected away from the second counter. If the magnetization of both bars is inverted, the lens concentrates the negative mesons and rejects the positive ones.

Placing an iron absorber under their lens, Conversi and his colleagues found that only positive mesons originated the delayed electrons characteristic of the spontaneous decay. Their conclusion was that practically all the negative mesons stopped in the iron were captured by the nuclei before having the possibility to disintegrate spontaneously [13,14], these results being in agreement with Tomonaga and Araki expectations.

Between 1946 and 1947 the same group repeated the experiment [15], this time utilizing several different kinds of absorbers, and they found that negative mesons, according to the kind of absorber employed, behave in a different way when stopped in the matter. While positive mesons decay more or less as if they were free, both in iron and carbon, the negative ones are attracted by nuclei and react with them in the case of heavy elements, while decay nearly as free in the case of light elements as carbon.

In the table shown in the figure the difference $M = (III) - (IV) - P$ between the threefold (III) and fourfold (IV) delayed coincidences, with a correction P, gives the number of mesons stopped in the absorber and ejecting a disintegration electrons which produce a delayed coincidence. In the table "Sign" refers to the sign of mesons focused by the magnetic field. The great yield of negative decay electrons from carbon shows a marked difference between it and iron as absorbers.

Shortly afterwards, G.E. Valley [16] using a cloud chamber in a magnetic field to separate positive from negative mesons, confirmed the results of the Italian experimenters and showed that in beryllium and water, as well as in carbon both positive and negative mesons undergo spontaneous decay.

This was not the behaviour predicted for the Yukawa particles. They should strongly react both with heavy and light elements, because they are subjected to the intense nuclear forces.

The Italian group result had an immediate international resonance. Fermi, E. Teller[2] and V.C. Weisskops[3] proved [17] that the disagreement between the exper-

---

[2]Edward Teller (1908–2003) was a physicist born in Budapest, who migrated to America after 1933. He was a physics professor at George Washington University. He worked at Los Almos on atomic bomb. After the war he came first at the Chicago University in 1946, and then since 1953 at California University at Berkeley. He is often named the "father" of the hydrogen bomb.

[3]Victor Weisskops (1908–2002), Austrian theoretician, was born in Vienna where he studied physics. Then went to Gottingen where he graduated in 1931. In 1942 he became a US citizen and joined the Manhattan atomic bomb project working in the theoretical section with H. Bethe. He was Director General at CERN from 1960 to 1965.

imental result and the theoretical expectations was by a factor 10^{12}. J.A. Wheeler[4] derived the capture probability for negative mesons, showing it should increase approximately as the fourth power of the atomic number [18]. Bruno Pontecorvo [19] and a shortly after, albeit independently, Gianpietro Puppi [20] (1917–2006) and other theoreticians [21–23] put forward the hypothesis of the universality of the Fermi interaction, that is that the same interaction govern the beta decay and the atomic K capture[5] as well as the spontaneous decay and the nuclear capture of the meson. H. Bethe and R.index[aut]Marshak, R. Marshak [24], anticipating the pion discovery, formulated the theory of the two-component meson according to which the mesons observed at sea level are secondary particles, not subjected to the strong interaction, deriving from the decay of primary mesons, that nobody had still seen, produced in the upper layers of the atmosphere and to be identified with the Yukawa particle. A two-component meson theory was also formulated by S. Sakata and T. Inoue ([25, 26] see also [27]). They suggested that the muons should not be considered responsible for the nuclear forces and estimated the mass of the Yukawa particle to be about twice the muon mass. The Yukawa particle should decay in a muon and a neutral particle with a lifetime of the order of 10^{-21} s.

The identification of the muons with the Yukawa mesons exhibited also other difficulties we do not consider here.

The inevitable conclusion of the theoretical discussions was that the mu mesons are not subjected to nuclear forces and therefore are not the particles predicted by Yukawa. As a consequence the concept of the universality of the week interaction emerged, with the recognition of the existence of a family of apparently elementary particles not subjected to the strong interaction, later called *leptons*. The muon belongs to this family and is not responsible for the nuclear interactions; as a consequence it was deprived of the name of meson, now reserved exclusively to particles with strong interactions, and was indicated simply with its name *muon* or μ. The Conversi, Pancini, and Piccioni experiment is today considered a milestone in the development of particle physics.

The week forces exerting among elementary particles constitute one of the four fundamental interactions to which all observed physical phenomena may be reduced (they are: strong, electromagnetic, weak, and gravitational). Weak interactions own their name to the intensity of the corresponding force, that is largely weaker than the strong forces which bind the nuclei and the electromagnetic forces that bind atomic and molecular electrons, being however much stronger than gravitational forces, which in any case do not play any role in nuclear phenomena. While the

---

[4]John Archibald Wheeler (1911–2008) was an American theoretical physicist. After a period (1933–1935) in Copenhagen under Bohr, he moved to the North Carolina University. In 1938 he was appointed physics professor at Princeton and then at the University of Texas (1976). He worked on unified theories and has given important contributes to nuclear physics. With N. Bohr he explained nuclear fission. He worked at Los Alamos on the H bomb.

[5]The name K capture describes the process of the capture by a nucleus of an electron from atomic K energy levels.

electromagnetic and gravitational interactions extend to infinity, with laws that are inversely proportional to the square of the distance, the strong and weak forces decrease exponentially with the distance, and therefore manifest themselves over a limited distance whose scale is connected with the mass M of the lighter carrier of the corresponding force, according to the relation $r \sim \hbar/Mc$. For the nuclear forces, $r \sim 10^{-13}$ cm, so the carrier (the pion, as we will see below) has a mass $M_\pi = 140\,\mathrm{MeV}/c^2$; for weak interactions $r \sim 10^{-18}$ cm and the carriers, the so-called *intermediate bosons*, have a mass of the order of $100\,\mathrm{GeV}/c^2$.

Weak interactions are responsible for beta radioactivity in nuclei and for all the decay processes of the free neutron, of *hyperons*[6], of charged pions, of K mesons[7], and of the charged leptons heavier than the electron (mu and tau). To the nuclear reactions induced by weak interactions, we owe the abundant energy production of the Sun and other stars (see Chap. 11).

9.2   A New Powerful Investigation Method: Nuclear Emulsions

At this point of our story, a new method for studying the cosmic ray particles entered the field and held the bench for several years providing a bounty of exceptional results.

The Wilson chamber is a wonderful instrument. It allows to "see" the elementary particles and the results of their collisions with atoms or nuclei. However, as any experimental device, it has its own limitations. Due to the low density of gases, very few of the particles entering the chamber are able to collide with nuclei or stop in it.

To improve the situation, the physicists of the 1930s started to build larger and larger chambers and to put solid material (like carbon or metals) plates in their interior. The technique of chambers with a multiple disposition of plates showed very useful in a number of studies. However it was not yet a satisfactory solution to the problem, because when a particle interact in a plate (that usually is a few millimetres or centimetres thick), it is impossible to know exactly what happens at the point where the collision takes place. Similarly when a particle stops in a plate and then decays, the decay products may be observed only if they have enough energy to come out of the plate. The direct observation of the interaction and decay requires a dense substance inside which the particles have a large probability to suffer collisions or to stop and in which they leave visible tracks.

---

[6]Elementary particles with a mass heavier that the neutron, null electric charge, or equal to one elementary positive or negative charge. They are unstable with decay processes that bear to the formation of mesons and of a proton or a neutron or another hyperon. They have a semi-integer spin.

[7]See later.

In the middle 1940s, the physicists set up a detector with the desired properties: the *nuclear emulsions*.

Photographic emulsions were largely used in the studies on electromagnetic radiations and radioactive materials. They were responsible for the discovery of X-rays by Roentgen, and played an important role in nuclear physics. The effect of high energy radiations upon photographic emulsions is similar to that of light. The ionizing particles sensitize the grains of silver bromide that they encounter along their path in the emulsion. A suitable developer solution reduces the grains to silver and under a microscope the individual trajectories of the ionizing particles appear as an irregular line of dark silver grains ([28]; further details may be found in Grilli and Sebastiano [29]).

A Japanese, S. Kinoshita [30], working with Rutherford before the First World War, demonstrated that ionizing particles, like alpha particles, darken photographic plates, but missed to realize, since he exposed the plate for too long a time, that they leave along the path a series of developed grains.

The track was observed by M. Reinganum [31] the following year and the fact was confirmed by many other researchers. Therefore, by using a microscope, the path of a particle is made visible. However, the emulsions used by Kinoshita and also by other people until about 1945 were sensible only to highly ionizing particles, as alpha particles; on the contrary, electrons were invisible.

The first works of Kinoshita ([30]), and Makower and Walmsley [32, 33] on alpha particles were done with photographic emulsion of a few microns thickness which were able to register tracks only if the particles were impinging at a grazing angle. To detect the full path of a low-energy alpha particle at all incidence angles, emulsions nearly 50 micron thick are needed. Moreover, the ordinary photographic emulsions are sensitive only to relatively slow particles which leave a very dense track of ions along their path. For these reasons the photographic emulsions were little used as means of detecting charged particles, even if they were used in researches on alpha particles [34–38].

Thicker emulsions were prepared in 1927 by Myssowsky [39]. They were used for studying the distribution of radioactive inclusions in rocks and plants [40]. Zhdanov [41, 42] continued their study in Russia at the State Radium Institute.

Eventually, in 1937 the photographic emulsion method was applied to the research on cosmic rays by Blau and Wambacher [43] at the Radium Institute in Vienna.

During those years, the effect of the silver bromide grain dimensions and density in the emulsion on the formation mechanism of tracks and on the dimensions and density of the developed grains was studied, in relation to the sensitivity and ability of the emulsions to discriminate among the different particles [44–48].

The emulsion technique had a quality jump due to a group of physicists at Bristol University, under the guidance of C.F. Powell and Occhialini.

Cecil Frank Powell was born on December 5, 1903 at Tonbridge, Kent, the son of a gunsmith and graduated in Cambridge. As a post-graduate student, Powell worked in Cavendish Laboratory under C.T.R. Wilson and E. Rutherford until 1927 when he gained his Ph.D. and moved to the University of Bristol as research assistant to

A.M. Tyndall. He remained in Bristol, being eventually appointed physics professor in 1948 and director of the Wills Physics Laboratory in 1964. He died in Bellano, Italy on August 9, 1969 while he was attending a Summer School organized by the Italian Physical Society[8] in Varenna.

Thanks to Powell, Bristol became the leader centre for the study of nuclear particles with photographic emulsions. In 1947, Powell in collaboration with Giuseppe Occhialini, published a fundamental work on the subject: *Nuclear Physics in Photographs* ([49]; another textbook is [50]).

In 1947, he discovered with this technique the presence of the π meson in cosmic rays and for this discovery he was awarded the physics Nobel Prize in 1950.

Working in collaboration with many researchers—in particular with L.V. Chilton and C. Waller—of the Ilford firm [51], a famous firm of photographic products, C.F. Powell, succeeded in producing an emulsion that contained much more silver than ordinary photographic emulsions and was sensible also to electrons. These emulsions resulted useful in a great variety of problems, and with time the technique was refined by increasing the gelatine thickness and introducing a series of improvements in its use and in the development techniques [52].

A short time later, also Kodak in England and Eastman Kodak in America started to work on the new emulsions and gave important contributions to the new technique with the development of even more specialized emulsions ([53–55] and also [56]). A refined technique was developed which allowed, with relatively simple means, to measure the charge, the velocity, the mass and other characteristics of the observed particles and for some time this technique dominated the study of elementary particles.

The emulsions had a nominal thickness from 50 to 100 μm and contained a large concentration of silver halides, nearly 10 times that one of ordinary plates, in the form of very small grains with dimensions of the order of 0.1–0.6 μm (in ordinary plates, the grain dimensions are of the order of a micron). Varying the mean dimension of the halide grains, it is possible to change the sensibility towards different kinds of ionizing radiations. Neutrons can be studied doping the plates with boron or lithium and by observing the tracks of the particles resulting from the reactions induced by neutrons: in this way it is possible to measure their momentum.

The stopping power in emulsion is approximately 1,700 times the one of air, that is one micron in the emulsion is equivalent to about 0.17 cm in normal air. This allows to shorten the path of the studied particles with respect to the one in air and therefore makes it possible to see the whole trajectory of a given particle in a single emulsion or in more emulsions suitably stacked one over the other.

New and ingenious developing techniques allowed to use emulsions even one millimetre thick, that is more than 100 times thicker than the emulsions available

[8]Since 1953 the Italian Physical Society (SIF) organized in Varenna, on Lake Como, in a former monastery, Villa Monastero, the International Summer School of Physics "Enrico Fermi" that has grown to an excellent standard: up to now, 45 Nobel laureates have been directors of courses. Enrico Fermi gave a memorable series of lectures on "Pions and nucleons" in the summer 1954.

before. Successively, the so-called *stripped* emulsions were developed, that is an emulsion without any support that could be piled a layer over the other to make what was called an *emulsion chamber*. In this way very long tracks could be observed and, in the case of nuclear reactions, it was possible to observe also the tracks of the produced particles.

The tracks of fast particles in photographic emulsions resemble in many aspects the tracks in the Wilson chamber. In fact the analysis of both tracks is very similar. In particular the density of grains produced by a particle is a monotonic function of its ionization rate. The more strongly the particles are ionizing, the more numerous are the grains and the greater is their initial energy, the longer are the resulting tracks. Therefore a relation between the grain density and the velocity exists which connect these quantities very accurately, enabling the identification of the involved particle and its energy under favourable circumstances. More elaborate methods were also put in operation. Moreover, with respect to the Wilson chamber, the nuclear emulsion has a sufficiently high density and is sensitive for the whole exposition time and therefore is a very efficient detector; particularly it is suited to investigate particles with short lifetime and range. Decay processes with a mean lifetime of 10^{-11} s may easily been detected by means of this simple and cheap technique.

On the other side the development technique is rather delicate. Observation must be made with microscopes and this is a long and tedious task.

In the case of cosmic rays that produce phenomena occurring with a low frequency, let us say a few events a day or even a month, the continuous sensitivity of emulsions is practically a must, in terms of economy of time and material. For researches with balloons or on mountains, the simplicity, robustness, and the low weight make the photographic plate the ideal instrument, to which we may add the continuous sensitivity. Some emulsions are insensitive to weakly ionizing particles, but for the study of heavily ionizing particles that present themselves very seldom, in the mid of fast electrons and mesons, it may be actually advantageous to suppress this background. However, for expositions of weeks or months the *fading*, that is the disappearing with the time of the latent image may create problems (for this see [57]).

The power of the technique is connected to the exceptional spatial resolution (about 1 μm) with which tracks are registered and with the possibility to deduce in general, by measurements of scattering and grain density, the direction in which a particle travels before coming at rest in the emulsion and the approximate value of its mass (good reviews are [53, 54]).

When a charged particle passes through a material medium, it is subjected to frequent small deviations in its direction of motion as a result of Coulomb scattering. The problem was first analysed by E.J. Williams [58, 59] and its solution may be used to find the mass of the particle. Another method is to use *straggling* to derive the energy of the particle. Fluctuations in the range of monoenergetic particles occur because of the discontinuous nature of the ionization processes, and these fluctuations (*straggling*) become of great importance in the emulsion due to the finite size and relatively small number of the grains making up the track. In addition, the halide grains of the emulsion are not homogeneously distributed in the gelatine, leading to regions of lower than average grain concentration. The uncertainty of range is proportional to the one in energy ($\Delta R/R \approx \Delta E/E$).

Particles of charge greater than two produce tracks with virtually no grain structure even at high energies, the track appearing as solid filaments of silver. Such energetic heavy ions are found in cosmic rays at high altitude. The rate of energy loss of these particles is so great that secondary electrons are produced with sufficient energies to have observable ranges in the emulsion. The number of such electrons or δ-rays is a function of the energy loss per path length (dE/dx) and, in conjunction with range determination, provides a means of estimating the charges and energies of these particles.

9.3   The Discovery of the π Meson

In 1944, Blackett invited Occhialini, at the time in Rio de Janeiro, to work in England on the atomic bomb project. Occhialini arrived to England on January 23, 1945; however, because of his Italian nationality, was not allowed to participate in military researches. He, therefore, remained in London to work with Edward Appleton (1892–1965) and then, for a short period at the General Electric Company. At the end he was invited by Tyndall, the laboratory director, to join the Powell group in Bristol.

Meanwhile, Blackett, through the Atomic Energy Cabinet Advisory Committee established a panel on emulsions, chaired by Joseph Rotblat (1908–2005)[9], to help and encourage the industrial firms, like Ilford and Kodak, to produce better and more sensitive emulsions and other firms to make microscopes with improved accuracy. Among the members of the panel there were Powell, Perkins of London Imperial College, and others. Occhialini was not a member but provided external contribution.

In this way, Occhialini, Powell and their collaborators developed a new kind of plates to detect the tracks of particles. In 1946, they were joined by César Lattes (1924–2005) from Sao Paulo.

During a speleological excursion on the Pyrenees, August 1946, Occhialini exposed the new silver bromine richer nuclear emulsions [52] to cosmic rays at Pic du Midi, in the French Pyrenees, at an elevation of 2,870 m where an observatory for cosmic rays had been built. A few weeks later, these were developed, and 65 tracks of mesons were found that stopped at the end of their range in the emulsion.

---

[9]Joseph Rotblat was a combination of distinguished scientist and leading peace advocate. He was involved with the Manhattan Project for the construction of the atomic bomb, but then decided to resign on ethical grounds. He received a PhD in nuclear physics from the University of Warsaw in 1938 and went to work under James Chadwick at Liverpool University. He worked as nuclear physicist and then moved into medical physics. He started the Pugwash Conference on Sources and World Affairs in 1957. In 1995, Rotblat was awarded the Nobel Peace Prize.

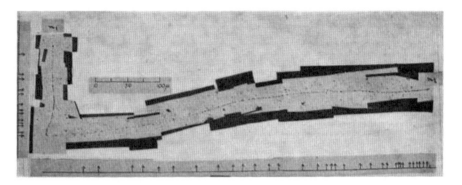

Fig. 9.2 From Lattes et al. [60]. The π meson is the *upper* track on the *left* that increases its thickness until it stops; and at that point, a long track starts that intensifies at the extreme right and represents the decay muon (Reprinted with permission from C.M.G. Lattes, H. Muirhead, G.P.S. Occhialini, C.F. Powell, Nature **159**, 694 (1947). Copyright 1947 by Nature Publishing Group)

Among these tracks, two showed particularly interesting features [60] (one of them is shown in Fig. 9.2).

When inspecting nuclear emulsions, sometimes a track is found that stops in the emulsion itself. This phenomenon has a characteristic aspect: as its velocity decreases, the ionization of the particle slowly increases, and at the end the track becomes very dense. In this way the direction in which the particle travels is found.

The two tracks that appear in Fig. 9.2 show a first particle that has stopped in the emulsion and from which a second particle emerges.

The authors decided to apply the term "meson" to any particle with a mass intermediate between that of a proton and an electron. Therefore, they described the event saying that the examination of grain density showed that the two particles were two mesons whose mass could not be determined with confidence because errors were very large.

A number of tentative explanations to interpret these two events were discussed but no final conclusion was taken.

Lattes, to find a confirmation of these results, exposed other plates in the meteorological station of Chacaltaya, in Bolivia, at 5,230 m of elevation, a few kilometres from La Paz. The results of this new experiment were published a few months after, with a discussion that considered the tracks obtained at Chacaltaya and on Pic du Midi. There were 644 tracks of mesons (193 in Bolivia and 451 in France) stopping in the emulsion [61]. In 40 cases secondary mesons were produced and in 11 cases the second meson stopped in the emulsion. However, only in two cases the track of the first meson was long enough to allow to determine the ratio between the mass of the two mesons. To do this, the number of developed grains was measured. In these 11 cases, the ionization of the first particle was always greater than that of the second one, showing that this was heavier than the second. In the two cases in which sufficiently long tracks were obtained both for the primary and the secondary particle, the ratio between the two masses was estimated to be 2.0 for one case and 1.8 for the other. Notwithstanding the large errors in the mass determination, the examination of all the tracks led to the conclusion that the secondary mesons had all the same mass and that they were emitted with constant kinetic energy. In this paper Powell introduced the denomination of π meson (we call it *pion*) for the heavier particle and μ for the other.

From these results and an accurate examination of all the possible explanations, the Bristol group, after discarding several interpretations, succeeded to demonstrate with certainty that the interpretation of an event like the one shown in Fig. 9.2 was that a heavy meson enters the emulsion and is slowed. Then it decays emitting a lighter meson and a neutral particle.

The heavier meson could not be the meson associated with the hard component of cosmic rays which decays giving rise to an electron and not to another meson. Therefore, the light meson could possibly be a muon and the heavier one a new particle.

If this explanation was correct, the secondary meson (mu), after stopping in the emulsion, should decay and produce an electron. The emulsions used in the 1947 experiment were not sensitive enough to detect particles of minimum ionization as fast electrons. When emulsions sensitive to those particles were available, in 1949, the decay of the light meson was detected [62].

The interpretation given by the Bristol group was unquestionable and the two mesons were named π the heavier, and μ, the lighter. The story has that when Powell wrote the paper describing the experiment, his typing machine had two Greek characters only: π and μ.

In part 2 of the former paper [61], the Bristol group [63] dealt with the origin of the slow mesons observed in photographic emulsions and their relation to the mesons forming the penetrating component of the cosmic rays, observed with the Wilson chamber. Photomicrographs were presented which showed that some of the slow mesons, ejected from nuclei during "explosive disintegrations" (stars) could enter the nuclei and produce a second disintegration. Most of the observations were made with boron-loaded emulsions, which had improved properties with respect to fading, a suggestion by Lattes.

Over 644 cases, they found 105 events in which mesons produce disintegrations with the emission of heavy particles.

Already 1 year after the first discovery, in 1948, 31 cases were found of muon production by the decay of the heavier meson which allowed a first determination of the masses [64]. Their exact values were the object of extended measurements. Y. Goldschmidt–Clermon and collab. (see for example [65]) gave $m_\pi = 272 \pm 12$ and $m_\mu = 202 \pm 8$ electron masses. A summary of the first measurements was given in a review paper by Powell on mesons [66] who gave for the ratio of the π mass to the μ mass the value 1.65 ± 0.15. The value today accepted is 1.32.

The discovery of the π meson clarified the meson problem. The quantum of nuclear forces, predicted by Yukawa, was the π meson and not the μ. The π mesons were the particles produced in the high-energy nuclear interactions. The fact that muons did not interact strongly with nuclei was no more surprising, because the muons were not produced in nuclear interactions, but were born from the π meson decay.

If π mesons are the Yukawa particle, then, when negative π mesons stop in the matter, they must be suddenly captured by nuclei and the released energy should produce an explosion. This is exactly what happens. The negative π mesons that stop in nuclear emulsions are never seen to decay into a muon. On the contrary a "star" appears, that is a number of tracks coming out from a common point at the end of their path, as was shown for the first time by D.H. Perkins [67], at Imperial College, London, and by Occhialini and Powell [60] in 1947. The star is produced by an atomic nucleus that explodes disintegrating.

The π mesons decaying in μ when stopping in the emulsion should be positively charged so to be repelled by nuclei and not to react with them. Stars were also observed in the paper by Lattes et al. [60].

In early days, Powell (see also [68]; also [66]) was used to classify the "mesons" detected in nuclear emulsions according to the effects they produced. He generically labelled the heavy mesons that produced stars at the end of their range as *sigma mesons*. Later they were identified with the negative pions. The tracks in which the meson stops without producing disintegrations were by Powell classified as due to *rho mesons*. At the beginning it was not possible to conclude from the Powell work that *rho* and *sigma* were with certainty positive and negative mesons, respectively, although this seemed very likely. It was not even sure the *sigma* mesons were heavier than *muons*, because there were not two simultaneous tracks and the sole grain counting does not give the absolute value of the mass. The *sigma* mass was obtained by measuring the scattering of the trajectory. Further experiments [65] showed that the masses of *sigma* and π were the same. Eventually, all the *rho* were found to have positive charge and be positive *pions* while all the *sigma* mesons were negative *pions*.

Because the mesons found in the cosmic radiation at sea level were mostly muons, the π mesons should decay so fast that no one survive the travel in the atmosphere. First calculations indicated a mean lifetime considerably shorter than $1\,\mu s$. The first determination of the pion decay was made in experiments in 1948 and 1950 with particle accelerators [71, 72] finding a mean life value of 2.55×10^{-8} s, that is nearly 100 times shorter than that of muons. This value is practically the one today accepted (26 ns). The accelerator experiments allowed also a very accurate measurement of the mass: 273 electronic masses.

The story is that Lattes left Bristol at the end of 1947 and went to the Radiation Laboratory at Berkeley, California, where there was a 184-in. cyclotron, the biggest in the world, producing a beam of alpha particles with 380 MeV energy looking for artificial production of mesons. Working with Eugene Gardner, Lattes discovered that the cyclotron had been producing mesons for over a year, which Gardner had not been able to detect [73, 74]!

We may ask if earlier evidences of π mesons in the cloud chamber were escaped to observation. Fretter [75] and Brode [76, 77] made accurate determinations of the meson masses in cloud chamber and their results excluded the presence of a large fraction of mesons with masses between 250 and 300 electronic masses (although some results suggested the possibility of these masses [78]). In any case, due to their short lifetime, it is understandable that pions are easily found at high altitudes and therefore are not easily observed at sea level with the Wilson chamber.

Once the lifetimes of π and μ were determined, the question arose about the nature of their decay products. In the decay of a meson some mass is transformed in energy according to the Einstein formula. The disappearing mass is the difference between the father meson mass and the sum of the masses of the particles which appear in the decay. This energy difference appears as kinetic energy of these particles.

Considering first the π meson decay, the only visible track that appears at the end of its path is that of a muon. However at least another particle should come out, due to momentum conservation, and because no other track is visible, this means that to be a neutral particle. If we assume that only an invisible particle is produced, then there is only one way the energy can be shared between it and the pion in order to have opposite and equal momenta (the decaying particle being at rest has null momentum). If instead more than two particles come out, they may share energy and momentum in many different ways.

Now, according to the experimental facts, every time a muon stops in the emulsion it travels always the same distance from the origin point: about 600 μm. Because the μ acquires always the same energy (4.17 MeV), the total energy is divided in only one way and therefore only an invisible particle is involved.

Who is this neutral particle? Energy balance and momentum conservation show its mass should be nearly zero. If one wishes not to involve a new type of neutral particle, it is simpler to assume it to have zero mass as a photon or a neutrino. An experiment performed by C. O'Ceallaigh [79] in order to decide whether the neutral particle is a photon gave negative results, and therefore it was decided that it was a neutrino.

Accordingly, a π^+ or π^- nearly always decays in a muon and a neutrino

$$\pi^{+(-)} \rightarrow \mu^{+(-)} + \nu_\mu (\underline{\nu}_\mu)$$

The positive pion decays in a positive muon and a neutrino, and the negative pion decays in a negative muon and an antineutrino. Due to momentum conservation, the muon and neutrino need to travel in opposite directions [60].

Other rare decay forms were found in which a charged pion decays in an electron (positron) and an electronic antineutrino (neutrino) or in a neutral pion, an electron (positron) and an antineutrino (neutrino).

Negative pions, instead of decaying, usually produce violent nuclear disintegrations observed already in the first Lattes et al. [60] work.

The discovery of the pion started a number of researches all performed with nuclear emulsions. This research was very inexpensive, but required a crowd of observers (typically women) who spent the whole day at a microscope searching for the tracks and measuring them. This was an activity rather suited for the European countries after the war, in particular for Italy that soon became one of the most important centres for this kind of research. This is one of the reasons for the prominence given in Italy to the study of elementary particles: at the beginning it was a rather inexpensive kind of research.

With their researches the Bristol group solved two big puzzles of the past: they established the existence of two kinds of mesons, the heavier π being associated with the Yukawa mesons which mediated the nuclear forces. The lighter one, resulting from the pion decay and observed essentially at low altitudes (as the Anderson and Neddermeyer mesons), does not exhibit strong interactions with nuclei (in agreement with the Conversi et al. experiment of 1947 in Roma). Many people asked which was the role of the muon. Isidor Rabi (1898–1988) is reported to have exclaimed "*Who ordered it?*"

9.4 The Stars

In 1936, C.D. Anderson and S.H. Neddermeyer [80], R.B. Brode and collab. [81], H. Fussell [82] and other people began to find, in their Wilson chamber, examples of apparently simultaneous emission from a nucleus of several strongly ionizing particles, looking like a *star*.

The following year (1937), in Austria, also Marietta Blau and H. Wambacher [43, 83] using photographic emulsions exposed on Hafelekar mount at 2,300 m, found disintegration stars. These stars appeared in the photographic emulsions exposed for a long time, and consisted in groups of tracks of particles (later called *prongs*) coming out from a single point. Blau and Wambachen were using emulsions sensitive only to strongly ionizing particles; therefore, they attributed the tracks to protons and alpha particles with energies of the order of 10 MeV and tried to see if there was radioactive contamination in the plate. However the particle energy was too high and so they concluded the stars to be the result of disintegrations caused by cosmic rays.

The most convincing proof that stars are produced by cosmic rays and are not the result of some radioactive transformation was given by G. Setter and H. Wambacher [84] who found that the rate of production of stars increased rapidly with altitude ascending from 200 to 3,400 m. Also the sum of the produced protons and the number of tracks per stars was increasing.

H. Wambacher [85] wrote a bibliography of investigations with nuclear emulsions and cloud chamber up to 1938; he studied in detail 154 stars of more than two particles and estimated the energy of single particles that he identified as protons. Other people [86, 87] found alpha particles and protons and sometimes even more heavy fragments.

With cloud chamber the research was very difficult due to the very small number of visible stars. Extensive researches were done later by W.M. Powell [88, 89] at 4,300 m. Among 20,000 photos he found that only 13 stars were created in the chamber gas, and 156 produced in the lead of the interposed plates. W.E. Hazen [90] reported, among 8,500 photos made at 3,000 m, only 2 stars in the chamber air and 56 in lead.

What produces stars? On this question a lively discussion arose. We omit here to detail all the proposals. E. Bagge [91] believed that they were produced by gamma rays.

The Heisenberg's theory on the origin of showers—which were attributed to explosive events resulting from the collision of a nucleus with an atom—was also considered.

After the discovery of the muon, the problem was from where this particle comes. Being instable, it should be produced in the atmosphere. Someone believed that muons could be produced in pairs through materialization of photons, in a process similar to the one that brings to the production of electron–positron pair. However, as proved later, this process is very rare.

The most probable hypothesis was that they were produced in multiple processes during nuclear collisions and somehow connected to stars. Gleb Watagin [92] (1899–1986), then in Brazil, and Janossy [93, 94], at Manchester, were the first to consider this hypothesis, around 1940. Janossy performed an experiment placing a number of counters with interposed thick lead plates and suitable electronic circuits to detect coincidences among non-aligned counters. The residual coincidences that could be found could not be produced by particles of ordinary showers which were not able to pass the shielding and would certainly be indicative of the production of muons by disintegrations of cosmic rays in the material above the counters.

W.M. Powell (quoted references) thought instead that neutrons were responsible.

Rossi and collab. [95] reported that the ionization pulses in an unscreened ionization chamber were mostly caused not by showers but by strongly ionizing particles produced in nuclear disintegration by cosmic rays.

Eventually, D.H. Perkins [67] and G.P.S. Occhialini and C.F. Powell [57, 70] demonstrated that in the photographic emulsions stars were formed at the end of slow mesons.

The use of nuclear emulsion proved to be the most suitable for the study of star formation. The nuclear emulsion allowed to "see" the nuclear interaction at the point in which it occurred and could cumulate in time these events, making easier the study of these rather rare processes.

With nuclear emulsions it was demonstrated that pions (especially the negative ones), besides decaying, may produce violent nuclear disintegrations. This discovery by C.M.G. Lattes et al. [63] gave raise to a new series of researches, all done with the nuclear emulsion technique.

Figures 9.3 and 9.4 show two early examples of stars in nuclear emulsions obtained by the group in Bristol [60].

In 1947 D.H. Perkins reported that the star number found in plates placed below a thick lead screen was not much smaller than the number found without screen. Because the screen was sufficiently thick to absorb practically all the electrons and photons, these particles could not be the cause of star production.

In 1948 John Tinlot [96], at MIT, showed that while the production rate of penetrating showers increased a factor 32 between sea level and 4,300 m on mount Evans, the number of muons increased only by a factor 2.5. Therefore, the particles producing the penetrating showers in nuclear collisions could not be muons.

Fig. 9.3 An example of a nuclear reaction which was tentatively interpreted as a nitrogen nucleus struck by a muon producing two alpha particles, two protons, and four neutrons (Reprinted with permission from G.P.S. Occhialini, C.F. Powell, Nature **159**, 186 (1947) Copyright 1947 by Nature Publishing group).

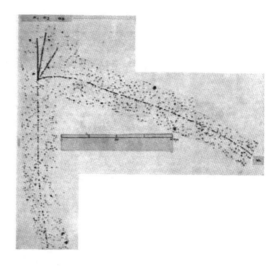

Fig. 9.4 A disintegration produced by a primary of high energy. Six tracks may be distinguished radiating from a common centre (Reprinted with permission from G.P.S. Occhialini, C.F. Powell, Nature **159**, 186 (1947) Copyright 1947 by Nature Publishing Group).

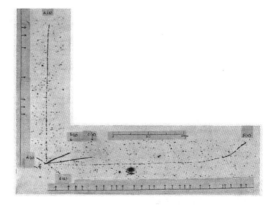

These results confirmed that high-energy electrons, photons, and muons, that form most of cosmic radiation at sea level, do not interact appreciably with atomic nuclei. The inability of muons to interact with atomic nuclei explains why these particles are slowly absorbed in the atmosphere and thus able to arrive at sea level.

The particles of cosmic radiation able to break nuclei, making stars and producing secondary mesons, very rare at sea level, but increasing rapidly with altitude, were therefore identified as protons, neutrons, pions (and the strange particles that were later discovered).

For several years, the cosmic-ray stars were the sole direct proof of interaction between cosmic rays and atomic nuclei. The energy required to produce these stars was modest on the cosmic-ray scale. However, stars were rather rare. Therefore the nuclear interactions did not play any fundamental role in the general scheme of cosmic-ray phenomena. Their study was, however, the first in which the effect of interaction of high-energy particles and nuclei could be observed. With the creation of powerful particle accelerators, these researches were continued with these machines that allowed to have a number of events enormously greater of the ones occurring in cosmic rays.

The fast increase of the number of protons, neutrons, and pions with altitudes made highly desirable for physicists interested to nuclear reactions produced by these particles to make experiments at the highest possible altitudes. The sounding balloons were perfect from this point of view. Also, locations on mountains were very suitable and a number of laboratories at high altitudes were built. We may mention the one on the Alps, near the Pic du Midi (2,877 m) in the central Pyrenees, 15 km from Bagneres-de-Bigorre and close to an already famous solar observatory. In Italy, the Laboratorio di Testa Grigia, near Cervinia, at 3,500 m, was operated by Gilberto Bernardini, Piccioni, and Carlo Castagnoli (1924–2005)[10]. In the United States, the Inter-university High-altitude Laboratory of Echo Lake, in the Rocky Mountains near Denver, Colorado, at an altitude of 3,240 m, was also used. In Bolivia, a laboratory was built at Chacaltaya, on a mountain of the Andinian Bolivian altopiano, near La Paz, at 5,200 m. After the discovery of the pion and the exposition of nuclear emulsion by Lattes, it assumed world relevance for cosmic-ray researches. The high altitude is such that the atmospheric pressure is only one-half than that at sea level. There were performed nuclear emulsion experiments and the experiment BASJE (Bolivian Air Shower Joint Experiment) which was the first experiment devoted to study gamma rays of very high energy through the study of showers [97].

Among the groups who studied most actively the star formation, in these laboratories or with balloons were Marcel Shein of Chicago University, Leprince-Ringuet of Ecole Polytechnique, Paris, and C.F. Powell, Bristol. The main result of the study of these nuclear interactions was the discovery of an enormous quite unexpected number of new particles that created great confusion.

9.5   The Neutral Mesons

Since the discovery of positive and negative muons, physicists asked themselves if neutral mesons could exist. To explain the experimental observations of the independence from the charge of the nuclear forces, N. Kremmer [98] pointed out that besides the Yukawa charged mesons, also a neutral pion should exist.

---

[10]C. Castagnoli graduated in physics at Pisa University in 1947 and started a collaboration with Roma where he was associated professor from 1948 to 1959 when he was appointed physics professor first in Parma (1959–1960) and then in Torino (1960–1987). He was president of the Italian Physical Society (1974–1981), created the Istituto di Cosmogeofisica of CNR at Torino in 1968. He performed the first observation with nuclear emulsions of an antiproton annihilation. Among its studies were the decay of pions and extended air showers (EAS).

In 1948, the problem was discussed by Oppenheimer [99] who suggested that neutral mesons should decay very rapidly into photons which could be responsible for the most of the photons and electrons found in cosmic rays at sea level.

The Oppenheimer suggestion proved to be correct.

The neutral pion was observed through its decay products (two photons) since the end of 1949, both in cosmic rays [66, 100] in Bristol and in experiments with the Berkeley cyclotron by W.E. Crandal, B.J. Moyer (1912–1973), and H.F. York [101–103].

The discovery of the neutral pion was a milestone in particle physics.

Its mean life was measured by A.G. Carlson [104] who found it to be $(1–5) \times 10^{-14}$ s. The experiment consisted in determining the spectrum of the gamma radiation in the atmosphere at 70,000 ft. (21 km) by the observation on the scattering of pairs of fast electrons recorded in photographic emulsions exposed in high-flying balloons. The detailed form of the spectrum was found consistent with the assumption that the gamma rays originated by the decay of neutral mesons. It was found that the mass of the neutral mesons was 295 ± 20 electronic masses, and that they were created in nuclear explosions with an "energy spectrum" similar to that of the charged pions. The up-to-date value of its lifetime is 83×10^{-14} s.

Photos with a cloud chamber using an accelerator were also obtained in 1950, by the group at MIT that studied cosmic rays [103, 105]. Confirmation of the decay into two gamma rays was obtained by Panofsky [106]. This experiment was a direct measurement of the parity of the negative pion, but we do not wish to deal with these problems.

9.6 The Lambda Meson

In 1948 Auger and collab. [107] suggested that a meson with a mass intermediate between the electron and the muon could exist. The mass of this λ meson was not determined because its existence was inferred indirectly from the properties of extended showers, but it was assumed between 3 and 10 electronic masses. However, later it came out it was not necessary to explain the properties of extended showers.

9.7 The New Particles

Between 1947 and 1962, the number of elementary particles increased from four to more than one hundred and after it exploded. At the beginning these "strange" particles were discovered in cosmic rays, but successively this kind of investigation was rapidly supplanted by the use of great particle accelerators. The identification of the different particles and their interpretation on the basis of theoretical models was very arduous and emerged gradually from a chaos of results. This is not the place to trace the story—rather complex—of these researches. Some references are given in the note [108–112]. We limit ourselves to give a short account of the beginning of this great epopee initiated by the studies of cosmic rays.

In fact, muons and pions were not the sole particles found in cosmic rays. Already in 1944, Leprince–Ringuet [113, 114][11] (1901–2000), in the Largentière laboratory on Alps at 1,000 m elevation, studying in a Wilson chamber the elastic scattering between penetrating charged particles (filtered by 10 cm of lead) and electrons, deduced the existence of a new particle of mass approximately 990 times the electron mass. In retrospective this seems to have been the first measurement of the mass of the charged K meson. His result was not immediately believed even if it was received with benevolence.

From September 27 to October 2, 1948, the eighth Solvay Conference [115] was held in Bruxelles under the general theme "Elementary particles". Powell presented a summary of the particles discovered in cosmic rays and R. Serber of Berkeley, California, presented results obtained by bombarding a target with high-energy alpha particles accelerated with the Berkeley cyclotron. The threshold for the production of the π meson was 260 MeV. At the conference, in the convivial meetings, some rhymes and songs were played. One of these, called the *meson song*, is an arrangement of a piece of classical American literature done by Childs, Marshak, McCreary with their wives and Valley of MIT. Here it follows

There are mesons pi, there are mesons mu
The former ones serve as nuclear glue
There are mesons tau, or so we suspect
And many more mesons which we can't yet detect

> *Can't you see them at all?*
> *Well, hardly at all*
> *For their lifetimes are short*
> *And their ranges are small.*

The mass may be small, the mass may be large,
We may find a positive or negative charge,
And some mesons never will show on a plate
For their charge is zero, through their mass is quite great

> *What, no charge at all?*
> *No, no charge at all!*
> *Or, if Blackett is right*
> *It's exceedingly small*

Some beautiful pictures are thrown on the screen,
Through the tracks of the mesons can hardly be seen,

---

[11]Louis Leprince-Ringuet was an engineer. In 1929, he worked in the X-ray laboratory of the physicist Maurice de Broglie and decides to work in nuclear physics. He was a Professor of physics at Ecole Polytechnique (1936–1969) and Collège de France (1959–1972).

Our desire for knowledge is mostly deeply stirred
When the statements of Serber can never be heard.

   What, not heard at all?
   No, not heard at all!
   Very dimly seen
   And not heard at all!

There are mesons lambda at the end of our list
Which are hard to be found but are easily missed,
In cosmic-ray showers they live and they die
But you can't get a picture, they are camera-shy.

   Well, do they exist?
   Or don't they exist?
   They are on our list
   But are easily missed

From mesons all manner of forces you get,
The infinite part you simply forget,
The divergence is large, the divergence is small,
In the meson field quanta there is no sense at all.

   What, no sense at all?
   No, no sense at all!
   Or, if there is some sense
   It's exceedingly small.

A summary of the conference, in the form of a French ballade, was presented by Casimir:

Les électrons et positrons
Mésons et forces nucléaires
Deuterons, protons, photons
Particules permanentes ou très éphémères
Particules bien connues ou trou trouvées naguère
Ou bien particules pas encore observées
Tout cela à été notre affaire
Et c'est là le but du Conseil Solvay.

Les grands corps en rotation
Le magnétisme de la terre
Et des étoiles ; les explosions,
Gerbes nées dans l'atmosphère
Et la doctrine complémentaire
Qui a clairement montré
Qu'il faut renoncer aux images vulgaires
Et c'est là le but du Conseil Solvay.

On a vu que le triton
N'est point un être légendaire,
Quant au problème de l'électron,
La situation théorique était claire
Quand Oppi a dit qu'il pourra se défaire
De toute divergence et de l'infinité,
Une fois qu'il sera devenu grand-père;
Et c'est là le but du Conseil Solvay.

Envoi

Leprince Ringuet quoiqu'on ne sache guère
Si la tau-particule est une réalité
Nous voulons bien le croire, seulement pour vous plaire
Et c'est là le but du Conseil Solvay.

The last strophe shows the general attitude at that time towards the particle hypothesized by Leprince-Ringuet.

In 1947, just a few months after the paper on π mesons, the first strange particles, the K or kaon, the Λ particle, etc., were discovered in cosmic radiation at high altitude.

The first announce of the discovery was given in the paper *"Evidence for the existence of new unstable elementary particles"* published on Nature [116] by George Rochester[12] (1908–2001) and Clifford Butler[13] (1922–1999), both from Manchester, who gave the first suggestion of the existence of heavy neutral and charged unstable particles, presenting results obtained in an investigation with the large cloud chamber built by Blackett [117] in 1936. Over about 50 photos of penetrating showers they observed two particular events. The first one shows two tracks, with the origin apparently coincident to form an inverted V, in the chamber gas under a thick metallic plate.

Both tracks of the two charged particles had minimal ionization and no proof was found of knock fragments in the vicinity of the apex. On the basis of a careful

---

[12]George Dixon Rochester studied at Armstrong College, Newcastle (then a constituent part of Durham University). He first worked in spectroscopy, then in 1937 was appointed to an assistant lectureship in Manchester under L. Bragg and had his PhD in spectroscopy. In 1938 P.M.S. Blackett arrived in Manchester and Rochester joined him to work on cosmic rays with the cloud chamber. After the war, together with C. Buttler they take over the 11-ton "Blackett magnet" to study penetrating showers and discovered V particles. After Blackett move to Imperial College, London, Rochester was named director of the Physical Lab until he was offered a chair in physics at Durham (1955) where he remained for the rest of his life.

[13]Clifford Charles Buttler worked first in electron diffraction at University of Reading and in 1937 joined Rochester in Blackett's group in Manchester. In 1953 he moved with Blackett to Imperial College where he became a full professor in 1957, heading the High-Energy Nuclear Physics group. Later he became head of Imperial College Physics Department. From 1970 to 1975 he was director of the Nuffield Foundation and from 1975 to 1985 he served as vice-chancellor of Loughborough University of Technology. He played important roles in IUPAP.

Fig. 9.5 The V-shaped track is in the lower right side (Reprinted with permission from G.D. Rochester and C.C. Buttler, Nature **160**, 855 (1947) Copyright 1947 by Nature Publishing Group)

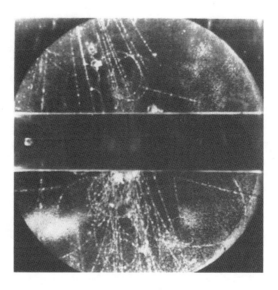

analysis, Rochester and Butler concluded the event could be explained in terms of the spontaneous decay in fly of a new type of heavy neutral particle ($V°$) which generated a pair of charged particles of opposite sign and notable magnetic rigidity.

This event is shown in Fig. 9.5. Just, a few millimetres below the lead plate, in the lower right-hand quadrant, two tracks are visible which form an inverted V with an opening angle of 67 ° extending to the lower right. These tracks were both at minimum ionization and were due to a positive and a negative particles. The V-shaped tracks could be explained only by assuming they were generated by an object of a mass about $1,115\,m_e$, which decayed into a couple of pions. The $V°$ event was obtained on October 15, 1946.

In the paper also another event with a V shape was presented which was interpreted as the decay of a charged particle (V^+). This event was obtained on May 23, 1947.

It was produced by a charged particle that decayed into one electron and one or more unidentified neutral particles (see Fig. 9.6).

These events remained unique for more than a year, but the certainty of the experiment and of its interpretation was such that there was no doubt on the existence of the new kind of particle that, on a suggestion of Blackett and Anderson, was called V particle. Today these particles are called Kaons. The name of K particle, or kaon, was proposed by Amaldi et al. [118, 119].

The two observed events were therefore the decay of kaons and represented, one a neutral particle, whose track obviously is not seen, which decayed into two particles (identified as a positive and a negative pion respectively) that were seen as a pair of V-shaped tracks (from which the name V particles). The other event was a charged particle that decayed into one electron and one or more unidentified neutral particles. The incident particle, shown in Fig. 9.6, is the straight line on the upper right that is apparently deflected through 19° in the gas of the chamber. It is

Fig. 9.6 The decay of a charged V particle into an electron and a neutral particle is shown on the upper right extremity in the photo (Reprinted with permission from G.D. Rochester and C.C. Buttler, Nature **160**, 855 (1947) Copyright 1947 by Nature Publishing Group)

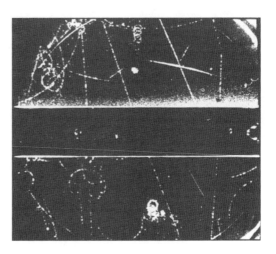

a positive particle. The successive positive particle penetrated the 3.4 cm lead plate. These tracks were interpreted as the decay in flight of a positive V particle into a secondary positive particle, probably a π meson, and one or more neutral particles.

The Bristol group [120], from the examination of electron-sensitive emulsions exposed at the Jungfraujoch, found another example of decay of a charged kaon into two positive and one negative pions, allowing a decisive measurement of its mass. This event was initially attributed to a meson called *tau*.

A confirmation of the events found by Rochester and Butler came from photos obtained with a cloud chamber operated at high altitude by Seriff et al. in California [121]. They found 34 examples of V-shaped tracks associated with penetrating showers in lead at sea level and at 3,200 m elevation. Of these 30 described neutral particle decays and 4 showed decays of charged particles. The lifetime of the neutral particle was found of the order of 3×10^{-10} s.

The decay of the neutral V particle was studied by Armenteros et al. [122] at Manchester.

The description of kaons, at the beginning, was very confused, due to the multiplicity of the decay modes and to the difficulty of measuring them.

At that time it was common to classify the observed particles not on the basis of their true nature, but on the basis of the effects they produced in the detector. An example is the sigma and rho mesons as classified by Powell in order to distinguish them through their effect in the emulsions. Similarly, according to their decay modes kaons received different names.

The neutral kaon was found to decay into two, one positive and the other negative, pions. Due to the equal masses of the two pions, the distribution of their track momenta was equal and so were the typical angles they formed with the direction of the incident kaon. However in 1950, an event in which the positive track appeared to be a proton was found [123]. In the decay of a neutral particle into one proton plus a negative pion, the proton, due to its larger mass, tends to keep most of the momentum, producing an asymmetric configurations of the tracks. The group at Pic du Midi of R. Armenteros [122] and the group with the Wilson chamber, in Indiana, lead by R. Thompson [124] were therefore able to distinguish these two decay processes. The first decay, symmetric, represents the decay of a neutral kaon into two pions according to the reaction

$$K^\circ \to \pi^+ + \pi^-$$

The other process was instead identified as bearing to a new particle of mass 1150 MeV which was named Λ particle, and decayed according the scheme

$$\Lambda \to p + \pi^-.$$

The Λ particle was the first example [122] of an *hyperon*, a particle heavier than the proton. Later C.F. Powell [125] wrote: "*An excited hydrogen atom, to use the simplest example, consists of a proton and an electron in a state of higher energy than in the normal atom. The analogy might then suggest that the excited nucleon consists of a proton and an associated π^-—that the Λ° is a composite particle. Such a view could not have been finally excluded while our knowledge was confined to the decay of the free Λ° particle... These considerations suggest that the Λ° is an excited nucleon in a different sense from that suggested by familiar analogies. We are entering a new field where basically new concepts remain to be established*".

The cosmic ray studies also produced the first evidence of K^+ [126, 127], and of the charged hyperons Ξ^- and Σ^+

The Ξ^- was called "cascade" particle because it decayed in one Λ plus a negative pion that in turn decayed in other particles [128]. The Σ^+ was found [129, 130] to decay into a neutron plus a positive pion or into a proton plus a neutral pion

$$\Sigma^+ \to n + \pi + \quad \text{or} \quad \Sigma^+ \to p + \pi^\circ.$$

At variance with the neutral kaon, the charged kaon was found to have many types of decay. To the two pions decay, discovered in 1947 was given also the name of *theta* particle. The decay into three pions was named *tau* particle. Nearly 10 years were needed to find that *theta* and *tau* were two different decay modes of the same particle: the kaon. The famous puzzle theta–tau ([131, 132], see also [133]) was so finally solved. In all these researches, the nuclear emulsions were of great help.

These initial discoveries made since the spring of 1950, opened the new era of the so-called *elementary particles* in physics. The Λ, Ξ^-, and Σ hyperons appeared, as the V particles of Rochester and Butler, not only unexpected, but also provided with unusual characteristics: from one side, they were relatively copiously produced in high-energy nuclear collisions, in approximately 1 % of all events. Thus, if the same mechanism was responsible for their production and decay, their lifetime should be of the order of 10^{-21} s. On the other side they decayed with relatively long lifetimes spanning from 10^{-10} to 10^{-8} s. Therefore these so-called "strange" particles were produced (always in pair) through the strong nuclear forces, but they decayed through the weak nuclear forces [134]. M. Gell-Mann suspected that some unknown property was conserved and that this explained the slow decay of kaons. He named the new property "strangeness".

A research started strictly interconnected to theory and experiment. It is a fascinating story that has not yet ended but we cannot deal with it here. A good review of the initial status of research has been given by Rochester and Butler [135–139].

At first it was a hard life. Gell-Mann[14] said [140]: "*Strange particles… were not considered respectable, especially among the theorists. I am told.. that when he wrote his excellent paper on the decay of the tau particle into three pions Dalitz was warned that it might adversely* affect *his career, because he would be known as the sort of person who worked on that kind of things*". And, Dalitz [141] remembered: "*Pion physics was indeed the central topic for theoretical physics in the mid 1950s, and that was that the young theoretician was expected to work on. The strange particles were considered generally to be an obscure and uncertain area of phenomena, as some kind of dirt effect which could not have much role to play in the nuclear forces, whose comprehension was considered to be the purpose of our research*".

The number of particles discovered in cosmic rays was still increasing when the great accelerators started to work, able to create in the lab first pions, and then all the other particles discovered in cosmic rays and even more.

The Brookhaven cosmotron, that started to operate in 1952, belonged to a new generation of accelerators, built to study fundamental interactions at energies yet never attained. It was soon employed to study strange particles. In a series of experiments with the cloud chamber Ralph Strutt and his group [142–145] demonstrated that the new particles were produced in pair as in the reaction

$$\pi^- + p \rightarrow \Lambda^\circ + K^\circ$$

confirming the hypotheses of several Japanese groups [146–148] and of Pais [134]. The group succeeded also in finding negative and neutral sigma hyperons. We stop here because we are entering another world.

References

1. S. Tomonaga, G. Araki, Phys. Rev. **58**, 90 (1940)
2. M. Conversi, in *The Bird of Particle Physics*, ed. by M.I. Brown, L. Hoddeson (Cambridge University Press, New York, 1983), p. 242
3. M. Conversi, Il Nuovo Saggiatore **1**(1), 33 (1985)
4. M. Conversi, O. Piccioni, Nuovo Cimentio **2**, 71 (1944)
5. M. Conversi, O. Piccioni, Phys. Rev. **70**, 859 (1946)

---

[14]Murray Gell-Mann (1929–) was an American Physicist and a professor at Caltech. He was mainly concerned with the study of elementary particles. He introduced the strangeness and the quarks. He was awarded the Physics Nobel Prize in 1969.

6. G. Bernardini, G.C. Wick, M. Conversi, E. Pancini, Phys. Rev. **60**, 535 (1941)
7. G. Bernardini, M. Conversi, E. Pancini, E. Scrocco, G.C. Wick, Phys. Rev. **68**, 109 (1945) and references wherein
8. B. Rossi, Nature **128**, 300 (1931)
9. L.M. Mott-Smith, Phys. Rev. **35**, 1125 (1930)
10. L.M. Mott-Smith, Phys. Rev. **37**, 1001 (1931)
11. L.M. Mott-Smith, Phys. Rev. **39**, 403 (1932)
12. C.F. Weizsaecker, Ann. Phys. **17**, 869 (1933)
13. M. Conversi, E. Pacini, O. Piccioni, Phys. Rev. **68**, 232 (1945)
14. M. Conversi, O. Piccioni, Phys. Rev. **70**, 874 (1946)
15. M. Conversi, E. Pacini, O. Piccioni, Phys. Rev. **71**, 209 (1947)
16. G.E. Valley, Phys. Rev. **72**, 772 (1947)
17. E. Fermi, E. Teller, V. Weisskopf, Phys. Rev. **71**, 314 (1947)
18. J.A. Wheeler, Phys. Rev. **71**, 320 (1947)
19. B. Pontecorvo, Phys. Rev. **72**, 246 (1947)
20. G. Puppi, Nuovo Cimentio **5**, 587 (1948)
21. O. Klein, Nature **161**, 897 (1948)
22. T.D. Lee, M. Rosenbluth, C.N. Yang, Phys. Rev. **75**, 905 (1949)
23. J. Tiomno, J.A. Wheeler, Rev. Mod. Phys. **21**, 1441 (1949)
24. R.E. Marshak, H.A. Bethe, Phys. Rev. **72**, 506 (1947)
25. S. Sakata, T. Inoue, Prog. Theor. Phys. **1**, 143 (1946)
26. S. Sakata, T. Inoue, Bull. Phys. Math. Soc. Jpn **16**, 232 (1942)
27. M. Taketani, Prog. Theor. Phys. **3**, 349 (1948)
28. H. Yagoda, *Radioactive Measurements with Nuclear Emulsions* (Wiley, New York, 1949)
29. M. Grilli, F. Sebastiano, Storia delle tecniche di rivelazione delle particelle elementari: le emulsioni nucleari, Il Nuovo Saggiatore **5**(3), 68 (1989)
30. S. Kinoshita, Proc. Roy. Soc. (Lond.), **A83**, 432 (1910)
31. M. Reinganum, Phys. Z. **12**, 1076 (1911)
32. W. Makower, Nature **99**, 98 (1917)
33. W. Makower, H.P. Walmsley, Proc. Phys. Soc. (Lond.) **26A**, 261 (1914)
34. W. Michl, Akad. Wiss. Wein. **121**, 1431 (1912)
35. S. Kinoshita, H. Ikeuti, Philos. Mag. **29**(6), 420 (1915)
36. H. Ikeuti, Philos. Mag. **32**(6), 129 (1916)
37. R. R. Sahni, Philos. Mag. **29**(6), 836 (1915)
38. E. Muehlestein, Arch. Sci. **4**, 38 (1922)
39. L. Myssowsky, P. Tschijow, Z. Phys. **44**, 408 (1927)
40. V.I. Baranov, S.I. Kretschmer, C. R. Acad. Sci. (URSS) **1**, 111 (1935)
41. A. Zhdanov, J. Phys. Radium **6**, 233 (1935)
42. A. Zhdanov, Proc. State Radium Inst. (URSS), **3**, 7 (1937)
43. H.M. Blau, H. Wambacher, Nature **140**, 585 (1937)
44. H.M. Blau, J. Phys. Radium **5**, 61 (1934)
45. A. Jdanoff, J. Phys. Radium **6**, 233 (1935)
46. H.J. Taylor, Proc. Roy. Soc. (Lond.) **A150**, 382 (1935)
47. C.F. Powell, Nature **144**, 155 (1939)
48. T.R. Wilkins, H.J. St. Helens, Phys. Rev. **54** (1938) 783
49. C.F. Powell, G.P.S. Occhialini, *Nuclear Physics in Photographs* (Oxford University Press, New York, 1947)
50. C.F. Powell, P.H. Fowler, D.H. Perkins, *The Study of Elementary Particles by the Photographic Method* (Pergamon, New York, 1959)
51. C. Waller, J. Photogr. Sci. **1**, 41 (1953)
52. C.F. Powell, G.P.S. Occhialini, O.L. Livesey, L.V. Chilton, J. Sci. Instrum. **23**, 102 (1946)
53. M.M. Shapiro, Rev. Mod. Phys. **13**, 58 (1941)
54. V.A. Beiser, Nuclear emulsion technique, Rev. Mod. Phys. **24**, 273 (1952)
55. Y. Goldschmidt-Clerman, Annu. Rev. Nucl. Sci. **3**, 141 (1953)
56. P. Demers, Phys. Rev. **70**, 86 (1946)

57. G.P.S. Occhialini, C.F. Powell, Nature **159**, 186 (1947)
58. E.J. Williams, Proc. Roy. Soc. (Lond.) **A169**, 531 (1939)
59. E.J. Williams, Phys. Rev. **58**, 292 (1940)
60. C.M.G. Lattes, H. Muirhead, G.P.S. Occhialini, C.F. Powell, Nature **159**, 694 (1947)
61. C.M.G. Lattes, G.P.S. Occhialini, C.F. Powell, Nature **160**, 453 (1947)
62. R. Brown, U. Camerini, P.H. Fowler, H. Muirhead, C.F. Powell, Nature **163**, 47 (1949)
63. C.M.G. Lattes, G.P.S. Occhialini, C.F. Powell, Nature **160**, 486 (1947)
64. C.M.G. Lattes, G.P.S. Occhialini, C.F. Powell, Proc. Phys. Soc. (Lond.) **A61**, 173 (1948)
65. Y. Goldschmidt-Clermont, D.T. King, H. Murihead, D.M. Ritson, Proc. Phys. Soc. **61**, 138 (1948)
66. C.F. Powell, Rep. Prog. Phys. **XIII**, 350 (1950)
67. D.H. Perkins, Nature **159**, 126 (1947)
68. C.F. Powell in Symposium, Bristol, Sept. 1948, Butterworths 1949, p. 83
69. G.P.S. Occhialini, C.F. Powell, Nature **160**, 453, 486 (1947)
70. G.P.S. Occhialini, C.F. Powell, Nature **159**, 93 (1947)
71. J.R. Richardson, Phys. Rev. **74**, 1720 (1948)
72. O. Chamberlain, R.F. Mozley, J. Steiberg, C. Wiegand, Phys. Rev. **79**, 394 (1950)
73. E. Gardner, C.M.G. Lattes, Science **107**, 270 (1948)
74. J. Burfening, E. Garner, C.M.G. Lattes, Phys. Rev. **75**, 382 (1949)
75. W.B. Fretter, Phys. Rev. **70**, 625 (1946)
76. R.B. Brode, Phys. Rev. **75**, 904 (1949)
77. R.B. Brode, Rev. Mod. Phys. **21**, 37 (1949)
78. T.C. Merkle Jr., E.L. Goldwasser, R.B. Brode, Phys. Rev. **79**, 926 (1945)
79. C. O'Ceallaigh, Philos. Mag. **41**(7), 838 (1950)
80. C.D. Anderson, S.H. Neddermeyer, Phys. Rev. **50**, 263 (1936)
81. R.B. Brode, H.G. MacPherson, M.A. Starr, Phys. Rev. **50**, 381 (1936)
82. H. Fussel, Phys. Rev. **51**, 1005 (1936)
83. H.M. Blau, H. Wambacher, Proc. Ac. Vienna **146**, 623 (1937)
84. G. Setter, H. Wambacher, Phys. Z. **40**, 702 (1939)
85. H. Wambacher, Phys. Z.. **39**, 883 (1938)
86. E. Schopper, E.M. Schopper, Phys. Z. **40**, 22 (1939)
87. A. Filippov, A. Jdanov, I. Gurevich, J. Phys. (USSR) **1**, 51 (1939)
88. W.M. Powell, Phys. Rev. **69**, 385 (1946)
89. W.M. Powell, Phys. Rev. **61**, 670 (1942)
90. W.H. Hazen, Phys. Rev. **65**, 67 (1944)
91. E. Bagge, Ann. Phys. **39**, 512 (1939)
92. G. Wataghin, M. de Souza-Santos, P.A. Pompena, Phys. Rev. **57**, 339 (1940)
93. L. Janossy, Proc. Roy. Soc. Lond. **A179**, 361 (1941)
94. L. Janossy, P. Ingleby, Nature **145**, 511 (1940)
95. H. Bridge, B. Rossi, Phys. Rev. **71**, 379 (1947)
96. J. Tinlot, Phys. Rev. **75**, 519 (1949)
97. K. Suga, I. Escobar, G. Clark, W. Hazen, A. Hendel, K. Muratami, J. Phys. Soc. Jpn **17**(Suppl. AIII) (1962)
98. N. Kremmer, Proc. Camb. Philos. Soc. **34**, 354 (1938)
99. H.W. Lewis, J.R. Oppenheimer, S.A. Wouthuysen, Phys. Rev. **73**, 127 (1948)
100. C.F. Powell, Edinburgh Conference Nov. 1949
101. W.E. Crandall, B.J. Moyer, H.F. York, Am. Soc. Meeting, Stanford Dec. 1949
102. W.E. Crandall, B.J. Moyer, H.F. York, Phys. Rev. **78**, 89 (1950)
103. R. Bjorklund, W.E. Crandall, B.J. Moyer, H.F. York, Phys. Rev. **77**, 213 (1950)
104. A.G. Carlson, J.E. Hooper, D.T. King, Philos. Mag. **41**(7), 701 (1950)
105. J. Steinberg et al.., Phys. Rev. **78**, 802 (1950)
106. W.K.H. Panofsky, R.L. Aamodt, J. Hdley, Phys. Rev. **81**, 565 (1951)
107. P. Auger, J. Daudin, A. Freon, R. Maze, C. R. Acad Sci. **226**, 169 (1948)

108. L.M. Brown, M. Dresden, L. Hoddeson, M. Riordan (ed.) *Third International Symposium on the History of Particle Physics: The Rise of the Standard Model (Stanford, CA, June 24–27, 1992)* (Cambridge University Press, New York, 1995)

109. A. Pais, *Inward Bound* (Oxford, Clarendon 1986)

110. L.M. Brown, L. Hoddeson (ed.) *International Symposium on the History of Particle Physics, The Birth of Particle Physics* (Cambridge University Press, New York, 1983)

111. M. Brown, L. Hoddeson (ed.) Colloque Int. Sur l'Historire de la Phys. des Particules, Paris 21-23 July, 1982. J. Phys. Coll. 43(supplement 12) C8

112. V.L. Fitch, J.L. Rosner, in *Twentieth Century Physics* vol II ed. by L.M. Brown, A. Pais, B. Pippard (IoP publishing Ltd, AIP Press Inc. New York 1995) p. 635–794 cap. 9

113. L. Leprince-Ringuet, M. Lheritier, C. R. Acad. Sci. **219**, 618 (1944)

114. L. Leprince-Ringuet, M. Lheritier, J. Phys. Radium, **7**, 65 (1946)

115. Les Particules Elementaires, Rapports et Discussions du Huitieme Conseil de Physique tenu à l'Universitè libre de Bruxelles du 27 septembre au 2 octobre 1948 R. Stoops, Brussels 1950

116. G.D. Rochester, C.C. Butler, Nature **160**, 855 (1947)

117. P.M.S. Blackett, Proc. Roy. Soc. **A154**, 564 (1936)

118. E. Amaldi et al., Nuovo Cimentio **11**, 213 (1954)

119. E. Amaldi et al., Nature **173**, 123 (1954)

120. R. Brown, U. Camerini, P.H. Fowler, H. Muirhead, C.F. Powell, D.M. Ritson, Nature **163**, 82 (1949)

121. A.J. Seriff, R.B. Leighton, C. Hsiao, E.W. Cowan, C.D. Anderson, Phys. Rev. **78**, 290 (1950)

122. R. Armenteros, K.H. Barker, C.C. Butler, A. Cachon, A.H. Chapman, Nature **167**, 501 (1951)

123. V.D. Hopper, S. Biswas, Phys. Rev. **80**, 1099 (1950)

124. R.W. Thompson, H.O. Cohn, R.S. Flum, Phys. Rev. **83**, 175 (1951)

125. C.F. Powell, Excited nucleons, Nature **173**, 469 (1954)

126. C. O'Ceallaigh, Philos. Mag. **42**(7), 1032 (1951)

127. P.H. Fowler et al., Philos. Mag. **42**(7), 1040 (1951)

128. R. Armenteros, Philos. Mag. **43**(7), 597 (1952)

129. C.M. York, R.B. Leighton, E.K. Bjornerud, Phys. Rev. **90**, 167 (1953)

130. A. Bonetti et al., Nuovo Cimentio **10**, 345, 1736 (1953)

131. R.H. Dalitz, Philos. Mag. **44**(7), 1068 (1953)

132. R.H. Dalitz, Proc. Phys. Soc. **A69**, 527 (1956)

133. H.B. Newman, T. Ypsilantis (eds.), *Dalitz Kaon Decays to Pions: the τ–θ Problem in History of the Original Ideas and Basic Discoveries in Particle Physics* (Plenum Press, New York 1994), p. 163

134. A. Pais, Phys. Rev. **86**, 663 (1952)

135. G.D. Rochester, C.C. Butler, Rep. Progr. Phys., **16**, 364 (1953)

136. W.D. Walker, Progr. Elem. Part Cosmic Ray Phys. **IV**, 73 (1960)

137. H.S. Bridge, Progr. Cosmic Ray Phys. **III**, 145 (1956)

138. R.W. Thompson, Progr. Cosmic Ray Phys. **III**, 255 (1956)

139. J.G. Wilson, Progr. Cosmic Ray Phys. **II**, 57 (1954)

140. M. Gell-Mann, *Colloque Int. sur l'Histoire de la Physique des Particules* (Paris, July 1982) J. Physique (Paris) **42**(Suppl.), C8–395 (1982)

141. R.H. Dalitz, *Colloque Int. sur l'Histoire de la Physique des Particules* (Paris, July 1982) J. Phys. (Paris) **42**(Suppl.), C8–406 (1982)

142. W.B. Fowler, R. Shutt et al., Phys. Rev. **90**, 1126 (1953)

143. W.B. Fowler, R. Shutt et al., Phys. Rev. **91**, 1287 (1953)

144. W.B. Fowler, R. Shutt et al., Phys. Rev. **93**, 861 (1954)

145. W.B. Fowler, R. Shutt et al., Phys. Rev. **98**, 121 (1955)

146. Y. Nambu et al., Progr. Theor. Phys. **6**, 615 (1951)

147. H. Miyazawa, Progr. Theor. Phys. **6**, 631 (1951)

148. S. Oneda, Progr. Theor. Phys. **6**, 633 (1951)

Chapter 10
The Extended Showers

10.1 Introduction

It was immediately understood that, because mesons are instable, they cannot be primary particles but should be produced in the interactions of primaries with the atmospheric nuclei. But which are the primaries? Millikan's gamma photons? The electrons used by Stoermer in his calculations?

In 1941, M. Schein and collab. [1], using a counter telescope carried by balloons, showed that the majority of particles near the top of the atmosphere are not electrons or gamma rays, but protons. Since then people accepted that the primary radiation consists essentially of protons.

Successive studies made after the war, at the end of the 1940s and early 1950s, with high-altitude balloons, aeroplanes, rockets, and satellites demonstrated that all the nuclei of the Periodic Table are present [2], and electrons, gamma rays, and neutrinos; however, protons are by far the most abundant component.

The discovery that the primary cosmic rays are mainly composed of protons imposed the replacement of the photon–nucleus collisions with the proton–nucleon interactions as the principal source of mesons in the high atmosphere. The paradox connected to the production and absorption of muons was solved in 1947 with the discovery of the pion. These discoveries and their subsequent analysis brought to a complete revision, accomplished in the early 1950s, of the events that occur in a shower.

When entering the atmosphere, the primary cosmic rays give rise to a chain of nuclear interactions whose result is the formation of the secondary radiation. Protons suffer the first collision with a nucleus of the atmospheric gases after having traversed a mean thickness of approximately $70 \, \mathrm{g/cm^2}$ of matter (that is 1/15 of the total air mass over the sea level). Alpha particles (helium nuclei) and the heavier nuclei collide even before. As a consequence, the probability that a primary cosmic ray escapes nuclear collisions and succeeds to reach the sea level is practically nonexistent. When a primary particle (for example, a proton) hits a nucleus, usually disintegrates it, starting a series of processes that consist in the possible production

M. Bertolotti, *Celestial Messengers: Cosmic Rays*, Astronomers' Universe,
DOI 10.1007/978-3-642-28371-0_10, © Springer-Verlag Berlin Heidelberg 2013

of neutrons, protons, pions, kaons, antiparticles, etc., that may have enough energy
to break other nuclei. The most numerous created particles are π mesons. The
neutral π mesons have a very short lifetime (8.3×10^{-14} s) and therefore, for
the greatest part, decay into photons without participating in other collisions. The
photons give place to a chain process: the ones which have energy greater than the
equivalent to the rest mass of a pair of positive and negative electrons (1.02 MeV)
in the electrostatic field of a nucleus may undergo pair production of electrons.
A photon may also produce fast secondary electrons by accelerating slow electrons
through the Compton effect. For photon energies of about 20 MeV in air and
5 MeV in lead, the probabilities for Compton scattering and for pair production are
equal; below these values Compton effect rapidly becomes predominant, and above
them pair production. The positron–electron pairs in turn irradiate new photons,
which start the process anew, contributing to the soft component. This way an
electrophotonic shower develops.

Charged π mesons, having a longer lifetime than the neutral ones, have a
reasonable probability to suffer nuclear collisions before decaying, originating *stars*
if they are negative, and muons and neutrinos if positive. Muons have a relatively
long lifetime (2.2×10^{-6} s) and a strong penetrating power because they have weak
interaction with nuclei and lose energy almost through ionization.

The slower muons decay into electrons and neutrinos before reaching the sea
level. The faster muons, taking advantage of the relativistic time dilation, are able
to reach the sea level before decaying, and form most of the hard component. They
are found to be able to traverse relevant lead thicknesses before producing electrons
in cascade or are absorbed through ionization and electromagnetic processes.

An appreciable part is also found underground.

The products of the interaction of the primary particles move practically all in
the same direction and give place to a shower of other interactions that, in the low
atmosphere, allow the production of several millions of secondary particles. All
these secondary particles arrive practically at the same time on a plane perpendicular
to the direction of motion of the original particle—the axis of the shower—but during their path along kilometres in the atmosphere are scattered around the axis over
distances of several hundreds of metres. Therefore, the shower covers a circular area
of many thousands of square metres with a maximum density in the central region.

It should be noted that the production of showers is the normal consequence of
the interaction of all the primary rays in the atmosphere. The low-energy cosmic
rays trigger showers that exhaust completely before reaching the low atmosphere.
As the primary energy increases, showers of a few particles able to reach the ground
are generated. All these showers are extended because the air scatters the secondary
particles away from the axis, but except the case in which a large number of them
reaches the level at which the observation is made, it is difficult to demonstrate
experimentally that the shower exists. Many single particles of cosmic rays detected
with a small apparatus near the ground belong to these extended but meagre
showers. When the number of secondary particles reaches a few thousands and the
centre of the shower is at a distance of a few tens metres from the apparatus, then it
is very probable that two or more particles of the shower arrive simultaneously over

a small area and a reasonable experimental proof of the existence of the shower may be obtained.

About 95% of the particles in a shower are electrons and photons, a few percent are muons, and only 1% is made of particles strongly interacting with matter (for example, pions or other mesons). The prevalence of electrons and photons gives the impression that these particles are the most important component of the shower and lead to the initial interpretation of the showers as a purely electromagnetic phenomenon.

If the primary particles have very high energies (larger than 10^{14} eV) the generated shower is well detectable, it covers a vast area and takes the name of Extended Atmospheric Shower (EAS). Such primary particles have such a low flux that it is practically impossible to detect them on board of spacecrafts. They can, on the other hand, be detected indirectly through the extremely numerous (from over one hundred thousand up to tens of billions) secondary particles of the showers. The experiments are preferably made on high peaks (where the showers reach their maximum development) placing the detectors over surfaces of 10 or more square kilometres and observing the flux. The collected data allow determining the number (N) of particles in the shower at the time of its maximum development. From this number the energy of the incident particle may be derived. The maximum value of N measured up to now is 10^{11} from which the primary energy was derived to be 10^{20} eV. During the first investigations, the nature of the primary particle could not be established; only later the simulation codes were able to identify them somehow.

Obviously to derive the information of the primary energy, simulations are made that describe the shower on the basis of a number of hypotheses on the interactions that may have occurred.

At energies lower than about 10^{14} eV, the cosmic ray flux is high enough and may be measured directly with apparatuses on board of balloons and satellites.

EAS contain everything: nucleons, nuclei, hard gamma rays, mesons (π, K,...), electrons, muons (leptons), neutrinos, etc.

Figure 10.1 shows schematically the development of a shower.

10.2   The Discovery of Extended Atmospheric Showers, EAS

In 1933, in Asmara, Eritrea, Bruno Rossi [3] completed his measurement campaign quantitatively ascertaining the east–west asymmetry that demonstrates that the primary radiation is composed for the greatest part of positive particles. He wrote:

> *"it looked as if...occasionally on the instruments very extended corpuscular showers came that produced coincidences between counters even rather far one from the other. Unfortunately I had no time to closer study this phenomenon".*

Perhaps this was the first observation of an EAS.

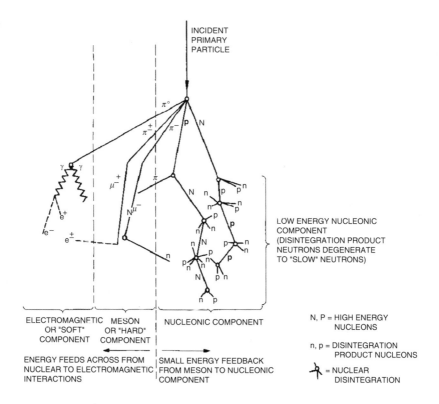

Fig. 10.1 Sketch of the development of a shower starting from an energetic primary. The different components are shown: the soft component made by the electrophotonic shower (electrons and photons), the hard component (essentially muons) and the nucleonic component, produced by violent interactions of fast nucleons with the atmosphere nuclei that produce besides pions, which feed the first two components, protons and neutrons that afterwards are slowed down. Neutrinos are also produced in the decay of muons and mesons.

A few years later, in Paris, using Geiger–Müller counters in coincidence, Pierre Auger and Ronald Maze [5], in 1938, observed again that counters placed several metres from one another detected simultaneously the arrival of cosmic ray particles, showing that they were secondary particles originated in a common source. They performed a detailed study and got credit of the discovery, apparently ignoring the Rossi's result.

Pierre Victor Auger (Fig. 10.2) was born in Paris in 1899, studied at Ecole Normale Superieure under J. Perrin (1870–1942), and earned his PhD in 1926. In 1936 he was appointed professor of physics at the University of Paris. In 1945 he was minister of education. From 1948 to 1960 he was a director of the science department of UNESCO, position that he left to become the president of the French Spatial Commission, and in 1964, he became the general director of the European Space and Research Organization, where he remained until his retirement in 1967.

Fig. 10.2 Pierre Auger

He discovered the effect that brings his name, in which an excited atom to jump back on a lower energy state, gives its energy to one of its electrons (instead of emitting a photon) that is expelled from the atom.

He also took an interest in science popularization and published poetry volumes. He died in Paris in 1993.

The re-discovery of extended showers by Auger and his group arose from a technical improvement in the resolving time of coincidence circuits made by Maze [6], that allowed detecting coincidences only if they occurred within 1 μs of each other. With this better time resolution, the group demonstrated that counts occurred, practically at the same time, in different places even if far apart one from each other.

Further experiments made by Auger [7] at the station of Pic du Midi at 2,877 m elevation and on Jungfraujoch, showed coincidences even between counters placed at a distance of 300 m one from another [8–11].

It was the definitive confirmation of the existence of EASs, showers of secondary particles and nuclei produced in collisions of high-energy primary particles with the air molecules.

Also K. Schmeiser and W. Bothe [12] had previously found coincidences between counters distant up to a half metre and Kolhoerster and collab. [13] measured coincidences between counters distant several metres.

On the basis of the Jungfraujoch measurements, Auger concluded he had observed showers with energies of 10^{15} eV (at that time the sole other source of fast particles was radioactivity and the higher registered energies from natural and artificial radioactivity reached only a few MeV). The energy value is, however, not quite correct as at those times the shower energy was calculated only on the basis of the electrophotonic shower theory.

The Auger experiments showed how wide could be the ground extension of showers. He also demonstrated that in showers some protons and neutrons exist, while the predominant part was electrons and photons and a small penetrating component able to pass more than 20 cm of lead. In his researches he also measured the strong variation with altitude of the shower composition and the barometric coefficient, and estimated the density of particles. Auger performed also measurements after the war [14] with three groups of counters separated by a few meters.

Confirmation of the extended shower existence came also from an experiment by A.C.B. Lovell and J.G. Wilson [15], in which they placed two cloud chambers triggered by Geiger–Müller counters at a distance of a few metres from each other at the Physics Department of the University of Manchester, finding that when the two chambers were operated in coincidence from the counters, parallel tracks were seen as it should be expected if many particles were travelling in the shower. Other experiments were done with cloud chambers [16] and ionization chambers [17–19]. Also Skobeltzyn [20] studied the extension of the showers finding that they spread over distances as long as 1,000 m.

Extended showers were soon interpreted as pure electrophotonic cascades of high energy and described theoretically by Euler [21] and Molière [22]. However, starting from the first results of Auger, a strong penetrating components was found which was definitely proved by detailed experiments near sea level by Wathagin ([23]; see also [24]) (1899–1986) and his co-workers. The results were confirmed by L. Janossi and P. Ingleby [25]. Surprisingly, Auger [26] and Janossy [27] suggested that the bulk of the penetrating particles in air showers could not be mesons or nucleons, suggesting instead tentatively that these particles might be "λ-mesons" we told of in Chap. 9.

The correct description as electro-nuclear showers emerged slowly and was completed in the early 1950s. In 1956, K. Greisen [28] (1918–2007), who was the first PhD student of Rossi, performing a vast review on extended showers, surprisingly did not quote Rossi as the first discoverer.

10.3   The Building of Large Detectors for the Extended Showers. Cosmic Ray Physics Goes "Big"

Because it was clear that some showers were the result of primary cosmic rays of enormous energy, their study was continued to get information on the origin of primary cosmic rays and to investigate the nature of the interactions of extremely energetic particles. After the Second World War, large arrays of detectors started to be built, based on different physical principles to measure at ground the shower dimensions and the number and the arrival times of its particles.

It was then, and it is still now, the only possibility to register the arrival of very high energy primaries that, otherwise, due to their rarity, would escape detection by apparatuses carried by balloons or space vectors. The story of the evolution of the measuring devices, that were built practically everywhere in the world, is complex and here we restrict to a short description of some of the fundamental steps.

Aside from the experimental implementation, a vast theoretical research started on the way showers evolve, that is well documented, up to 1956, in the paper by K. Greisen ([28]; see also [29]) and that in more recent years has given rise to numerical simulation models of the greatest utility. Much of our understanding of how showers develop and grow comes from detailed studies made with Monte Carlo simulations of both electromagnetic and hadronic[1] cascades. The electronic cascades are simulated using the Heitler theory [30] while hadronic cascades are treated extending the Heitler's treatment and using for the hadronic interaction models the data extrapolated from low-energy accelerators. Examples of these calculations may be found in [31–34].

The group that Bruno Rossi established at MIT was very active. Robert Williams [35] measured the dimensions of extended showers with a new method suggested by Rossi that consisted in the sampling of the density of particles in the shower with an array of fast ionization chambers.

The experiment was performed on a mountain (between 3,050 and 4,300 m) using four fast ionization chambers with electronic coincidences and photographic recording of the height of the pulses. Twenty-seven showers each one with a total energy of more than 10^{15} eV were registered.

An ionization chamber gives a signal that is directly linked to the created ion pairs and it is therefore particularly suited to study the particle density in the shower. The first series of experiments was performed at Climax, Colorado, at an elevation of 3,050 m during the winter 1946–1947. Four chambers were placed along a line at one metre from each other. Another configuration, mainly used at Doolittle Ranch (near Echo Lake), Colorado, elevation 3,050 m, in the summer 1947, used instead two pairs of chambers each one formed by two superposed chambers, each pair separated by a variable distance. A third configuration used three chambers placed on the sides, 1.22 m long, of an equilateral triangle, with a fourth chamber at its centre. In a nearby experiment there was also a Wilson chamber. This last configuration was used at Doolittle and on the top of Mount Evans (4,300 m). The principal objective was to obtain information on the structure of showers by finding the particle density in different points of the shower. The study was made by changing the distance between counters. By measuring the density of particles on the counters, the distance from the axis may be derived, making some reasonable hypothesis about the lateral distribution of particles. It was demonstrated that the shower had a single centre with a high particle density, and that the density decreased going away from the centre, as predicted by theory (even if it was not just right, the theory was not too sensitive to the single processes which occur in the shower). It was found that the shower was mainly formed by very high energy electrons, even if other particles and photons were present.

The problem of what initiated the shower was rather controversial, but it was soon understood (see for example, G. Cocconi [36]) that any primary particle, whose energy undergoes any degenerative process giving rise in the high atmosphere to several electrons of high energies, may be responsible for an extensive shower.

---

[1]Pions, strange particles, protons, and neutrons belong to the class of particles known as the hadrons, which are distinguished by their participation in the strong nuclear interactions. Hadrons are all composites of quarks and antiquarks.

This is due to the fact that the main characteristics of the multiplicative theory principally depend on the behaviour of the less-energetic electrons and photons.

Williams' experiment measured an event due to a particle of energy 10^{17} eV. However, the inability of Geiger–Müller counter arrays to give measurements on the arrival direction of showers was a big limitation.

Cocconi et al. [37] performed an experiment at Echo Lake, Colorado, elevation 3,260 m, using an ensemble of counters to measure the particle density in the shower as a function of the separation between counters. They showed that there was no evidence of multiple cores in the shower; that is, the shower had a single core. The lateral distribution of the electrons was found to agree well with that calculated by G. Molière [31,38,39], taking into account only Coulomb scattering of the electrons.

The group lead by Rossi [40] eventually, in 1953, overcame the inability of Geiger–Müller counters to give information on the arrival direction of showers by developing a technique that determined the arrival direction through a measurement of the arrival time of the signals in scintillation counters separated by a few tens of metres.

If the primary hits perpendicularly to the atmosphere, the produced shower axis remains perpendicular and the particles found at ground on a horizontal plane (that in this case in perpendicular to the axis) arrive simultaneously on the counter array. On the other hand, if the shower axis forms an angle with the vertical, the orthogonal plane on which all the particles arrive contemporarily, is inclined with respect to the horizontal plane, and from the difference of the times with which the particles arrive on its different points, the incidence angle of the primary ray may be derived.

The time differences are extremely short, of the order of one microsecond or less, and therefore very fast detectors must be used. Rossi realized that the scintillation counters had the right characteristics.

Scintillation counters were developed ([41], for liquid scintillation counters: [42,43]) in the 1940s, even if the spintariscope, successfully used by E. Rutherford in his researches on radioactivity around 1910, may be considered a counter of this kind. The working principle is based on the property (scintillation) of some substances (scintillators) to emit light when they are traversed by ionizing particles. The counter is made by a scintillator that may be a solid, liquid, or very often a plastic material, connected via a light guide or an optical fibre, to a photomultiplier able to transform the light pulse in an electrical current pulse that may be amplified and measured. The electrical signal is proportional to the amount of detected light which in turn depends on the number of charged particles that traversed the active material of the detector.

In the summer of 1952, a young visitor from the University of Padua, Piero Bassi joined the Rossi group and together with George W. Clark, another Rossi student, they verified the Rossi idea that the arrival direction of a shower could be determined by measuring the differences in the arrival times on a great number of detectors of the shower particles. Rossi had understood that the very short decay time of the fluorescence of the liquid scintillators, recently discovered, made possible the construction of fast detectors of great area. The liquid scintillators used in the experiment were soon replaced by home-made plastic scintillators [44].

Fig. 10.3 From Clark et al. [45]. Schematic diagram of the detector configuration. The four detectors in the circle C were just used sometimes (from G. Clark, J. Earl, W. Kraushaar, J. Linsley, B. Rossi, F. Scherb, Nuovo Cimento **8**, 623 (1958))

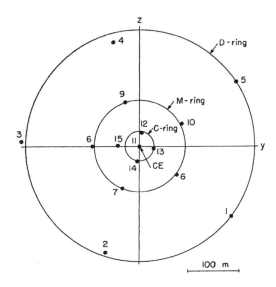

So, in 1955, the group lead by Bruno Rossi at MIT set up the new technique for the determination of the energy and arrival direction of the primary cosmic ray that originated the extended shower. The density distribution of secondary particles, observed on different positions in a counter array, was used to localize the centre of the shower and to obtain the primary energy (*density sampling* method). The arrival direction of the primary (that coincides with the shower axis) was determined by the arrival time differences of the shower front on the different counters (*fast timing*).

In the following 10 years, fast timing and sampling were employed in a series of experiments at MIT, in India, and in Bolivia [45].

At Kodaikanal in South India, at 2,800 m above the sea level, in collaboration with V. Sarabhai and E. Chinis of the Physical Research Laboratory of Ahmedabad, the MIT group put into operation, in the August 1956, a station with four detectors at the vertex of a square with side 35 m long. Thousands of showers per day were detected [45] with average dimensions of around 100,000 particles, presumably produced by a primary of energy around 10^{14} eV.

The MIT group put also into operation, between 1954 and 1957, a detecting station in Massachusetts which was named Harvard Agassiz Station [46]. Liquid scintillators were installed, each one containing approximately 1,000 l of toluene, a liquid with a good scintillation efficiency, but highly inflammable. One counter was set on fire by a lightning and so by 1956 all the liquid scintillators were substituted by plastic ones. There were 15 plastic scintillators (11 plus 4 that were used sometimes, all of 105 cm diameter) in a double circular configuration (for a diameter of about 500 m) (Fig. 10.3)

The scintillators were coupled to photomultipliers, oscillographs, and a photographic recording system.

The researchers were able to accurately determine the dimensions, the location of the centre, the arrival direction, and the lateral distribution of particles in each of the many measured showers.

The analysis of results showed very high energies of the primary cosmic rays in the range between 10^{15} and 10^{18} eV, allowing the derivation of the energy spectrum.

A particle of energy close to 10^{19} eV was observed. Cosmic rays of such energy, that is about 1 joule[2], are called Ultra High Energy Cosmic Rays (UHECRs). Above 10^{19} eV, the flux is only about 1 particle per square kilometre per year.

In this experiment the good performance of the *fast time method* (Bassi) to find the arrival direction and of the *density sampling* (Williams) to determine the location of the shower centre was demonstrated.

A report on the operation of the station from 1954 to 1957 was presented in 1961 [46]. The primary purpose of the experiment was to study the spectrum and the arrival directions of very high energy cosmic rays to gain information on their origin. A second purpose was to obtain accurate data on the shower structure to give criteria to judge the validity of the shower theories based on very high energy interaction models.

After Agassiz, a bigger experiment was built, at Volcano Ranch, New Mexico, under the direction of John Linsley, Livio Scarsi, and Bruno Rossi [47, 48]. In the desert, 19 plastic scintillator counters were installed, each one with a surface of 3.3 m^2. The scintillators were coupled to photomultipliers and oscillographs, with a photographic recorder. In 1962 the diameter of the Volcano Ranch array was enlarged to 3.6 km. It was the first gigantic array detector. With it, Linsley [49] discovered the first event that he estimated was produced by a particle with energy above 10^{20} eV, generating a shower of 50 billions of particles. This was the first indirect observation of a primary cosmic ray of such high energy, and remained the only one for several years.

The success of these experiments marked the start, everywhere in the world, of the construction of large arrays of detectors to study extended showers.

A few years before, in 1960 the experimental setup [50] BASJE (Bolivian Air Shower Joint Experiment) was built, dedicated to the study of gamma rays through the produced shower. The detecting system was provided with a muon detector of 60 m^2, which at the time was the largest in the world. The research consisted in studying the showers deprived of muons and therefore presumably initiated by primary gamma rays. The experiment was realized, also thanks to the contribution of Bruno Rossi. The gamma rays entering the atmosphere produce a purely electromagnetic shower of only electrons, positrons, and gamma rays that develops without generating muons and hadronic components[3]. It is therefore possible to distinguish it from showers initiated by primary charged particles which, besides the electromagnetic component, give also rise to a nuclear component.

---

[2]One joule is approximately the energy of a tennis ball travelling at a speed of 85 km/h.

[3]If they are very energetic, however, also gamma rays may produce mesons by photoproduction.

Moreover, cascades generated by incident cosmic ray nuclei develop differently from purely electromagnetic cascades. Interactions of high-energy hadrons replenish the electromagnetic component via π° production as the shower develop. As a consequence, both the longitudinal development and lateral structures are more complex than in photon-induced showers.

The results of the BASJE experiment were discussed in 1965 [51]. However, no gamma rays were found. After the discovery of the microwave cosmic background at 2.7 K, the reason was understood. The reaction between a high-energy photon and one of cosmic background produces electron–positron pairs

$$\gamma + \gamma_{2.7\,K} \rightarrow e^+ + e^-$$

and suppresses the gamma rays with energies above 10^{15} eV. Although the cross-section for a process of this kind is extremely small, after a path of the order of 10 kiloparsec[4], the high-energy photon flux reduces to half its value and therefore, because of the very large distances that these photons must travel to come to us, practically very few of them arrive and the BASJE experiment was too small to detect them.

The technique of counters arrays was realized on scales larger than 8 km^2 in many stations, besides Volcano Ranch (USA). We mention some of them: Haverah Park (UK), Narribri (Australia), Yakutsk (Russia), AGASA (Japan), and KASCADE (Germany).

In 1967, Alan A. Watson and collaborators installed in England, 220 m above sea level, on the Haverah Park site, in North Yorkshire Moors, 17 miles from the Reeds University (that was operating it) and 3 miles from Harrogate, an array of 200 Cerenkov detectors in water [52], coupled to photomultipliers and grouped into 48 stations in coincidence placed over an area of 12 km^2 and distributed in groups at variable distance from 150 m up to 2 km from the centre of the site.

The Cerenkov effect gets its name from Pavel Alekseyevich Cerenkov (1904–1990), the Russian physicist who discovered it in 1934. It consists in the light emission by charged particles travelling in a medium at a speed greater than that of the light in that medium. In water, the light travels at approximately 220,000 km/s, nearly 70% its velocity in vacuum (circa 300,000 km/s). When a charged particle travels in water at a speed greater than that, faint light flashes are emitted in a narrow cone along the propagation direction of the particle that, due to relativistic effect, narrows as the particle velocity approaches that of the light in vacuum.

Tanks of water of suitable depth equipped with photomultipliers may therefore measure the energy released in the form of Cerenkov light by the particles of the shower. The walls of the tank are internally covered by a reflecting material to direct as much Cerenkov light as possible towards the photomultiplier [53, 54].

Two kinds of Cerenkov detectors were operated; the majority of them, used for all the time of the experiment, employed tanks of dimensions 2.25 m^2 × 1.2 m.

---

[4]1 Kiloparsec = 1Kps = 3,260 light years = 3.09×10^{19} m.

A minority of detectors, operated only during 4 years, had dimensions 1 m^2 × 1.2 m. The experiment took place from July 1968 to July 1987. In 20 years of activity in the site, also eight events of energy larger than 10^{20} eV were observed. In 1987 the site was closed (a report of activity is M.A. Lawrence et al. [52]).

At Narribri, Australia, in 1968 the experiment SUGAR (Sydney University Giant Air-Shower Recorder) started to operate. In this experiment 56 stations each one consisting of a pair of scintillators of 6 m^2 buried 1.7 m under ground were operated [55]. Although the covered area was more than 60 km^2, the spacing between stations, typically 1.61 km, was too large to determine the location of the centre of the shower and the arrival direction of the primary. Also, the photomultipliers created problems. In any case nine events with approximate energy of 10^{20} eV were reported [56].

In Tokyo, K. Suga and collaborators built an array (INS-LAS), in 1968, and reported one event at first evaluated 4 × 10^{21} eV [57]. Such a large value was later subject to criticism and lowered to a more probable value of 2.5 × 10^{20} eV [58].

In the then Soviet Union, the study of extended showers started under the supervision of D.V. Skobeltzyn, in 1944 and in 1946, near the Pamir station. In 1956, two groups were operating: one in Moscow at the Lebedev Physical Institute of the USSR Academy of Sciences and the University of Moscow with G.T. Zatsepin and the other in Georgia at the Georgian Science Academy, organized by professor Andronikasvili. People in Moscow were making studies on the Pamir Mountains and at sea level near Moscow, where Georgi Zatsepin and collaborators built the first Geiger–Müller counter array connected to circuits suitable for the observation of extended showers. The most energetic observed shower resulted produced by a primary of an energy 10^{16} eV.

In 1960, a station at an elevation of 3,340 m near Almaty (Alma Ata) was open ([59], a review of some Russian research is in [60]).

Another array was built at Yakutsk, Siberia, operated by the Institute of Cosmophysical Research and Astronomy that started to acquire data in 1970 and was developed to cover an area of 18 km^2 in 1974 [61]. It was by far the most complex of the giant arrays [62]. There were detectors on surface and underground and 35 photomultiplier systems to measure the Cerenkov radiation in air, with a technique we will describe soon. The site was later enlarged to 86 scintillation detectors and 43 Cerenkov detectors coupled to photomultipliers grouped into 43 stations in coincidence placed over an area of 25 km^2. The distances between stations, some of which were installed underground, varied from 100 to 500 m. On May 7th, 1989, an event of 1.2 × 10^{20} eV was observed. In 1995 the site was reconfigured over 10 km^2 and so a detailed study of the structure of showers around 10^{19} eV [63] was possible.

In Japan, near the observatory of Akeno, a village about 120 km west of Tokyo, AGASA (Akeno Giant Air-Shower Array) was built [64–66] with 111 scintillation detectors, with an area 2.2 m^2 each, placed about 1 km from each other. Moreover, 27 screened detectors of varied area for muon observation were assembled. AGASA covered an area of 100 km^2. The construction began in 1987 and events were measured since 1991. On December 3rd, 1993, a shower that arrived nearly vertically and fell almost completely in the detector array was revealed, which was produced by a primary of 2×10^{20} eV. This was the most energetic event ever detected.

AGASA detected about ten events of energy greater than 10^{20} eV. The site was closed in 2004.

KASCADE (KArlsruhe Shower Core and Array Detector) [67,68] gives data limited to energies below 10^{17} eV. Another array of KASCADE is larger (KASCADE-Grande) and covers an area of 1 km^2. With it, the energy may be extended to 10^{18} eV. Kascade is able to measure the shower dimension at ground separately for protons and muons with energy of the order of GeV. Information on the composition of the primaries may be obtained by using simulations of showers through the atmosphere.

The experiment KASCADE-Grande [69] is located on the site Forshungzentrum Karlsruhe (Germany) to measure showers with energy between 0.5 PeV and 1 EeV [68]. The site [70] consists of the original KASCADE experiment and an extension by the Grande array, covering an effective area of 0.5 km^2.

The last station to be constructed and at present in full operation is the Auger array in Argentina that covers 3, 000 km^2, which we will discuss later.

10.4 The Atmosphere May Be used as a Detector. Why Not?

Basically three methods may be used to detect extended showers. The generic method consists in the distribution of a number of particle counters over a large area and in the direct detection of the particles that survive at the level at which detection is made. This was essentially the method used by Auger in his work in 1938. From the number of detected particles an idea of the energy of the primary cosmic ray that generated the shower may be gained, and this was the method used in the first built stations, as we have described.

With this technique, by evaluating the number of secondary particles that arrive on the measurement plane, the energy of the primary may be inferred, by applying suitable models of the interactions that develop in the shower. Obviously this estimate depends on the calculation model that is applied.

Alternative methods could be devised. Two have been developed with success. We may in fact ask why to employ big detectors at ground instead of using the whole atmosphere as a detector. The application can be made through two distinct effects. One method is the study of the Cerenkov light generated in the atmosphere by the fast electrons which travel at relativistic speed in the shower. The other method consists in studying the fluorescence light that these electrons may produce by exciting the nitrogen molecules of the atmosphere, that subsequently decay by emitting fluorescent light in the band 300–400 nm.

The idea to use the Cerenkov light emitted in air by fast particles can be dated back to 1947. In that year, P.M.S. Blackett [71] calculated that approximately 10^{-4}

of the total intensity of the night sky was due to the Cerenkov radiation produced by the passage of cosmic ray particles through the atmosphere. A few years later, W. Galbraith and J.V. Jelley [72] discovered that intense light pulses associated to showers in air of primary energy larger than 10^{14} eV could be detected with a very simple instrumentation. The possibility to detect the emitted light was also studied in Italy, using in the laboratory single particles in air [73, 74]. A number of further experiments [75–77] revealed that the light pulses were effectively due to the Cerenkov light.

The light receiver is very simple and cheap to build. The light is collected by a parabolic mirror and sent to a photomultiplier. The most severe limitation is that clear moonless nights are needed in locations without lights from nearby towns. In the experiments, the contributions of fluorescence from the one of Cerenkov emission have to be separated, according to the method that one wishes to select.

In the second method the terrestrial atmosphere is used as a big scintillator, as suggested at the Norikura Symposium in 1958. The first discussions about this possibility were made by K. Suga ([78] quoted by [79]) and A.E. Chudakov [78] at the Fifth Interamerican Seminar in La Paz, Bolivia.

The fast shower electrons which travel in the atmosphere may excite the nitrogen molecule that subsequently decay emitting fluorescent light in the wavelength region of 300 450 nm. The mechanism is the same as the one which produces the aurora borealis. Experiments with alpha particles and protons of 57 MeV provided the data necessary for the project. Also an experimental measurement of fluorescence efficiency produced by a deuteron beam in air was made [80].

The fluorescence technique allows deriving the primary energy transferred to electrons and positrons. The output is comparable with that produced in the Cerenkov processes in air, but the small lateral widening of the Cerenkov emission limits its use to detect showers. In fact, the light emitted by a relativistic particle by Cerenkov effect is emitted in a narrow cone whose axis is along the particle direction of motion. On the other hand, fluorescence light is emitted from the excited molecules in all directions and therefore may be seen laterally. In the fluorescence method the light is detected by photomultipliers and the profile of the shower in the atmosphere may be derived directly. An extended shower produced by a primary cosmic ray of energy greater than 10^{19} eV forms in the atmosphere a fluorescence band long 10–100 km or more, depending on the nature of the primary and of the incidence angle with respect to the vertical.

The light emitted is isotropic and showers produced by primaries of 10^{19} eV were predicted to be detectable at distances of about 20 km.

The time duration of the flash is 30–300 ms. Although the nitrogen fluorescence accounts only for a small part of the energy losses of the shower, that are dominated by the losses for the ionization of the air molecules, laboratory experiments have shown that the fluorescence is nearly proportional to the total ionization losses.

The first station was built on the hill of Mount Pleasant, near Ithaca, in 1964, by K. Greisen and his group [81]. With two additional stations that formed a triangle with sides of 11, 16, and 12 km the experiment continued until 1967. In 1965, Greisen proposed a new optical system to collect and register the emitted light, which was built in 1966 [82]. However, due to Ithaca's harsh winter and to the low probability to have showers of sufficient energy, nothing was observed. Subsequently, Greisen [83] built a prototype instrument at Cornell, again without success.

In 1968, Tanahashi, returning from a visit to the Cornell experiment, repeated the experiment on the mount Dodaira, near Tokyo, collecting the light with special Fresnel lens with a diameter of 1.6 m and placing 27 photomultipliers in the focal plane. Over 6,000 events registered during 90 h, his group eventfully succeeded to detect the first fluorescence light by an EAS of energy above 5×10^{18} eV [84]. This event was clearly distinguished by the other events of Cerenkov light. In the euphoria of this result, two more telescopes were built, without success.

However, neither Cornell, nor the Japan sites were climatically suitable for the employment of the method. Eventually, in 1976, a prototype was built at Volcano Ranch that was shown to work properly by studying the coincidences that were obtained with the detection made with the Linsley array. This way, an unambiguous detection of the fluorescence emission from showers was obtained, checking the coincidence with a detector array on ground [85]. This result brought to the development of an apparatus that is operating with great success, called Fly's Eye [86], designed and operated since 1981 by the group led by Keuffel at Utah University, in the Dugway desert, Utah, 160 km southwest of Salt Lake City.

There were two detectors, Fly's Eye I (FEI) and Fly's Eye II (FEII), separated by a distance of 3.4 km. FE-I consisted of 67 mirrors of 1.6 m diameter and 880 photomultipliers to cover the whole sky. In front of each photomultiplier was a hexagonal "Winston" light cone to extend the collection area of each phototube from a circular to a hexagon. Because of the light cone, the actual shape of the field-of-view for each pixel was hexagonal, leaving no gaps between photomultipliers. The coverage of the sky with hexagonal pixels was reminiscent of the compound eye of a fly. For this reason, the Utah experiment was named "The Fly's Eye". FEII consisted of 36 mirrors and 464 tubes, and it observed half the sky in the direction of FEI. It began to operate in November 1986 [87]. The two detectors were also located in the middle of the array called CASA-MIA.

The several dozens of mirrors of each one of the two telescopes focus different sky portions on an array of fast photomultipliers that allow realizing a sort of movie of the shower while it propagates through the atmosphere. The study may be done treating separately the data from the two telescopes (monocular method) or treating them together (a stereo analysis is possible in this case) [88].

In November, 1991, an event produced by a primary particle of 3.2×10^{20} eV was detected [89]. Both FEI and FEII were operated until 1993.

In mid-1998 a successor to Fly's Eye, known as HiRes, started regular operation at the Dugway site [90]. This detector is a collaboration among University of Utah, University of Adelaide, Columbia University, The University of Illinois, and University of New Mexico. The facility has a pair of atmospheric-fluorescence telescopes (HiRes-1 and HiRes-2) separated by 12.5 km.

These telescopes were built under the guidance of P. Sokolisky and Eugene Loh of the Utah University and registered the showers produced by primary cosmic rays

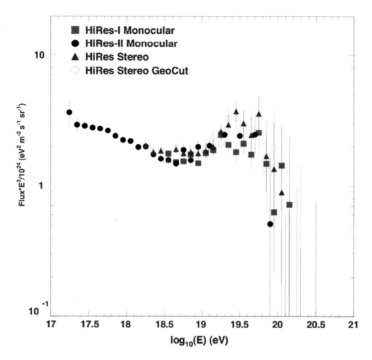

Fig. 10.4 Cosmic ray flux as a function of energy (Reprinted from R.U. Abbasi et al., Astro particle Physics **32**, 53 (2009). Copyright 2009, with permission from Elsevier)

with energy above 10^{17} eV, since 1997. Also, this system allows a stereoscopic vision of the shower.

Figure 10.4 (from Abbasi et al. [91]) shows the cosmic ray flux as a function of energy, measured since 1997, with the exception of an interval of seven months after September 11, 2001, during which the access to the area was denied to civilians.

The fluorescence method allows a measurement of the ionization density in a single shower at different altitudes and therefore the registration of its longitudinal development (a recent study is [92]). One of the advantages of the fluorescence method is its ability to measure directly the depth (X) at which the shower reaches its maximum. The X distribution has given important results concerning primary composition. However, the production of fluorescence is small: its faint dazzle may be compared to that one of a few watts lamp emitting in the ultraviolet that travels at the speed of light. Therefore, a very fast and sensitive detection electronics, moonless night observations, clear sky, and absence of light pollution are needed [79].

In the 1990s, the air fluorescence detector of HiRes was used in connection with the Michigan muon array (MIA) [93]. The HiRes is located at a vertical atmospheric depth of 860 g/cm^2, overlooking the MIA array which is 3.4 km away and 150 m lower. The HiRes viewed the night sky over an elevation range from

$3°$ to $70°$ with an array of 14 optical reflecting telescopes and imaged the EAS as it progressed through the detection volume. By using a suitable model [94] of the shower development one can measure both the primary particle energy and the depth at which the shower reaches its maximum size. MIA [95] consisted of over $2,500$ m^2 of active area distributed in 16 patches of 64 scintillation counters, and measured the EAS muons arrival times. With this combined measurements the precision of the geometrical reconstruction of the shower was greatly improved. Measurements were done between 1993 and 1996 obtaining evidence of a changing cosmic ray composition between 10^{17} and 10^{18} eV.

While the last generations of arrays have all been designed to measure primary particles of very high energy, the study of showers produced by primaries of low energy remained somehow uncovered. To study showers at relatively low energies, of the order of 10^{15} a 10^{16} eV, great emulsion chambers were used, built with some layers of lead spaced out by photographic emulsions [96, 97]. The chambers had surfaces of the order of 100 m^2 and are exposed for months. With this kind of chambers experiments were done on mounts Kanbala [98], Fuji [99], Pamir [100], and Chacaltaya [101].

The flux above 10^{15} eV is of the order of 10^{-10} cm^{-2} sr^{-1} s^{-1}. Using 100 m^2 at an equivalent altitude of 550 g/cm^2, approximately ten events per year are obtained.

In 2001, Kieda et al. [102] proposed a new method for the cosmic ray measurement, using only the Cerenkov light emitted by the only primary entering the atmosphere, separated by that one of the subsequent shower. This light may be measured on the ground and, being proportional to the square of the charge of the particle, may allow identifying it. The method has been demonstrated using the High Energy Stereoscopic System H.E.S.S. [103] in 2007 [104].

10.5 The Auger Observatory

In 1999, in Argentina, an International group of about 300 people, under the guidance of James W. Cronin (1931–) of the University of Chicago—Nobel Prize 1980 together with V.L. Fitch (1923–) for their research of the decay process of the K meson that brought them to the discovery of a small asymmetry between matter and antimatter, known as CP violation—and Alan A. Watson,of Leeds University, England, started the construction of the south site of the gigantic AUGER (Observatory Pierre Auger) covering an area of $3,000$ km^2, about ten times the extension of the town of Paris, on the Andine plateau, west Argentina, near the small town of Malargue, in the Pampa Amarilla (yellow prairie) 1,000 km west of Buenos Aires (see Fig. 10.5). The project had been proposed in 1992 [105].

The observatory is a "hybrid" system employing two different methods to detect and study the high-energy cosmic rays. One technique detects the shower particles through the Cerenkov effect in enormous water tanks on the ground. The other technique studies the shower development by observing the fluorescence and Cherenkov light produced by charged particles along the shower trajectory in the atmosphere [105–107].

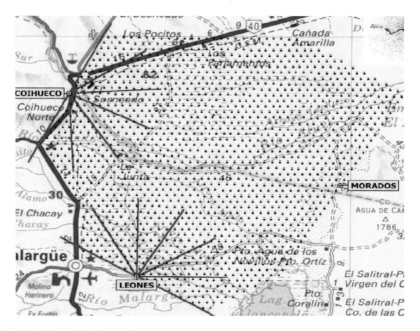

Fig. 10.5 Map of the Auger site

Fig. 10.6 A photo of one of the water tanks of the Auger system. On the horizon the Andes mountains can be seen

There are 1,600 water Cerenkov detectors, each one of 12,000 L (see Fig. 10.6) at a distance of 1.5 km from each other and four stations with 24 fluorescence telescopes each one distributed inside the array.

The water tanks are used to measure the energy flux of electrons, photons, and muons in the shower, while the weak light emitted isotropically from the shower moving through the atmosphere can be detected by the fluorescence telescopes. The configuration combines the potentiality of a high statistics due to a practically 100% time operation of the water tanks, with the power of the calorimetric determination of energy from the fluorescence telescopes.

The atmospheric depth, X_{max}, at which the longitudinal development of a shower reaches its maximum may be approximately expressed as a function of the logarithm of the energy E and mass A of the primary particle as [107]

$$\langle X_{max} \rangle = \alpha \left(\ln E - \langle \ln A \rangle \right) + \beta$$

The coefficients α and β depend on the nature of hadronic interactions. The shower maximum can be observed directly with fluorescence detectors and the energy is derived by the number of secondary particles. The Auger facility is able to measure both quantities and so to derive information on both the energy and mass of the primary particle, by using suitable models of simulation.

The principal task is to study the spectrum of cosmic rays above 10^{19} eV.

The AUGER project considered also the installation of a second site of 3,000 km^2 in the north hemisphere in Utah, USA, but at the moment its financial support has been cut.

The first results were reported in 2005. In 2007, 1,400 of the total 1,600 detectors were working. Its results are very useful to understand the nature and origin of cosmic rays.

10.6   Is There a Limit to the Energy of the Primary Cosmic Ray?

According to the big-bang theory, the Universe came into existence in a sudden explosion and has been expanding and cooling ever since. Nowadays, almost 14 billions year later, radiation has attained equilibrium and is distributed as that of a black body at a temperature of 2,725 K so that it exists essentially in the microwave range (cosmic microwave background radiation). Physicists Arno Penzias (1933–) and Robert Wilson [108] (1936–) discovered those microwaves by accident in 1965 and won the Nobel Prize 13 years later.

After this discovery, Kenneth Greisen [109] (1918–2007) at Cornell University and Georgi T. Zatsepin and Vadem A. Kuz'min [110, 111] in Moscow, independently, pointed out that if the high-energy particles were protons and if their sources were uniformly distributed in the Universe, the interaction with the microwave background radiation would influence the particle distribution in the region of 10^{20} eV.

In fact, a proton with that energy, due to the relativistic Doppler effect, "sees" a photon of the background radiation as if it was a high-energy gamma ray, and above 10^{18} eV it loses its energy in slightly less than one Hubble time[5] due to pair production colliding with it. At higher energies, just under 10^{20} eV for a proton, photo-meson production becomes the dominant energy loss mechanism throughout the reaction

$$p + \gamma_{2.7\,K} \rightarrow p + \pi^\circ$$

So the proton produces a neutral π meson losing approximately 15% of its energy. On great distances this process eliminates the high-energy protons. The ones we detect should have been originated nearby. Stecker [112], a couple of years later, calculated the mean energy loss time for protons propagating in space as a function of energy. For example, protons of 5×10^{19} eV energy have only 50% probability to arrive from distances larger than 100 Megaparsec[6].

For heavier nuclei at high energies, photodisintegration in collision with dilute optical radiation becomes important [113].

From the initials of its authors the effect has been dubbed GZK effect. The sudden decay expected in the number of protons above 6×10^{19} eV is called the GZK cut.

In 2003, the AGASA collaboration reported having observed 11 events above 10^{20} eV and no proof of the GZK cut during 10 years of measurements [114]. Later the Fly's Eye collaboration has instead reported the observation of a cut [115]. The problem is still very controversial.

10.7   What About the Future?

To detect very high energy primaries, space-based observatories may allow reaching the necessary instantaneous aperture and integrated exposure.

Two programs have been studied, both of which use the fluorescence light emitted by the shower particles in the atmosphere as viewed at night from space. In one project dubbed S-EUSO (Super Extreme Universe Space Observatory) [116] a satellite is used, as shown in Fig. 10.7.

In the second project, studied by NASA, two satellites are employed as to have a stereoscopic vision (Orbiting Wide-angle Light-collector or OWL).

An increase by a factor approximately 30–50 is expected in collecting power compared with the Auger system.

---

[5]This time is of the order of 2×10^{10} years.

[6]1 Megaparsec $= 1\,\text{Mps} = 3.26 \times 10^6$ light years.

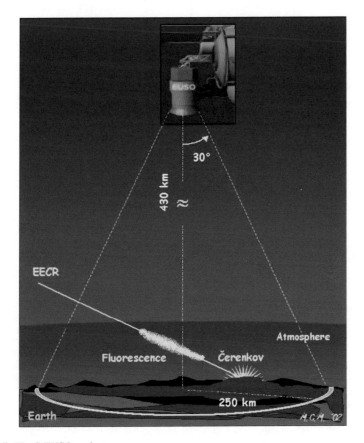

Fig. 10.7 The S-EUSO project

References

1. M. Schein, W.P. Jesse, E.O. Wollan, Phys. Rev. **59**, 615 (1941)
2. P. Freier, E.J. Lofgren, E.P. Ney, F. Oppenheimer, H.L. Bradt, B. Peters, Phys. Rev. **74**, 213 (1948)
3. B. Rossi, Ric. Sci. Suppl. **1**, 579 (1934)
4. B. Rossi et al., Ric. Sci. **5**(1) 559 (1934)
5. P. Auger, R. Maze, T. Grivet-Meyer, C. R. Acad. Sci. **206**, 1721 (1938)
6. R. Maze, J. Phys. Radium **9**, 162 (1938)
7. P. Auger, R. Maze, C. Robley, C. R. Acad. Sci. **208**, 164 (1939)
8. P. Auger, R. Maze, C. R. Acad. Sci. **207**, 228 (1938)
9. P. Auger, R. Maze, P. Ehrenfest Jr., A. Freon, J. Phys. Radium **10**, 39 (1939)
10. P. Auger, P. Ehrenfest Jr., R. Maze, J. Daudin, C. Robley, A. Freon, Rev. Mod. Phys. **11**, 288 (1939)
11. P. Auger, R. Maze, C. Robley, C. R. Acad. Sci. **208**, 1641 (1939)
12. W. Bothe, W. Gentner, H. Maier-Leibnitz, W. Mauser, E. Wilhelmy, K. Schmeiser, Phys. Z. **38**, 964 (1937)
13. W. Kolhoerster, I. Matthes, E. Weber, Naturwissenschaften **26**, 576 (1937)

14. R. Maze, A. Freon, J. Daudin, P. Auger, Rev. Mod. Phys. **21**, 14 (1949)
15. A.C.B. Lovell, J.G. Wilson, Nature **144**, 863 (1939)
16. J. Daudin, J. Phys. Radium **7**, 302 (1945)
17. L. Lewis, Phys. Rev. **67**, 228 (1945)
18. R. Lapp, Phys. Rev. **69**, 328 (1946)
19. K.L. Kingshill, L. Lewis, Phys. Rev. **69**, 159 (1946)
20. V.D. Skobeltzyn, G.T. Zatsepin, V.V. Miller, Phys. Rev. **71**, 315 (1947)
21. H. Euler, Z. Phys. **116**, 73 (1940)
22. G. Molière, Naturwissenschaften **30**, 87 (1942)
23. G. Wataghin, M.D. De Souza Santos, P.A. Pompeia, Phys. Rev. **57**, (1940) 61, 339
24. G. Wataghin, Suppl. Nuovo Cimento **6**, 539 (1948)
25. L. Janossy, P. Ingleby, Nature **145**, 511 (1940)
26. P. Auger, J. Daudin, A. Freon, R. Maze, C. R. Acad. Sci. **266**, 169 (1948)
27. D. Broadbent, L. Janossy, Proc. Roy. Soc. **A192**, 364 (1948)
28. K. Greisen, The extensive air showers, in *Progress in Cosmic Ray Physics*, vol. **III** (North-Holland Pu.Co., Amsterdam, 1956), p. 3
29. K. Greisen, Ann. Revs. Nuclear Sci. **10**, 63 (1960)
30. W. Heitler, *The Quantum Theory of Radiation* (Oxford University Press, New York, 1954)
31. G. Molière, in *Kosmische Strahlung*, ed. by W. Heisenberg (Springer, Berlin, 1953)
32. E.J. Fenyves et al., Phys. Rev. **D37**, 649 (1988)
33. S. Mikocki et al., J. Phys. G Nucl. Part. Phys. **13**, L85 (1987)
34. S. Mikocki et al., J. Phys. G Nucl. Part. Phys. **17**, 1303 (1991)
35. R. Williams, Phys. Rev. **74**, 1689 (1948)
36. G. Cocconi, Rev. Mod. Phys. **21**, 26 (1949)
37. G. Cocconi, V.C. Tongiorgi, K. Greisen, Phys. Rev. **76**, 1020 (1949)
38. G. Molière, Cosmic radiation, ed. by W. Heisenberg (Dover Publications, New York, 1946)
39. G. Molière, Phys. Rev. **77**, 715 (1950)
40. P. Bassi, G. Clark, B. Rossi, Phys. Rev. **92**, 441 (1953)
41. F. Marshall, J.W. Coltman, Phys. Rev. **72**, 528 (1947)
42. G.T. Reynolds, F.B. Harris, G. Salvini, Phys. Rev. **78**, 488 (1950)
43. H. Kallmann, Phys. Rev. **78**, 621 (1950)
44. G.V. Clark, F. Scherb, W.B. Smith, Rev. Sci. Instrum. **28**, 433 (1957)
45. G. Clark, J. Earl, W. Kraushaar, J. Linsley, B. Rossi, F. Scherb, Nuovo Cimento. **8**, 623 (1958)
46. G.W. Clark, J. Earl, W.L. Kraushaar, J. Linsley, B.B. Rossi, F. Scherb, D. Scott, Phys. Rev. **122**, 637 (1961)
47. J. Linsley, L. Scarsi, B. Rossi, Phys. Rev. Lett. **6**, 485 (1961)
48. J. Linsley, L. Scarsi, B. Rossi, Suppl. J. Phys. Jpn. **17**, 91 (1962)
49. J. Linsley, Phys. Rev. Lett. **10**, 146 (1963)
50. K. Suga, I. Escobar, G. Clark, W. Hazen, A. Hendel, K. Muratami, J. Phys. Soc. Jpn. **17**(Suppl. AIII), (1962)
51. Y. Toyoda, et al., *Proceedings of the 9th International Cosmic Ray Conference*, vol. **2** (London, 1965), p. 708
52. M.A. Lawrence, R.J.O. Reid, A.A. Watson, J. Phys. G **17**, 733 (1971)
53. S.C. Lillicrap, R.D. Wills, K.E. Turver, Proc. Phys. Soc. **82**, 95 (1963)
54. R.M. Tennant, Proc. Phys. Soc. **92**, 622 (1967)
55. C.J. Bell et al., J. Phys. **A7**, 990 (1974)
56. M.M. Winn et al., J. Phys. G **12**, 653 (1986)
57. K. Suga, H. Sukuyama, S. Kawaguchi, T. Hara, Phys. Rev. Lett. **27**, 1604 (1971)
58. V.M. Kagano, J. Phys. Soc. Jpn. **70**(Suppl. B), 1 (2001)
59. S.A. Slavatinskii, Tr. Fiz. Inst. Akad. Nauk SSSR **46**, 140 (1970)
60. G.T. Zatsepin, T.M. Roganova, Phys. Uspekhi **52**, 1139 (2009)
61. B.N. Afanasiev et al., *Proceedings of the Tokyo Workshop on Techniques for the Study of the Extremely High Energy Cosmic Rays*, ed. by M. Nagano (ICRR, University Tokyo, Tokyo, 1993)

62. A.A. Ivanov, S.P. Knurenko, I.Ye. Sleptsov, New Jour. of Physics 11, 065008 (2009)
63. A.V. Glushkov et al., Astropart. Phys. **4**, 15 (1995)
64. N. Chiba et al., Nucl. Instrum. Methods Phys. Res., Sect. **A311**, 338 (1992)
65. H. Ohoka et al., Nucl. Instrum. Methods Phys. Res. Sect. **A385**, 268 (1997)
66. M. Takeda et al., Phys. Rev. Lett. **81**, 1163 (1998)
67. T. Antoni et al., Astropart. Phys. **24**, 1 (2005)
68. T. Antoni et al., Nucl. Instrum. Methods. **A513**, 490 (2003)
69. G. Navarra et al., Nucl. Instrum. Methods. **A518**, 207 (2004)
70. W. Apel et al., Astropart. Phys. **31**, 86 (2009)
71. S. Blackett, in *Rep. Conf. Gassiot Comm. Roy. Soc.* 1947, London 1948 p. 34
72. Galbraith, J.V. Jelley, Nature, **171**, 349 (1953)
73. A. Ascoli Balzanelli, R. Ascoli, Nuovo Cimentio **10**, (1953) 1345
74. A. Ascoli Balzanelli, R. Ascoli, Nuovo Cimentio **11**, 562 (1954)
75. W. Galbraith, J.V. Jelley, Journ. Atmos. Terrestial Phys., **6**, 250 (1955)
76. J.V. Jelley, W. Galbraith, , Journ. Atmos. Terrestial Phys., **6**, 304 (1955)
77. N.M. Nesterova, A. Chudakov, Zu. Eksper. Teor. Fiz. **28**, 384 (1955)
78. K. Suga, *Proceedings of the 5th Interamerican Seminar on Cosmic Rays* 1962, vol. **2**, p. XLIX
79. M. Nagano, A.A. Watson, Rev. Mod. Phys. **72**, 689 (2000)
80. A.N. Bunner, thesis Cornell University 1964
81. K. Greisen, *Proceedings of the 9th ICRC*, vol. **2** (London, 1965) p. 609
82. A.N. Bunner, K. Greisen, P. Landecher, Can. J. Phys. **46**, 266 (1967)
83. G. Tanahashi, *Early fluorescence work Cornell and Japan, AIP Conference Proceedings* **433**, 54 (1998), see also J. Linsley, ibidem page 1
84. T. Hara et al., Acta Phys. Acad. Sci. Hung. **29**(Suppl. 3), 361 (1970)
85. H.E. Bergeson et al., Phys. Rev. Lett. **39**, 847 (1977)
86. R.M. Baltrusaitis et al., Phys. Rev. **D31**, 2192 (1985)
87. R.M. Baltrusaitis, et al., Nucl. Instrum. Methods Phys. Res. **A240**, 410 (1985)
88. R.U. Abbasi et al., Astrophys. J. **622**, 910 (2005)
89. D. Bird et al., Astrophys. J. **441**, 144 (1995)
90. P. Sokolsky, in *Proceedings of Workshop on Observing Giant Cosmic Ray Showers from >1020 eV Particles from Space, AIP Conf. Proc. No. 433, ed. By J.F. Krizmanic, J.Ormes, R.E. Streitmatter* (AIP, Woodbury, NY) 1998
91. R.U. Abbasi et al., Astropart. Phys. **32**, 53 (2009)
92. F. Arqueros, F. Blanco, J. Rosado, New J. Phys. 11, **065011** (2009)
93. T. Abu-Zayyad et al., Phys. Rev. Lett. **84**, 4276. (2000)
94. T. Gaisser, A.M. Hillas, in *Proceedings of the 15th International Cosmic Ray Conference, Plovdiv 1977*, vol. **8** (Bulgarian Academic Sciences, Plovdiv, 1977), p. 353
95. A. Borione et al., Nucl. Instrum. Methods Phys. Res. Sect. **A346**, 329 (1994)
96. C.M.G. Lattes et al., Suppl. Progr. Theor. Phys. **47**, 1 (1971)
97. T. Gaisser, G.B. Yodh, Annu. Rev. Nucl. Sci. **30**, 475 (1980)
98. J.R. Ren et al., Phys. Rev. **D38**, 1404, 1417, 1426 (1988)
99. J.R. Ren et al., Nuovo Cimentio **10C**, 43 (1987)
100. G.S. Bayburian et al., Nucleic Phys. **B191**, 1 (1981)
101. C.M.G. Lattes et al., Phys. Rep. **65**, 198 (1980)
102. D.B. Kieda, S.P. Swordy, S.P. Wakely, Astropart. Phys. **15**, 287 (2001)
103. J.A. Hinton et al., New Astron. Rev. **48**, 331 (2004)
104. F. Aharonian et al., Phys. Rev. **D75**, 042004 (2007)
105. J. Abraham et al., Nucl. Instrum. Methods **A523** 50, (2004)
106. I. Allekotte et al., Nucl. Instrum. Methods Phys. Res. **A586**, 409 (2008)
107. J. Abraham et al., Phys. Rev. Lett. **104**, 091101 (2010)
108. A.A. Penzias, R.W. Wilson, Astrophys. J. **142**, 419 (1965)
109. K. Greisen, Phys. Rev. Lett. **16**, 748 (1966)
110. G.T. Zatsepin, V.A. Kuz'min, J. Exp. Theor. Fiz. Pisma **4**, 14 (1966)
111. G.T. Zatsepin, V.A. Kuz'min, Sov. Phys. JETP Lett. **4**, 78 (1966)

112. F.W. Stecker, Phys. Rev. Lett. **21**, 1016 (1968)
113. F.W. Stecker, Phys. Rev. **180**, 1264 (1969)
114. M. Takeda et al., Astropart. Phys. **19**, 447 (2003)
115. R.U. Abbasi et al., Phys. Rev. Lett. **100**, 101101 (2008)
116. A. Santangelo, A. Petrolini, N. J. Phys. **11**, 065010 (2009)

Chapter 11
The Neutrino or, More Precisely, the Neutrinos: Elusive and Weird Particles, Able to Arrive from Very Far Away

11.1 Introduction

Although neutrinos are the most abundant component of cosmic rays at sea level, due to their very weak interaction with matter it has been possible to detect them only in the last 50 years, after large underground detectors became available.

We have already discussed how the idea of neutrino existence came to Wolfang Pauli, in 1930, in connection to beta decay. He was very upset by his hypothesis, so much so that he apparently declared: "*This night I slipped very badly because I have invented a particle experimental physicists will never succeed to see*".

And, indeed, the absence of an electric charge and the quasi null mass of the neutrino he postulated make its detection very difficult.

Nearly 30 years later, Bruno Pontecorvo[1] suggested that the neutrino associated with the muon could be different from the beta radioactivity neutrino (which comes with the electron). To tell apart the two, we call them nowadays *electronic neutrino* and *muonic neutrino*. After the discovery of the third lepton, the tau particle [3] and the appreciation that the *tauon neutrino* is different from the other two neutrinos, we may look at the neutrino as an elementary particle, electrically neutral, with spin $\frac{1}{2}$ and a mass so low that for long time it was assumed that it was actually zero. Three kinds of neutrinos exist (we say there are three *flavours*), usually indicated with the symbols ν_e, ν_μ, and ν_τ (electronic, muonic, and tauonic neutrino, respectively) and three corresponding kinds of antiparticles (antineutrinos) $\underline{\nu}_e$, $\underline{\nu}_\mu$, and $\underline{\nu}_\tau$.

Neutrinos and antineutrinos are distinct particles.[2]

Intense beams of muonic neutrinos and antineutrinos are obtained by decay of high energy pions and kaons produced by proton beams colliding in the big accelerators.

---

[1] B. Pontecorvo [1] (in Russian); translated in [2].

[2] F. Reines and R. Davies experiments discussed later here and B.P. Lazarenko and S. Yu Likanov [4]; G. Bernardini [5].

M. Bertolotti, *Celestial Messengers: Cosmic Rays*, Astronomers' Universe, DOI 10.1007/978-3-642-28371-0_11, © Springer-Verlag Berlin Heidelberg 2013

Experiments performed in 1989 at SLAC-SLC with linear e^+e^- collider [6] and with the collision ring [7–9] LEP of CERN, Geneva have shown that only three kinds of neutrino may exist, and we should not expect to find anymore.[3]

11.2   The Neutrino Detection

The neutrinos interact with matter only via the weak interactions and therefore they may traverse undisturbed huge thicknesses of any material. Occasionally, a neutrino may hit a particle (a nucleus or an electron) on its way. If the available energy is enough, this interaction produces one of the three leptons, that is an electron, a muon, or a tauon according to the kind of neutrino, and through the detection of the electron, muon, or tauon the neutrino may be detected and identified in an indirect way. Other interactions are possible and will be discussed later.

How much the Pauli intuition in 1930 was ahead of time can be understood noting that only as late as 1955 Frederick Reines and Clyde L. Cowan ([10] see also [11]) succeeded in experimentally proving the existence of the neutrino thanks to a research made on the particles produced in nuclear reactors. Actually, not the neutrinos but the antineutrinos were detected.

In the experiment, a delayed coincidence technique was employed to observe the reaction

$$\underline{\nu}_e + p \rightarrow e^+ + n, \tag{11.1}$$

in which an electronic antineutrino interacts with a proton in a large tank filled with a liquid scintillator doped with cadmium.

Frederick Reines was born in Paterson, New Jersey, on 16 March 1918 and entered Stevens Institute of Technology continuing his graduate studies at New York University earning a PhD in theoretical physics in 1944. From 1944 to 1959 he was Group Leader in the Theoretical Division at the Los Alamos Scientific Laboratory to study the physics and the effects of nuclear explosions. In 1951, he took a sabbatical period and decided to attempt the observation of the neutrino starting a collaboration with Clyde Lorrain Cowan (1919–1974), another Los Alamos staff member. They correctly believed that to solve the problem of the extremely low interaction of the neutrino with matter it was necessary to use a very intense neutrino source and a detector of very large dimension.

Initially they considered the use of neutrinos emitted by the decay of the fission products resulting in a nuclear bomb test. Reines remembers [12] that at the time he was considering this experiment, in the Summer of 1951, he knocked timidly on the door of Enrico Fermi who happened to be at Los Alamos, and said: *"I'd like*

---

[3]Theorists have been thinking about a fourth kind of neutrinos, called sterile neutrinos, since the late 1960s. These hypothetical neutrinos are called sterile because they do not interact at all with known particles. A sterile neutrino would not participate in weak interactions and would arise only from ordinary neutrinos oscillating into a sterile form.

to talk to you a few minutes about the possibility of neutrino detection". Fermi was very pleasant: *"Well, tell me what's on your mind?"* Reines expounded his idea: *"First off as to the source, I think that the bomb is best"*. After a moment's thought Fermi agreed: *"Yes, the bomb is the best source"*. Very good thought Reines and added: *"But one needs a detector which is so big. I don't know how to make such a detector"*. Again Fermi pondered the question and than agreed, and Reines registered in his mind the positive reaction of Fermi, until a chance of talking with Clyde Cowan. They were on their way to Princeton, when the airplane was grounded in Kansas City because of engine trouble. Just to let time go by, they started to discuss what could be done that was interesting in physics. *"Let's do a real challenging problem"* Reines said. Cowan suggested: *"Let's work on positronium"* (a hydrogen-like atom whose "nucleus" is a positron instead of a proton). *"No—* Reines replied—*already very able people are working on"* and added: *"Let's work on the neutrino, instead"*. Cowan agreed at once. Why to work on neutrino? Reines replies *"Because everybody said, you couldn't detect it. So we were attracted by the challenge"*. Bethe and Peierls [13] immediately after the Fermi paper on beta emission theory, concluded that there was no way to observe neutrinos. Twenty years later, after Reines succeeded in detecting them, he reminded Bethe of this statement, to which he replied *"Well, you shouldn't believe everything you read in the papers"*.

Going back to the Reines and Cowan idea of how to detect neutrinos, the two men pondered some time on the question and finally decided to use the reaction (11.1) written earlier. Some time later, they realized that an atomic bomb was not actually needed and that neutrinos emitted during the decay of the fission products in a nuclear reactor would be suited to the purpose. They decided to use the Hanford reactor, Washington. The reactor was built during the Second World War to produce plutonium for the atomic bomb. They wrote a letter to Fermi about their change of program on 4 October 1952, briefly explaining how they planned to do the experiment and received an immediate positive reply by Fermi (Fig. 11.1).

The first experiment, in 1953, gave still uncertain results [14,15], but they insisted choosing a new and more powerful reactor: the new reactor of Savannah River in the South Carolina. This time they succeeded to detect neutrino[4] (1955).

In principle, the used technique was very simple: one has to detect reaction (11.1), that should be considered as the inverse of the beta decay, in which an electronic antineutrino hits a proton (an hydrogen nucleus) producing a positron and a neutron. In the experiment, an antineutrino from fission products of the reactor is incident on a big tank filled with water containing cadmium chloride. Hitting a water proton it produced a neutron and a positron. The positron slows and at the end annihilates with an electron, producing two 0.5 MeV gamma rays, each one emitted in opposite directions that are detected by two large scintillation detectors placed on opposite sides of the tank. The neutron is slowed down by the water and captured by the cadmium, that has a high cross section for this process,

---

[4]see F. Reines and C.L. Cowan [16] for a complete description of the Savannah River experiment.

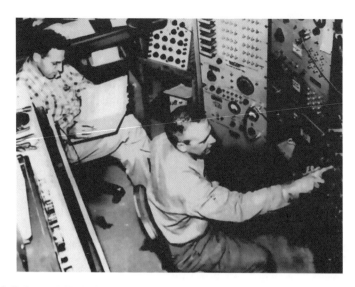

Fig. 11.1   Reines and Clyde Cowan at the control centre of the Harford Experiment in 1953

producing multiple gamma rays, which are also observed in coincidence by the two scintillation detectors. This particular reaction may be individuated among the many that may occur because the two photons emitted in the annihilation of the positron must be in coincidence and must be in a delayed coincidence with the photons produced microseconds after the neutron capture in cadmium.

The Savannah River reactor was well suited for these studies because of the availability of a well-shielded location 11 m from the reactor centre and some 12 m underground in a massive building. The high antineutrinos flux, estimated $1.2 \times 10^{13}/\text{cm}^{-2}$s, and the reduced cosmic ray background were essential to the success of the experiment which involved a running time of 100 days over the period of approximately 1 year.

After that, experimental observation of the reaction produced by antineutrinos was obtained, in 1955, Reines and Cowan sent a telegram to Pauli (14 June 1956):

We are happy to inform you that we have definitely detected neutrinos from fission fragments by observing inverse beta decay of protons. Observed cross section agree well with expected six times ten to minus forty four square centimetres" [17].

Pauli later said: *"Everything comes to him who knows how to wait"*.

In 1959, Reines was appointed professor and head of the Physics Department of the, at the time, Case Institute of Technology at Cleveland, Ohio. He then moved to the University of California at Irvine in 1966 and started an ambitious experiment in a gold mine in South Africa which allowed one of the first observations of the neutrinos produced in the atmosphere by cosmic rays. The apparatus was operated at a depth of 3,288 m below the surface of the earth in the East Rand Proprietary Mine, near Johannesburg, South Africa. A special tunnel was excavated at some distance from the gold-bearing reef to accommodate the scintillation detectors. Earlier experiments [18], in fact, had demonstrated the feasibility of studying cosmic-ray neutrino interactions with large scintillation detectors. The experiment operated from 13 December 1967 to 28 October 1971 to measure the muons produced by neutrinos, see [19].

In 1995, he was awarded the Nobel Prize in Physics for his detection of neutrinos. He died on 26 August 1998.

There are essentially three sources of neutrinos: the sun and the other stars, the nuclear reactors, and the decay in the atmosphere of mesons produced by cosmic rays.

11.3 The Solar Neutrinos

The Sun is a natural source of neutrinos that sends on Earth approximately 60×10^9 electronic neutrinos per cm^2s.

The origin of the energy dissipated by the Sun was for many years the object of study until, in 1929, Robert Atkinson and Fritz Houtermans [20] suggested that, at the very high temperatures existing inside the Sun, the kinetic energy of the thermal movement is such as to make it possible the existence of nuclear reactions able to liberate the necessary energy. The idea was successively elaborated and Hans Bethe [21], in a very beautiful paper nearly 10 years later, pointed out which could be the nuclear reactions that may produce inside stars the enormous quantity of energy needed for their subsistence.

In the Sun, according to the standard theory, the energy production comes out through the fusion of four protons in a helium nucleus of mass four, ^4He, with the release of electronic neutrinos and of an energy of about 26 MeV. The chain of the thermonuclear reactions is articulated in two cycles: the proton–proton cycle (pp) which represents the 98 % of the entire chain, and the carbon–oxygen–nitrogen (CNO) cycle which bears for somehow <2%.

The pp cycle produces electronic neutrinos in four types of reactions: pp, ^7Be, pep, ^8B, besides the reaction hep which, however, gives a very small contribution. The reactions are shown below.

At the beginning two protons collide, coupling together to give a deuterium nucleus (^2H), a positron and an electronic neutrino

$$p + p \rightarrow\, ^2\text{H} + e^+ + \nu_e \quad \text{(pp reaction)}. \tag{11.2}$$

This reaction occurs in 99.77 % of cases and the neutrinos are emitted with low energy (between 0 and 0.42 MeV). With a probability of about 0.23 % it may also occur that two protons and one electron collide among them producing again a deuterium nucleus and an electronic neutrino

$$p + p + e^- \rightarrow\, ^2\text{H} + \nu_e \quad \text{(pep reaction)}. \tag{11.3}$$

In this case the produced neutrino is monoenergetic with an energy of 1.44 MeV.

The deuterium produced in these two reactions interacts with a proton-producing helium of mass 3 and a gamma photon

$$^2H + p \rightarrow ^3He + \gamma. \tag{11.4}$$

Successively, in the 84.92 % of cases, the produced ^3He interacts with another ^3He, according to the reaction

$$^3He + ^3He \rightarrow ^4He + 2p \quad \text{(ppI reaction)}, \tag{11.5}$$

which produces ^4He regenerating the two protons. This chain is called ppI and the final result is the production of one ^4He and two neutrinos.

Two more reactions are possible; in one, occurring with a probability $10^{-5}\%$, a ^3He nucleus interacts with a proton to give a ^4He nucleus, a positron and an electronic neutrino

$$^3He + p \rightarrow ^4He + e^+ + \nu_e \quad \text{(hep reaction)}. \tag{11.6}$$

The emitted neutrino has a maximum energy of 18.773 MeV. In 15.08 % of cases the reaction is instead the following:

$$^3He + ^4He \rightarrow ^7Be + \gamma \tag{11.7}$$

and the produced beryllium gives place to two different reaction chains. In 99.9 % of cases it interacts with a negative electron to give ^7Li and an electronic neutrino

$$^7Be + e^- \rightarrow ^7Li + \nu_e. \tag{11.8}$$

This neutrino has an energy 0.86 MeV in 89.7 % cases and 0.38 MeV in the residual 10.3 % cases.

The lithium in turn interacts with a proton to give two ^4He nuclei

$$^7Li + p \rightarrow ^4He + ^4He. \tag{11.9}$$

This chain is called ppII. In it, two ^4He nuclei are produced and two neutrinos.

For 0.1 % of cases the produced beryllium originates the following chain in which high energy, between 0 and 14.1 MeV, neutrinos are produced:

$$^7Be + p \rightarrow ^8B + \gamma, \tag{11.10}$$

$$^8B \rightarrow ^8Be^* + e^+ + \nu_e, \tag{11.11}$$

the asterisk indicating that the beryllium of mass 8 is produced in a very unstable state. Accordingly, it spontaneously disintegrates producing two ^4He nuclei

$$^8\text{Be}^* \rightarrow\, ^4\text{He} +\,^4\text{He}. \tag{11.12}$$

This reaction chain is named ppIII.

The three possible processes all have in common the fact that in any case each ^4He nucleus is accompanied by two electron neutrinos.

The CNO cycle occurs with a probability of 2 % and also in it the emission of neutrinos is involved.

The father of this model—called the standard model—is John N. Bahcall[5] (1934–2006) who started its elaboration in the 1960s and worked on it until his death.

In summary, there are five separated sources of neutrinos in the proton–proton chain. Three sources produce neutrinos with a continuous energy spectrum: the pp reaction (maximum energy 0.420 MeV), the hep neutrinos (maximum energy 18.773 MeV), and neutrinos from beryllium of mass 8 (maximum energy <15 MeV). There are then two sources of monoenergetic neutrinos: the pep neutrinos with energy 1.442 MeV and the neutrinos from ^7Be that in the 89.7 % of cases are emitted with an energy of 0.862 MeV and for 10.3 % have an energy 0.384 MeV.

Also in the CNO cycle neutrinos are emitted with a maximum energy that is never greater than something <2 MeV.

The flux of pp neutrinos is $6 \times 10^{10}/\text{cm}^{-2}\text{s}$, while that one of neutrinos from ^7Be is $4.9 \times 10^7/\text{cm}^{-2}\text{s}$ and the flux of the other channels is even lower.

The nuclear fusion reactions occurring inside the Sun produce a great quantity of energy which arrives to us prevalently as photons, that is electromagnetic radiation. However, according to the theoretical calculations we have just outlined, 2–3 % of the energy should come as neutrinos.

Similar mechanisms operate in the stars.

The study of neutrinos coming from Sun is a way to verify if the above reactions really occur. Raymond Davis Jr., was the first to study solar neutrinos.

Davis was born in Washington, DC on 14 October 1914 and was educated in the Washington public schools, attending the University of Maryland and graduating in 1938 in chemistry. After working for the Dow Chemical Co. in Midland, Michigan for a year, he returned to the University of Maryland to take a Master's degree, before going on to Yale where he received his PhD in physical chemistry in 1942. Immediately after he entered the Army as a reserve officer. Most of his war years were spent at Dugway Proving Ground in Utah, observing chemical weapons tests. Upon his discharge from the Army in 1945, he went to work at the Monsanto Chemical Co., Ohio, doing applied radiochemistry of interest to the Atomic Energy Commission. In the spring of 1948 he joined the newly created

---

[5]John Norris Bahcall was born on 30 December 1934 at Shreveport, Louisiana, USA and died at the Institute for Advanced Studies, Princeton on 17 August 2005. He studied at Louisiana State University first philosophy, then physics, and eventually astronomy. He then moved to the University of California, at Berkeley where he earned an AB in physics in 1956. Later, in 1957 he had a MS at Chicago University and a PhD at Harvard, in 1961.After some appointments in several universities, in 1971 he was appointed Professor at the Institute for Advanced Studies, Princeton. He worked in several fields of astrophysics.

Brookhaven National Laboratory, which was dedicated to finding peaceful uses for atomic energy. He remained there from 1948 to 1985. Since 1985 he was appointed astronomy Professor at the Pennsylvania University.

In 2002 he was awarded the Nobel Prize for physics together with Masatoshi Koshiba (1926–) and Riccardo Giacconi[6] (1931–). Davis and Koshiba divided one half of the prize *"for pioneering contributions to astrophysics, in particular for the detection of cosmic neutrinos"*. We describe later the Koshiba contribution. Giacconi received the prize *"for pioneering contributions to astrophysics, which have led to the discovery of cosmic X-ray sources"*.

Davis remembers [22][7] that his first act, on arriving at Brookhaven, was to report to the chairman of the Chemistry Department and ask him what he was expected to do. To his surprise he was advised to go to the library, do some reading, and choose a project of his own. In the library he read a review paper on neutrinos [23] and decided to make an experiment in neutrino physics.

In early experiments, he attempted to detect the neutrino by studying the recoil of a nucleus of ^{7}Li resulting from the decay by electronic capture of ^{7}Be. In this process an electron is captured by a proton of the beryllium nucleus. The proton turns to a neutron emitting, in nearly 90 % of cases, a monoenergetic neutrino of 0.862 MeV and the resulting ^{7}Li nucleus recoils with an energy of 57 eV. Through a measurement of this process, a proof of the existence of the neutrino was obtained and Davis by measuring the energy of the recoiling Li obtained results in agreement with what was expected [24]. However he was preceded by P.B. Smith and J.S. Allen [25] and he decided to make a different experiment aimed at directly detecting neutrinos from a nuclear reactor using the method suggested by Bruno Pontecorvo in 1946 known as the *chlorine–argon method*.

In this method, a chlorine atom of mass 37 if struck by an electronic neutrino transforms to an argon atom of mass 37. The reaction may be written as

$$\nu_e + {}^{37}Cl \rightarrow {}^{37}Ar + e^-. \tag{11.13}$$

Argon is a noble gas and is easy to separate chemically from a large amount of chlorine-rich solvent. It is radioactive with a half-life of 35 days and can be counted with a gas-filled proportional counter. The experiment is sensitive only to electronic neutrinos of energy >0.814 MeV.

A first attempt, exposing a 3,900 l (1,000-gallon) tank of carbon tetrachloride at the Brookhaven Graphite Research Reactor, failed to detect any signal. A reactor emits antineutrinos, and the method only detects neutrinos. At that time, it was not known if the two particles were identical or not. In 1955 he published these negative

---

[6]Riccardo Giacconi was born in Genua, Italy, in 1931. He studied at University of Milano earning a degree in Physics and working on cosmic rays. In 1956, he obtained a Fulbright Fellowship to work in US and after some time he received an offer to work on space sciences and started a successful research in X-rays astronomy. He held several academic positions.

[7]The story is taken from this lecture.

results that showed that the antineutrinos from the reactor are not absorbed in the chlorine–argon reaction [26]. Reines using reaction (11.1), which is just sensitive to antineutrinos, was on the contrary able, in the same year, to obtain a positive outcome, so that the Davis results, which used reaction (11.13) only sensitive to neutrinos, gave the first experimental proof that the antineutrino is different from the neutrino.

Davis continued his research measuring radioactive atoms in meteorites and moon rocks and began thinking about detecting neutrinos coming from the Sun.

At the end of his paper [26] he wrote: *"Since the 3900-liter tank with its associated counters was more sensitive for detecting neutrinos than any previously reported device, it is of interest to consider the possibility of detecting neutrinos from the sun. Two processes are now considered important for energy production in the sun, the proton-proton chain and the carbon-nitrogen cycle. The neutrinos from the proton-proton chain have a maximum energy (0.41 MeV) below the threshold of 0.816 MeV for the $Cl^{37}(v, e^-)A^{37}$ reaction and therefore neutrinos from this chain could not be detected by this method. The neutrinos from the carbon-nitrogen cycle arise from the positron decay of N^{13} and O^{15} which have a maximum neutrino energies of 1.24 and 1.68 MeV respectively"*.

An early experiment was made in a laboratory 2,300 ft. underground in the Barberton Limestone Mine, near Akron, Ohio. Observing neutrinos from the Sun had the potential of testing the theory that the hydrogen–helium fusion reactions are the source of the Sun's energy. In the 1950s, however, the proton–proton chain emitting only low-energy neutrinos was believed to be the principal neutrino source, and these neutrinos had energy below the threshold of the chlorine–argon reaction.

A new measurement of the nuclear reaction ^{3}He $+^4$ He $\rightarrow^7$ Be $+ \gamma$ (the reaction (7) at page 234) by H.D. Holmgren and R.L. Johnston [27] in 1958 suggested that this reaction, initially believed to be very rare, had on the contrary a 15 % probability to occur. This reaction may produce energetic neutrinos that can be measured with the chlorine–argon method.

Figure 11.2 from Bahcall and Ulrich [28] shows the spectrum of solar neutrinos predicted by the standard solar model. It gives the neutrino flux per cm^{2}s at the earth distance. The solid curves refer to the pp chain and the dotted ones to the CNO chain. With the support of the Brookhaven National Laboratory, Davis built a much larger experiment in the Homestake Gold Mine in Lead, South Dakota. The detector itself consisted of a 400,0001 (100,000 gallon) tank filled with perchloroethylene (C_2 Cl_4), a solvent most commonly used for dry clearing of clothing, placed 1,500 m underground for shielding from spurious effects due to cosmic rays.

After many years of preparation, eventually, in 1968, Raymond Davis Jr., and his collaborators began the experiment to measure the flux of neutrinos coming from the Sun and compare with the theoretical expectations. A collaborator of Davis was Bahcall, the big expert of the solar model (the one we today call the standard model).

The most energetic neutrinos emitted by the Sun are those of the ^{8}Be reaction and they have enough energy to change ^{37}Cl into ^{37}Ar. The method is not sensitive to the lower energy pp neutrinos, so that only a very small fraction of neutrinos interact.

Traversing the big tank containing a fluid with chlorine, the solar neutrinos interacted with the chlorine atoms producing radioactive argon. An average of 0.4

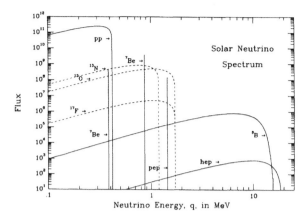

Fig. 11.2 The solar neutrino energy spectrum predicted by the standard solar model. The neutrino fluxes are given in number/cm^2/s/MeV at 1 AU. (*Solid curves*) Spectra from pp chain and (*dotted curves*) CNO spectra (Reprinted with permission from J.N. Bahcall and R.K. Ulrich, Rev. Mod. Phys. **60**, 297 (1988). Copyright 1988 by the American Physical Society)

Fig. 11.3 From right to left, Davis, Bahacall, and Don Harmer (from Mercury, March/April 1990)

argon atoms per day was produced. This number makes very clear the experimental difficulties of the experiment (Fig. 11.3).

Since the very beginning [29], the experiment showed a clear deficit in the solar neutrinos number: only one-third of the expected neutrinos were detected and this result was confirmed for all the duration of the experiment that ended in 1994. The solar neutrinos problem was born. We will present later the solution.

In Physical Review Letters, the Davis paper was followed by that of Bahcall [30] with the theoretical calculations of the expected neutrinos' flux.

The first Davis paper was entitled "*Search for neutrinos from the Sun*". The higher found value was 3SNU.[8] In the following Bahcall's paper, the prediction was 7.5SNU. Also after improving the system, the production rate of solar neutrinos resulted lower than that predicted by the solar model by a factor between 2 and 3. Davis himself believed that something was wrong with the standard model; many physicists thought there was something wrong with the experiment. A number of hypotheses were considered and a re-examination of the reactions that take place in the Sun was made. In 1988, Bahcall and R.K. Ulrich [28] reconsidered the standard solar model and re-calculated the neutrino flux that could be expected using different detection systems, confirming the existence of a strong difference between the measured and the theoretical values.

Today we know that the reason for this discrepancy was not an error in the Bahcall calculations. When Bahcall knew of the explanation, he reacted by saying he was feeling "*like that of a man falsely accused of a crime, whose innocence was vindicated 30 years later by a DNA test*".

Raymond Davis died on 31 May 2006.

For many years the Davis experiment remained the only one. Only in 1987 a new experiment started: Kamiokande. In Japan, the Institute for Cosmic Ray Research, University of Tokyo, started construction, in 1982, of the Kamioka underground observatory, located 1,000 m underground of the Mozumi Mine of the Kamioka Mining and Smelting Co., in Kamioka.cho, Gifu. It was completed in April 1983. The detector, named Kamiokande (Kamioka Nuclear Decay Experiment), was a tank which contained 3,000 tons of pure water and had about 1,000 photomultiplier tubes attached in the inner surface. The size of the tank was 16 m in height and 15.6 m in diameter. The photomultipliers collected the Cerenkov light emitted by the fast particles travelling in the water.

An upgrade of the detector was started in 1985 to observe the leptons produced by the interaction of the neutrino with water. The experiment was able to detect not only the solar neutrinos but, more in general, neutrinos coming from every direction due to the interaction of cosmic rays with the terrestrial atmosphere, that is the so-called *atmospheric neutrinos*. Neutrinos were detected by observing the Cerenkov radiation emitted by the recoil electrons in the scattering reaction

$$\nu + e^- \rightarrow \nu^* + e^*,$$

where the asterisk indicates the particle has a different energy. Both the direction and total energy are so measured. The reaction may occur with all the three kinds of neutrinos, but the cross section is higher for electronic neutrinos. The advantage is that the experiment is not limited to count down the neutrinos number but measures also their energy and provenience direction. However it is able to detect only neutrinos of energy >5 MeV.

---

[8] SNU is the shortname of solar neutrino unit and is 10^{-36} captures per target atom per second.

Kamiokande was supervised by Masatoshi Koshiba[9] and Yoji Totsuka[10] and was followed by Super-Kamiokande, supervised by Y. Totsuka and Yochiro Suzuki, an experiment based on the same technology but with an enormously bigger tank. Kamiokande had a surface coverage of 20 % by photomultipliers and cost about three millions US dollars. Super-Kamiokande cost 100 million US dollars. The detector consisted in an inner and an outer volume that contained 32,000 and 18,000 tons, respectively, of pure water. The external volume shields the inner volume in which the neutrinos reactions are studied. The inner volume is surrounded by approximately 11,000 photomultipliers that detect the Cerenkov light emitted by the neutrino struck electrons covering 40 % of surface. The construction started in 1991 and the observation began on 1 April 1996. On 12 November 2001 about 6,000 photomultipliers imploded. The system was partially repaired.

Kamiokande and Super-Kamiokande confirmed the Davis result [31–37]. However the solar neutrinos that they detected were still of relatively high energy. Also in this case the pp neutrinos escaped observation.

In 1991 a new experiment, Gallex [38, 39], performed in Italy inside the INFN Laboratorio del Gran Sasso of INFN started to take data using again a radiochemical detection scheme. This time gallium was used through the reaction

$$\nu + {}^{71}\text{Ga} \rightarrow {}^{71}\text{Ge} + e^-.$$

The produced germanium is radioactive. The reaction has a threshold for neutrinos of relatively low energy, >0.235 MeV. It is therefore able to measure the pp neutrinos. The tank contained 30 tons of gallium chloride. It was designed in 1991 as a collaboration among Italy, France, Germany, and Israel. Gallex measured the whole solar neutrino spectrum and confirmed the deficit observed by Davis [40]. The results were confirmed also by another Russian experiment, sensible to low energy neutrinos, called SAGE [41–45] and directed by Vladimir Gavrin of Moscow.

Meanwhile, in 1997, precise measurements of the sound velocity propagating inside the Sun were done, observing the periodic fluctuations of the light emitted by the solar surface. The measured sound velocity was in agreement with an accuracy of 0.1 % with the one calculated with the solar standard model, thus showing that Bahcall calculations were correct. So, why the flux of neutrinos did not agree with calculations?

---

[9]Masatoshi Koshiba was born in Toyohashi, Aichi on 1926. He graduated from the University of Tokyo in 1951 and received a PhD in physics at the University of Rochester, New York, in 1955. He was Associate Professor in 1963 at the University of Tokyo and then Professor in 1970 and Emeritus Professor in 1987. He had several other academic duties. He received Nobel Prize for Physics in 2002.

[10]Yoji Totsuka was born in Fuji, Japan in 1942. He studied in University of Tokyo and accepted there an Associate Professorship in 1979. In 1981, started work on Kamiokande experiment. Shortly after the supernova neutrino detection, Koshiba retired and Totsuka became leader of the Kamiokande project. In 2003, he became the Director General of Japan's high-energy physics organization, KEK. He died in 2008.

11.4 Atmospheric Neutrinos and Their Detection

The decay of pions, muons, and kaons produced in the collisions of the primary cosmic rays with the nuclei of the atmosphere originates a great number of neutrinos that to be distinguished by the ones of solar origin are called *atmospheric neutrinos*.

Electronic and muonic neutrinos are produced in the ratio 1–2. In fact, when cosmic ray particles enter the atmosphere, they interact with the N and O nuclei to produce π and K mesons. These mesons decay into μ and ν_μ. When later the secondary μ decays, an additional ν_μ and ν_e are produced. So at the end, one has two ν_μ against one ν_e. The number ratio $N(\nu_\mu)/N(\nu_e)$ is thus two. At higher energies the μ has time to reach the detector and in this case one misses one ν_μ and one ν_e. So at high energies the ratio becomes larger than two. In the early 1960s, many authors [46, 47] calculated their flux. These neutrinos are typically produced at altitudes between 10 and 20 km.

The experiments on atmospheric neutrinos started in the 1960s with big underground detectors. An experiment was done at Kolar Gold Field in southern India [48, 49]. Other experiments, that we already mentioned, were done by Reines in the East Rand Proprietary Mine in South Africa ([18, 50, 51], see also [52]). In these experiments, due to the high energy of the atmospheric neutrinos, a reaction in which a neutrino hits a nucleon and produces a muon plus something else may be employed. If the direction of arrival of the muon can be assessed, one may distinguish the muons arriving vertically from above—that are practically all produced by the decay of the cosmic ray pions—from the ones coming from below. Some muons may in fact be produced in the rocks near the detector and directed upward entering the mass of the detector before decaying. The neutrino that has produced the reaction may even have traversed the earth coming from the antipodes. No other particle is able to cross the earth to produce muons directed upward. Muons coming out grazing (horizontally) or directed upward are therefore produced by neutrinos.

These muons have a relatively long range in rocks, so that the detector volume is effectively expanded accepting events originated outside its sensitive volume. A muon emerging from the rocks below the detector must be produced by a muonic neutrino that has interacted at most 1 km from the detector. The neutrino itself may have been produced 12,000 km far by a cosmic ray hitting the atmosphere on the other side of the globe or may come from an astrophysical source 10,000 light years far.

The detectors used for the earlier observations [51, 53] were not able to distinguish between muons travelling downward or upward, and therefore, in order to avoid to identify the muons coming from above that are produced by cosmic rays with muons produced by neutrinos, people limited themselves to detect only the muons near the horizon: these particles were certainly all originated by atmospheric neutrinos.

More refined experiments revealed a handful of muons travelling upward, produced by atmospheric neutrinos originated on the other side of the earth.

In order to reduce the background made up by the muons produced directly in the atmosphere, it is necessary to use big deep underground detectors.

To have an idea of the difficulties encountered in the experiment one may consider that for every 100 billions of neutrinos that go through the earth just one interact with the matter forming the earth. A neutrino with an energy 100 GeV has a cross section for interaction with a nucleon of only $6.7 \times 10^{-6}\,\mu b$[11]; this means, e.g. that in water it may traverse on the average a thickness of 24 millions kilometres before suffering an interaction.

11.5   The Solution of the Solar Neutrinos Mystery

Bruno Pontecorvo,[12] in 1957, suggested neutrinos and antineutrinos could oscillate, that is transform continuously one into the other.

When Davis found the deficit in the number of solar neutrinos, the hypothesis of the oscillation among different kinds of neutrinos was resumed by Pontecorvo and Gribov [56–58] taking into account the two different kinds of neutrinos known at the time.

Already in 1962, Z. Maki, N. Nagakawa, and S. Sakata [59, 60] had suggested the possibility of transitions among leptons of different generations; Pontecorvo and Gribov suggested that the oscillations take place among the different kinds of neutrinos, a process that is possible only if neutrinos have a mass and more precisely different masses for different kinds of neutrinos. Since in those years the dominant idea was that neutrino has no mass, the idea was difficult to digest.

About the neutrino oscillation, what one means is that, given for example a beam of mu neutrinos produced by the decay in flight of pions, the composition of this beam should change as a function of the distance from the place of production. If, for example, initially only mu neutrinos were present, they should transform while travelling into electron and tau neutrinos and they in turn successively ones into the others and into mu neutrinos and so on. The theoretical calculations predict that these oscillations should occur with a sinusoidal law as a function of the travelled distance, with a spatial periodicity of the oscillations that should be proportional to the reciprocal of the neutrino mass difference squared and directly proportional to their energy.

Earlier experiments looking for these oscillations were performed using electronic neutrinos of low energy ($\leq$10 MeV) produced in a nuclear reactor. One experiment was made in USA at Palo Verde [61] and another, named Chooz [62,63], from the Ardenne place where it was located, gave negative results.

The CHOOZ experiment—operating at the end of 1990s—was located 1,050 m from the double unit of the Chooz nuclear reactor. The detector was placed in an underground laboratory under 100 m of rocks to reduce the cosmic ray background. A homogeneous detector was filled with 5 tons of a liquid scintillator doped with

---

[11]μb is a microbarn. The barn is an international unit of cross section, 1 barn $= 10^{-24}\,cm^2$.

[12]B. Pontecorvo [54] (in Russian); translated in [55].

gadolinium surrounded by thick scintillators. A more complex project with two identical neutrino detectors one near the reactor and the other some km far away is under construction [64].

Experiments were performed also with accelerators [65]. The first attempts, Chorus [66] and Nomad [67], were made at the beginning of 1990s in the CERN laboratory in Geneva, using the mu neutrinos artificially produced in an accelerator. The distance between the beam source and the detector, that is crucial to detect the oscillation, was, however, too short as was appreciated later, and no positive result was obtained. All these experiments made with neutrinos produced in the lab did not allow the observation of oscillations but were useful to better define the characteristics of the phenomenon.

Meanwhile, in an unexpected way, oscillations were actually observed with atmospheric neutrinos. As already mentioned, Kamiokande and Super-Kamiokande are able to "see" atmospheric neutrinos and their arrival direction. In particular they can detect neutrinos coming upward, that is neutrinos produced in the opposite hemisphere to that in which the detector is placed that have passed the earth before being detected. The Super-Kamiokande collaboration measured the ratio between events which could be attributed to muon neutrinos (muon-like) and events which could be attributed to electron neutrinos (e-like). The ratio between the muon neutrinos and the electron neutrinos had been calculated in great detail over a broad range of energies from 0.1 to 10 GeV [68, 69]. The Super-Kamiokande collaboration studied the ratio $R = (\mu/e)_{DATA}/(\mu/e)_{MC}$, where μ and e are the number of muon-like and electron-like events observed in the detector for both data and Monte Carlo simulation. This ratio R should be $R = 1$ if the physics in the Monte Carlo simulation accurately models the data. In the presence of neutrino oscillations, this ratio should depend on the travelled distance L. For detectors near the surface of the earth, the neutrino's flight distance is a function of the zenith angle of the neutrino direction. Vertically down-going neutrinos travel about 15 km, while vertically upward-going neutrinos travel about 12,000 km before interacting in the detector. The measurements found a zenith angle dependence of R which decreased with increasing the travelled distance. Considering the number of upward-going events U over the number of down-going events D it was found that in the case of μ-like data the ratio exhibited a strong dependence with the zenith angle, while no significant effect was observed in the e-like data. This result and the one of the R parameter could be interpreted as a diminution of the number of muon neutrinos by increasing L, while the number of electron neutrinos remains unchanged. Therefore, this result was interpreted as a change due to $\nu_\mu \rightarrow \nu_\tau$ oscillation.

Figure 11.4 shows what essentially may be interpreted as the parameter R we defined previously vs. the reconstructed L/E, where E is the neutrino energy. The points show the ratio of observed data to Monte Carlo simulation expectation in the absence of oscillations. The dashed lines[13] show the expected shape for $\nu_\mu \rightarrow \nu_\tau$ at $\Delta m^2 = 2.2 \times 10^{-3} \, eV^2$ and $\sin^2 2\vartheta = 1$.

---

[13]We explain in the next paragraph these numbers.

Fig. 11.4 The ratio of the observed data to Monte Carlo simulation expectation in the absence of oscillations versus reconstructed L/E. (Reprinted with permission from V. Fukuda et al, Phys. Rev. Lett. **81**, 1562 (1998). Copyright 1998 by the American Physical Society)

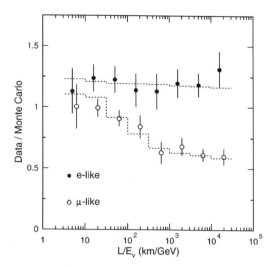

The result was first announced in 1998, during a conference held in Takayama, Japan. Thus, the Super-Kamiokande showed, in an unquestionable way, that the anomaly of atmospheric neutrinos had to be interpreted as the disappearing of mu neutrinos, the more as the path is longer through the earth [70, 71]. Therefore mu neutrinos oscillate along their path through the earth with a period of the order of the earth radius. Moreover, because the number of electron neutrinos does not change, the oscillation has to be interpreted as a transition of mu neutrinos into tau neutrinos. This result was confirmed by MACRO which performed similar measurements.

The MACRO (Monopole, Astrophysics and Cosmic Ray Observatory) experiment is the first big experiment installed at the Gran Sasso laboratories, L'Aquila, Italy, in 1989 as an International collaboration for the study of the existence of magnetic monopoles (particles that present only one magnetic pole, predicted in some theories but not yet found experimentally till now), cosmic rays, and neutrinos. The MACRO experiment is long 76.6 m. It measures the neutrinos produced in the atmosphere above the device and in the opposite hemisphere which therefore come from below after travelling 12,000 km through the Earth from one side to the other. The experiment found that muon neutrinos coming upward after traversing the Earth from one side to the other are approximately one-half the expected number, while the ones produced above the device are nearly the expected number. In 10 years MACRO detected approximately 400 neutrinos and the results were in favour of the oscillation of neutrinos [72].

Mostly contemporarily the Liquid Scintillator Neutrino Detector (LSND) collaboration [65] at the Los Alamos Meson Physics Facility claimed positive results for $\nu_\mu \rightarrow \nu_e$ oscillations using a ν_μ flux from π^+ decay in fly obtained by LAMPF 800 MeV proton linear accelerator. The experiment was very controversial, however.

Also, the Japan–American collaboration, KamLAND, lead by Atsuto Suzuki gave confirmation of the existence of oscillations by studying the antineutrinos emitted by nuclear reactors in Japan and nearby countries [73, 74].

Fig. 11.5 KamLand neutrino detector (Reprinted with permission from K. Eguchi et al. Phys. Rev. Lett. **90**, 021802 (2003). Copyright 2003 by the American Physical Society)

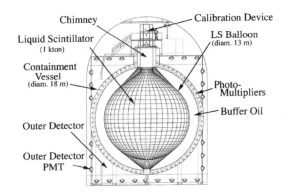

The purpose of the collaboration was a search for oscillations of the electron antineutrinos emitted from distant power reactors. In this experiment neutrinos were detected through the inverse beta decay

$$\underline{\nu}'_e + p \rightarrow e^+ + n,$$

in a liquid scintillator detecting both the positron and the delayed 2.2 MeV gamma rays produced by the neutron capture on a proton. KamLAND was placed on the site of the earlier Kamiokande under the equivalent of 2,700 m of rocks. Figure 11.5 shows the neutrino detector which consists of 1,000 tons of ultrapure liquid scintillator contained in a 13 m diameter transparent nylon-based balloon suspended in nonscintillating oil. The balloon is surrounded by 1879 photomultiplier tubes mounted on an inner surface of an 18 m diameter spherical stainless steel vessel. A 3,200 tons Cherenkov detector surrounds the containment sphere, absorbing gamma rays and neutrons from the surrounding rocks and tagging cosmic-ray muons.

The antineutrinos were provided by many nuclear reactors but the flux was actually dominated by a few powerful reactors at an average distance of about 180 km. The ratio of measured to expected neutrinos flux showed a marked decrease at the 180 km distance so demonstrating a disappearance of electron antineutrinos at long baselines. Further measurements confirmed the results [74].

Unfortunately, the transition of muon to tauon neutrinos, as observed by Super-Kamiokande and KamLAND could not be detected directly, because the experiments were insensitive to tau neutrinos. The OPERA (Oscillation Project with an Emulsion tRacking Apparatus) experiment installed in the Laboratori Nazionali del Gran Sasso is able to detect these neutrinos. The experiment is very impressive; a beam of neutrinos is sent from CERN, Geneva, over a baseline of 730 km towards the Gran Sasso Laboratories. Due to oscillations some of the muonic neutrinos sent by Geneva will change in tau neutrinos that may be detected through the tau particles produced by their interactions in a gigantic stack of nuclear emulsions. The experiment, proposed in 1997, is now working [75].

The beam produced at CERN is mainly composed of ν_μ with no ν_τ. If a tau neutrino comes to be detected this is an indication of the oscillation phenomenon.

Experimentally the two types of neutrinos can be distinguished because v_μ produces a muon and the v_τ produces a tauon. In a tracking detector the muon appears as a long almost straight track, while the tauon, which has a very short lifetime (picoseconds), decays already after a few hundreds micrometers. The detector should therefore have a micrometer resolution and the solution that was chosen was to build an emulsion cloud chamber based on sandwiches of thin $(50\,\mu m)$ emulsion sheets, providing the $1\,\mu m$ resolution tracking, interleaved with $1\,mm$ thick Pb sheets, providing the mass necessary to detect the neutrino. After $730\,km$ only a very small fraction of neutrinos, 1–2 % are expected to change their flavour. In OPERA 150,000 sandwiches are used including about $110,000\,m^2$ emulsion films and $105,000\,m^2$ lead plates for a total of about 1,250 tons. The experiment has started in 2008 and in 2010 the first event identified as a probable tau neutrino was found [76]. The claim made by the researchers that neutrinos were found able to travel faster than light has raised an outcry and has been later disproved.

11.6 Back to Solar Neutrinos

In a nickel mine at Sudbury, Ontario, Canada, is the site of an observatory for neutrinos: SNO (Sudbury Neutrino Observatory). It is a Canada–America–England collaboration led by Arthur McDonald, utilizing for the detection heavy water in a 1,000 tons detector. The heavy water allows to "see" simultaneously several reactions produced by solar neutrinos.

In earlier measurements, the SNO collaboration used the detector in such a way to detect only the solar neutrinos produced by ^8B. They found about one-third of the expected neutrinos produced by Sun [77]. Super-Kamiokande that is sensible to electron neutrinos but has also some sensitivity for the ones of the other kinds observed about one-half of the number calculated by the standard model.

Because the solar neutrinos are only of the electron type one should have obtained the same number. Combining the SNO and Super-Kamiokande[14] measurements, the SNO collaboration determined the number of solar neutrinos of all kinds (electron, muon, and tau) as well as the number of the sole electron neutrinos. The total number of neutrinos of all kinds was in agreement with that predicted for the Sun, while the electron neutrinos were only about one-third of the total.

This result was announced on 18 June 2001 and was confirmed by other experiments showing that the Sun emits the right number of neutrinos that, however, during the travel change partly in the other two species.

Now that the solar neutrinos anomaly was solved, solar neutrinos became a quantitative tool for astronomy. Indeed, the measurements of Super-Kamiokande and Sudbury had already yielded one remarkable constraint on the Sun, a direct

---

[14]T. Toshito et al., quoted in Ahmad [77].

determination of the core temperature to high precision, through measurement of the ^{8}B neutrino flux [78].

Critical analysis of the nuclear reactions important to energy generation in the Sun and other hydrogen-burning stars and to solar neutrino production was discussed in two papers one before the solar neutrino anomaly was solved [79] and a more recent one after the solution in 2011 [80].

11.7 The Neutrinos' Oscillation and Its Implications

Let us consider for the sake of simplicity the case of neutrinos oscillation only between the two flavour electronic and muonic. According to the Pontecorvo's suggestion the quantum mechanical wave function describing these neutrinos may be considered a superposition of the wave functions of two massive neutrinos ν_1 and ν_2, with masses m_1 and m_2, respectively, called Majorana's neutrinos.

During their travel in space the mixing of these wave functions may change. For a two-neutrino oscillation hypothesis, the probability for a neutrino produced in flavour state a to be observed in flavour state b after travelling a distance L through vacuum is

$$P_{a \to b} = \sin^2 2\vartheta \, \sin^2 \left[\frac{1.27 \Delta m^2 (\text{eV}^2) L(\text{km})}{E(\text{GeV})} \right],$$

where E is the neutrino energy, ϑ is called "mixing angle", and Δm^2 is the mass squared difference of the neutrino mass eigenstates.

Extending these considerations to three flavours of neutrinos, the results of the Super-Kamiokande and Sudbury experiments allowed to put some preliminary number as [70, 81]

$$\Delta m_{21}^2 = m_2^2 - m_1^2 \approx 8 \times 10^{-5} \, \text{eV}^2$$

and

$$\Delta m_{32}^2 = m_3^2 - m_2^2 \approx 3 \times 10^{-3} \, \text{eV}^2$$

for the mass squared differences of the three kinds of neutrinos.

The oscillation phenomenon implies therefore that neutrinos have non-null mass. However, the experiments do not allow to measure their mass but only the differences between the masses squared, and current data do not allow to discriminate between the two possibilities

$$m_1 < m_2 < m_3 \quad \text{and} \quad m_3 < m_1 < m_2.$$

An upper limit on the mass of the most massive neutrino, $mc^2 \leq 2 \, \text{eV}$, is set by measurements of radioactive decay of tritium [82].

The discovery that neutrinos have a non-zero mass is very important for astrophysics and cosmology. Massive neutrinos in the early 1980s were believed could help solving the problem of the so-called missing mass in the Universe, that is the circumstance that the mass present in the form of "visible" matter in the Universe is not enough to explain the observed evolution. The obscure mass of neutrinos could help to "close" the Universe. However, G. Blumenthal et al. [83] proved that dark matter had to consist of slow-moving, "cold" particles instead of fast neutrinos. The debate on dark matter and dark energy is still open.

11.8 Neutrinos from Stars: The Birth of Neutrino Astronomy

In the Universe, neutrinos are produced in a number of processes, as thermonuclear reactions in the interior of the Sun and other stars and in the stellar collapses that originate the supernovae explosions. The interest in astrophysics for the detection of extraterrestrial neutrinos is principally because they may reach us from regions in the Universe that otherwise would be inaccessible to observations. While, for example, electromagnetic radiations produced in the nuclear reactions in a star are not able to filter through the outer layers of the star, neutrinos, that interact only very weakly with matter, are able to escape easily going into space, telling us what happened.

The detection of high energy atmospheric neutrinos demonstrated that the technique may be applied to the high energy neutrino astronomy, provided that a large enough detector is built to detect the low rate of events.

As we will discuss later, an old massive star which has exhausted its nuclear fuel may collapse under its gravitational field, exploding and giving place to what we call a *supernova*. Two kinds of supernova exists. According to the conventional theory [84–86], in a supernova of the second type (II), about 90 % of its gravitational energy, that is nearly 3×10^{53}erg is irradiated in a few seconds during the explosion under the form of 10^{58} neutrinos of all kinds with an energy around 10–15 MeV.

On February 1987, a spectacular supernova, the first visible to the naked eye since the seventeenth century, exploded in the Large Magellano Cloud (Supernova SN1987A), a small satellite galaxy of our own Galaxy, at only 180,000 light years from us.

When the researchers who operated the Kamiokande and Irvine-Michigan-Brookhaven stations knew of the explosion, they executed an accurate analysis of the detected neutrinos and found a burst of 19 neutrinos simultaneously detected in about 10 s on 23 February. Kamiokande II detected 11 neutrinos in a time interval of 13 s [87]. The Irvine-Michigan-Brookhaven, in a Ohio salt mine, detected a burst with 8 neutrinos in the range of 20–40 MeV in 6 s [88]. The usual background was approximately 1 neutrino per day.

The result confirmed that supernovae are actually a nuclear explosion, and stimulated the development of two second-generation detectors, Super-Kamiokande that was a copy of its father, scaled to 50,000 tons and the Sudbury Neutrino Observatory in a nickel mine in Ontario with 1,000 tons of heavy water.

Some claim was also made of a detection of five neutrinos by the observatory on Monte Bianco but its reliability has been criticised[15].

Because of the proximity of supernova SN1987A, it has been possible to witness its evolution from explosion to remnant. SN1987A has thus become one of the most extensively studied extragalactic objects, with ground, airborne, and space observatories covering a wide range of the electromagnetic spectrum. The detection of neutrinos confirmed that the event marked the explosive death of a massive star. Examination of plates of the region obtained before the SN explosion allowed the detection of its progenitor, a blue supergiant (Sk-69 202) that was believed to have had an initial mass of 18–20 times the Sun mass [89].

Neutrino astronomy requires kilometre-scale neutrino detectors. Early efforts concentrated on transforming large volumes of natural water into Cherenkov detectors that catch the light produced when neutrinos interact with nuclei in or near the detector. In 1980s, a collaboration settled at the Hawaii University called DUMAND (Deep Underwater Muon and Neutrino Detector) considered for the first time to built a detector deep in the ocean [90] in a site about 40 km off the main island of Hawaii in 4,800 m of water. The project was subsequently withdrawn because the technology was not yet mature for a project of that kind. However, it paved the road for later efforts by developing many of the detector technologies in use today, and by inspiring the development, with a Russian–German collaboration, of a smaller instrument in Lake Baikal [91] in Siberia as well as efforts to commission neutrino telescopes in the Mediterranean.

In later 1980s a collaboration AMANDA (Antarctic Muon And Neutrino Detection Array) in the Antarctic glaciers was established at the Amundsenn-Scott Pole Station. After thousands of problems, four stripes of photomultipliers at a depth between 1,500 and 2,000 m in the ice were positioned, in 1995–1996, and AMANDA detected its first neutrino in 1996 [92]. The system was completed in 2000 and was in operation up to 2009.

AMANDA represented a proof of concept for the kilometre-scale IceCube, directed by Francis Halzen from University of Wisconsin, whose construction began in the 2004–2005 season and was completed 18 December 2010. IceCube transforms a cubic kilometre of deep and ultra-transparent Antarctic ice into a particle detector. It is located at the Amundsen-Scott research centre at the South Pole and is paid for by US National Science Foundation and collaborators in Germany and Sweden. A total of 5,160 optical sensors are embedded into the ice to detect the Cherenkov light emitted by the secondary particles produced when neutrinos interact with nuclei in the ice. The light patterns reveal the flavour of neutrino interaction and the energy and direction of the neutrino, making neutrino astronomy possible. The detector observes neutrinos with energies above 100 GeV. A deepcore infill array may identify a smaller sample with energies as low as

---

[15] See, for example, note 5 of the paper K. Hirata et al. [87].

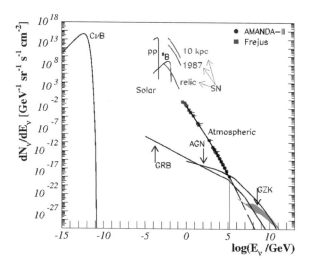

Fig. 11.6 The cosmic-neutrino spectrum (Reprinted with permission from V.F. Halzen and S.R. Klein, Phys. Today **61**, 29 (2008). Copyright 2008 by the American Institute of Physics)

10 GeV. The cosmic neutrino spectrum is shown in Fig. 11.6. There is a large variety of astrophysical sources. Neutrinos were produced in the big-bang (curve labelled CνB in the figure), and are now emitted by the Sun, supernovae (SN), active galactic nuclei (AGN), and originate also as a result of the GZK cutoff. In the figure also atmospheric neutrinos are indicated. The data points are from detectors at the Frejus underground laboratory [94] (red) and from AMANDA [95] (blue).

An Italian project, ANTARES has been completed off the cote d'Azur in France. The detection of muons and their arrival direction is made through the Cerenkov light generated in the sea water 2,400 m deep between Corsica and the southern cost of France.[16] Like AMANDA, ANTARES is a proof of concept for KM3NeT, a kilometre-scale detector in the Mediterranean Sea complementary to IceCube at the South Pole.

These experiments are designed to look for neutrinos possibly coming from the farthest points of the Universe [93].

Other collaborations are underway and new detection techniques are also under study, such as the use of acoustic pulses or radio waves emitted by the charged particles produced in a collision with a neutrino.

References

1. B. Pontecorvo, Zh. Eksp. Teor. Fiz. **37**, 1751 (1959) (in Russian)
2. B. Pontecorvo, Sov. Phys. JETP **37**, 1236 (1960)
3. M.L. Perl, Phys. Rev. Lett. **35**, 1489 (1975)
4. B.P. Lazarenko, S. Yu Likanov, Zh. Eksp. Teor. Fiz. **49**, 751 (1965)

---

[16]More information may be found in [96].

5. G. Bernardini, in *International Conference on High-Energy Physics*, Dubna, 1964
6. G.S. Abrams et al., Phys. Rev. Lett. **63**, 2173 (1989)
7. B. Adeva et al., Z. Phys. C **51**, 179 (1991)
8. G. Alexander et al., Z. Phys. C **52**, 175 (1991)
9. P. Abreu et al., Nucl. Phys. B **367**, 511 (1991)
10. F. Reines, C.L. Cowan Jr., Detection of the free neutrino. Phys. Rev. **92**, 830 (1953)
11. C.L. Cowan Jr., E. Reines, E.B. Harrison, H.W. Kruse, A.D. McGuire, Science **124**, 103 (1956)
12. F. Reines, in *Nobel Lectures. Physics 1991–1995*, ed. by G. Ekspong (World Scientific Publishing Co., Singapore, 1997)
13. H.A. Bethe, R.E. Peierls, Nature **133**, 532 (1934)
14. F. Reines, C.L. Cowan, Phys. Rev. **90**, 492 (1953)
15. F. Reines, C.L. Cowan, Phys. Rev. **92**, 830 (1953)
16. F. Reines, C.L. Cowan, Phys. Rev. **113**, 273 (1959)
17. C.P. Enz, No time to be Brief, Oxford University Press (2002), p. 488
18. F. Reines et al., Phys. Rev. D **4**, 80 (1971)
19. M.F. Crouch, P.B. Landecker, J.F. Lathrop, F. Reines, W.G. Sandie, H.W. Sobel, H. Coxell, J.P.F. Sellschop, Phys. Rev. D **18**, 2239 (1978)
20. R. Atkinson, F.G. Houtermans, Zeits. Phys. **54**, 656 (1929)
21. H.A. Bethe, Phys. Rev. **55**, 434 (1939)
22. J. Davis, in *Nobel Lectures. The Nobel Prizes 2002*, ed. by T. Frangsmyr (Nobel Foundation, Stockholm, 2003)
23. M.R. Crane, Rev. Mod. Phys. **20**, 278 (1948)
24. R. Davis Jr., Phys. Rev. **86**, 976 (1952)
25. P.B. Smith, J.S. Allen, Phys. Rev. **81**, 381 (1951)
26. R. Davis Jr., Phys. Rev. **97**, 766 (1955)
27. H.D. Holmgren, R.L. Johnston, Phys. Rev. **113**, 1556 (1958)
28. J.N. Bahcall, R.K. Ulrich, Rev. Mod. Phys. **60**, 297 (1988)
29. R. Davis Jr., D.S. Harmer, K.C. Hoffman, Phys. Rev. Lett. **20**, 1205 (1968)
30. J.N. Bahcall, N.A. Bahcall, G. Shaviv, Phys. Rev. Lett. **20**, 1209 (1968)
31. K.S. Hirata et al., Phys. Rev.Lett. **63**, 16 (1989)
32. K.S. Hirata et al., Phys. Rev. Lett. **65**, 1297 (1990)
33. K.S. Hirata et al., Phys. Rev. D **44**, 2241 (1991)
34. Y. Fukuda et al., Phys. Rev. Lett. **77**, 1683 (1996)
35. Y. Fukuda et al. Phys. Rev. Lett. **81**, 1158 (1998)
36. S. Fukuda et al., Phys. Rev. Lett. **86**, 5651 (2001)
37. Y. Fukuda et al., Phys. Rev. Lett. **82**, 1810 (1999)
38. M. Altman et al., Phys. Lett. B **285**, 16 (1999)
39. W. Hampel et al., Phys. Lett. B **447**, 127 (1999)
40. P. Anselmann et al., Phys. Lett. B **327**, 377(1994)
41. N. Abdurashitov et al., Zh. Eksp. Theor. Phys. **95**, 181 (2002)
42. J.N. Abdurashitov et al., Phys. Lett. B **328**, 234 (1994)
43. J.N. Abdurashitov et al., Phys. Rev. Lett. **77**, 4708 (1996)
44. J.N. Abdurashitov et al., Phys. Rev. C **59**, 2246 (1999)
45. J.N. Abdurashitov et al., Phys. Rev. C **60**, 055801 (1999)
46. M.A. Markov, I.M. Zheleznykh, Nucl. Phys. **27**, 385 (1961)
47. G.T. Zatsepin, V.A. Kuzmin, JETP **14**, 1294 (1962)
48. C.V. Achar et al., Phys. Lett. **18**, 196 (1965)
49. C.V. Achar et al., Phys. Lett. **19**, 78 (1965)
50. F. Reines et al., Phys. Rev. Lett. **15**,429 (1965)
51. M.F. Crouch et al., Phys. Rev. D **18**, 2239 (1978)
52. J.G. Learned, K. Mannhein, Annu. Rev. Nucl. Part. Sci. **50**, 679 (2000)
53. M.R. Krishnaswamy et al., Pramana **19**, 525 (1982)
54. B. Pontecorvo, Zh. Eksp. Teor. Fiz. **33**, 549 (1957) (in Russian)
55. B. Pontecorvo, Sov. Phys. JETP **6**, 429 (1958)

56. B. Pontecorvo, V. Gribov, Phys. Lett. B **28**, 493 (1969)
57. B. Pontecorvo, Zh. Eksp. Teor. Fiz. **53**, 1717 (1967)
58. B. Pontecorvo, Sov. Phys. JETP **26**, 984 (1968)
59. Z. Maki, M. Nakagawa, S. Sakata, Progr. Theor. Phys. **28**, 870 (1962)
60. M. Nakagawa, H. Okonogi, S. Sakata, A. Toyoda, Progr. Theor. Phys. **30**, 727 (1963)
61. F. Boehm et al., Phys. Rev. Lett. **84**, 3764 (2000)
62. M. Apollonio et al., Phys. Lett. B **420**, 397 (1998)
63. M. Apollonio et al., Phys. Lett. B **466**, 415 (1999)
64. T. Lasserre, Europhys. News **38**(4), 20 (2007)
65. C. Athanassopoulos et al., Phys. Rev. Lett. **81**, 1774 (1998)
66. E. Eskut et al., Phys. Lett. B **497**, 8 (2001)
67. P. Astier et al., Nucl. Phys. B **611**, 3 (2001)
68. M. Honda et al., Phys. Lett. B **248**, 193 (1990)
69. G. Barr et al., Phys. Rev. D **39**, 3532 (1998)
70. V. Fukuda et al., Phys. Rev. Lett. **81**, 1562 (1998)
71. S. Fukuda et al., Phys. Rev. Lett. **86**, 5651 (2001)
72. M. Ambrosio et al., Phys. Lett. B **517**, 59 (2001)
73. K. Eguchi et al., Phys. Rev. Lett. **90**, 021802 (2003)
74. T. Araki et al., Phys. Rev. Lett. **94**, 081801 (2005)
75. E. Pennacchio, Nuovo Cimento **33**, 77 (2010)
76. N. Agafonova et al., Phys. Lett. B **691**, 138 (2010)
77. Q.R. Ahmad et al., Phys. Rev. Lett. **87**, 71301 (2001)
78. B. Aharmin et al., Phys. Rev. Lett. **101**, 111301 (2008)
79. E.G. Adelberger et al., Rev. Mod. Phys. **70**, 1265 (1998)
80. E.G. Adelberger et al., Rev. Mod. Phys. **83**, 195 (2011)
81. The KamLAND Collaboration, Phys. Rev. Lett. **94**, 081801 (2005)
82. C. Weinheimer, Phys. Scripta T **121**, 166 (2005)
83. G. Blumenthal, S. Faber, J. Primack, M. Rees, Nature **311**, 517 (1984)
84. A. Colgate, R.H. White, Astrophys. J. **143**, 626 (1966)
85. J.N. Bahcall, A. Dar, T. Piran, Nature **326**, 155 (1987)
86. S.E. Woosley, T.A. Neaver, Annu. Rev. Astron. Astrophys. **24**, 205 (1986)
87. K. Hirata et al., Phys. Rev. Lett. **58**, 1490 (1987)
88. R.M. Bionta et al., Phys. Rev. Lett. **58**, 1494 (1987)
89. W.D. Arnet et al. Annu. Rev. Astron. Astrophys. **27**, 629 (1989)
90. A. Roberts, Rev. Mod. Phys. **64**, 259 (1992)
91. V.A. Balkanov et al., Nucl. Phys. Proc. **118** (suppl.) 363 (2003)
92. E. Andres et al., Nature **410**, 441 (2001)
93. V.F. Halzen, S.R. Klein, Phys. Today **61**, 29 (2008)
94. W. Rhode et al., Astropart Phys. **4**, 217 (1996)
95. A. Achterberg et al., Phys. Rev. D **76**, 042008 (2007)
96. S. Giordeno, Il Nuovo Sagg. Boll. Soc. Ital. Fis. **23**(5–6), 58 (2007)

Chapter 12
Which are the Primary Cosmic Rays?

12.1 Introduction

After the Second World War, now that the existence of cosmic rays was universally accepted, there were plenty of problems to solve: one was connected to the nature of mesons that brought to the discovery—after the muon—of the pion, the K mesons, and started in the 1950s the research on "elementary" particles. The latter was principally made using particle accelerators which allow studying easily in the laboratory of the different interactions by creating a number of events much greater than those present in cosmic rays. This kind of research was very expensive because to study the ever increasing number of particles with their large masses, greater and greater energies were needed and therefore bigger and bigger accelerators of ever increasing cost were built.

In 1953, in France, at Bagnères de Bigorre, one of the most important conferences on cosmic rays was held. Groups from all over the world were present who engaged in lively discussions on the new particles discovered in cosmic rays. The conference was closed by Louis Le Prince Ringuet: "*We know—he said—that in this moment Brookhaven is producing V° particles and will produce ever more. One may anticipate without doubt that the future of cosmic radiation as a field of particle physics will depend on accelerators... We* [cosmic ray people] *have the exclusive of rare phenomena, whose energies are much greater, oh yes, and surmise that for some time we will be protected by the fast growth of the machine energy as far as technique will be more and more developed... However we are in an unhappy situation; but it is not this perspective full of interest?*"

And in fact even the latest accelerator, the Large Hadron Collider (LHC) of CERN, Geneva, that may attain a final energy of 14 TeV (that is 1.4×10^{13} eV), is still seven orders of magnitude below the most energetic cosmic rays!

Beside the problem of the study of "elementary" particles, was that one of their mutual interactions and of the produced reactions. This research, started before the war, principally studying the *disintegration stars* produced by cosmic rays in nuclear

M. Bertolotti, *Celestial Messengers: Cosmic Rays*, Astronomers' Universe,
DOI 10.1007/978-3-642-28371-0_12, © Springer-Verlag Berlin Heidelberg 2013

emulsions, had some ten years of great activity immediately after the end of the conflict before being superseded by the use of the big accelerators.

Another problem was to decide what was the nature of primary cosmic radiation, whose principal component had been individuated in the protons already during the conflict, but that now could be studied with greater accuracy through high altitude measurements with planes, rockets and balloons, and then outside the Earth with spacecrafts.

With the instrumentation brought by these crafts, the study of primary radiation received an enormous impulse because on one side it was eventually possible to have direct measurements of which were the cosmic rays hitting the top of our atmosphere coming from outside, and on the other side it was possible to know the state of space outside the Earth. The particles with very high energy were, however, too rare to be studied with detectors on board of spacecrafts, but could be studied through the *extended showers* described in Chap. 10.

The exploration of space far from the Earth paved the way for a great number of discoveries, part of which were crucial to understand the nature of primaries and provide elements to study their origin.

Let us begin to see, very briefly and only for what is pertaining to the specific theme of cosmic rays, what was found relatively to the space surrounding our Earth and the magnetic field existing in the solar system.

12.2   A Short Digression on Space Researches

During the International Geophysics Year 1957–1958, the United States presented a project—denominated Vanguard—for the creation of artificial satellites to make scientific measurements in the high atmosphere. The first artificial satellite, the Sputnik I, was however put into orbit by the then Soviet Union on 4 October 1957, causing a great sensation, followed about 1 month later by the Sputnik II with the dog Laika on board. The first American satellite was Explorer I, launched on 31 January 1958.

Since then, thousands of satellites have been launched for various purposes: telecommunications, navigation help, surveying, meteorology, tracking, etc., by a great number of nations all over the world. With a further effort, special spacecrafts, with suitable scientific instrumentation, were sent everywhere in our solar system and the Moon was the first extraterrestrial body to be explored. After a first unsuccessful attempt of Pioneer I launched by the Americans on 11 October 1958, the Russians with Lunik I, launched on 2 January 1959, succeeded to pass at about 5,000 km from the moon surface. With the Apollo program, started in 1961, Americans, after a number of preparatory flights, also with human crew and an accident that cost the life to three astronauts, eventually in the mission of 16–24 July 1969, on 20 July landed on Moon in the zone of the Tranquillity Sea, and N.A. Armstrong (1930–) and E.E. Aldrin (1930–) were the first men to set foot on our satellite.

A great number of both Soviet and American spacecrafts since then explored the Moon, sometime moon-landing and also bringing back material.

The list of missions is by now endless. The solar system has been mostly explored in various ways by the Russians with the Lunik and Venera series, and others, and by the Americans with Pioneer, Mariner, Ranger, Viking, Voyager, and more. Pioneer X, launched on 3 March 1972, after passing near Jupiter, was the first to abandon the solar system. Europe developed its own program and expeditions were made also by China, Japan, and India, just to mention some.

All these spacecrafts, in various ways, have brought apparatuses for measurements and even scientific experiments, reporting a great deal of data, not only on the observed planets but also on the surrounding space, cosmic rays, space magnetic fields, etc.

12.3   The Solar Wind, The Magnetic Field, and The Heliosphere

One of the results of the exploration of space outside the Earth was the understanding of the solar influence on the primary cosmic rays and the discovery of the so-called *solar wind.*

The solar wind is a plasma supersonic flux (essentially ionized hydrogen and electrons) produced by the expansion of the solar corona in the interplanetary space. In its radial motion (with a velocity approximately between 250 and 800 km/s) the solar wind carries with it the Sun magnetic field, originating an interplanetary magnetic field. The region surrounding the Sun, with dimensions varying between 100 and 1,000 AU[1] permeated by the solar wind and the interplanetary magnetic field, is named *heliosphere.*

The hypothesis that the Sun emitted, at least occasionally, plasma clouds, had already been advanced early in the twentieth century by some geophysicists [K. Birkeland (1867–1917), S. Chapman (1888–1970), and V.C.A. Ferraro (1907–1974)] to explain geomagnetic storms. Chapman and Ferraro [1] proposed in 1931 that burst of particles emitted from the Sun would cause brief compression of the Earth's magnetic field (sudden storm commencement), often preceding large geomagnetic disturbances called *magnetic storms.* According to their model (now known to be erroneous), solar wind would only occur temporarily in connection with flares or other specific solar phenomena.

In 1951, L. Biermann [2–4] (1907–1986), studying the motions of CO^+ clouds in the tail of comets, proposed that they were due to the interaction of comet gases with a continuous plasma flux coming from the Sun. He estimated that the continuously blowing solar wind had a velocity around 500 km/s, which is a fairly close value to

---

[1]The Astronomic Unit (AU) is equal to the mean distance of Earth from the Sun. 1AU = 148.000.000 km.

what is known today. Biermann's proposal is now considered to sign the beginning of the modern view of solar wind, a name coined by E.N. Parker (1927–) only in 1958, when developing the theory of the continuous solar wind.[2]

One important discovery which was essential in the first studies of the effects of the magnetic field existing in the space surrounding the Earth was made in the late 1940s, concerning the neutron flux present in the atmosphere. Neutrons' presence had been observed since 1933 [6–8]. Experiments from the sea level up to about 11,000 m (35,000 ft) showed that the neutron number increases rapidly with height.

Due to their short lifetime (15 min) neutrons are mostly produced in the atmosphere by primary cosmic ray particles interacting with oxygen and nitrogen nuclei. The slowing down of these neutrons in the atmosphere was studied by Bethe et al. [9]. Through a series of measurements covering the period between December 1947 and November 1949, J.A. Simpson [10] from the University of Chicago showed that the neutron component has the largest latitude effect of any secondary component in the cosmic rays and that the neutrons are directly produced by the primary nucleonic component of low energy hitting the top of the atmosphere and therefore is most sensible to any flux variation of primary radiation. Simpson argued that more than 90% of the neutrons in the atmosphere are the disintegration product of evaporation type stars produced by nucleons of energy of the order of 300 MeV. On the basis of these results and of other studies, a neutron intensity monitoring network was already developed in 1948, through which it was possible to measure the time response and the intensity of nucleons produced by the low-energy portion of the primary cosmic ray spectrum, which is the portion that is sensible to changes in the magnetic space field.

The network consisted of six neutron monitors (Climax, Colorado; Sacramento Peak, New Mexico; Mexico City; Huancayo, Peru; Wellington harbour; Chicago).

On 23 February 1956, a solar flare[3] occurred that produced a dramatic increase in the cosmic ray flux. Large increases of cosmic ray intensity had already been observed in association with solar flares since 1942 [11,12]. The increase associated with the solar flare that took place on 23 February 1956, was the largest of all intensity increases.

At the time of this event, the sixth neutron monitor of the previously mentioned network, identical to the units in Chicago and Climax, was returning with the US Antarctic Expedition and was operating in the harbour of Wellington, New Zeeland. In addition, a neutron detection apparatus was carried by a balloon over Chicago during the cosmic ray increases [13–15]. The results of the analysis allowed deducing the primary-particle intensity spectrum as a function of particle rigidity

---

[2]A story of the solar wind has been described by E.N. Parker [5]. See also A.J. Dessler, Solar Wind and Interplanetary Magnetic Field, Rev. Geophys. **5**, 1 (1967).

[3]A solar flare is a sudden release of energy from a localized region on Sun, in the form of electromagnetic radiation and usually also of energetic particles. Flares occur in active Sun regions, especially at the boundary zones between solar spots of opposite magnetic polarity; their frequency changes during the solar cycle in a similar way as the spots.

for different times during the time of enhanced intensity. The study with neutron monitoring of this and some of the previous events showed that the particles producing the intensity increase occurred predominantly in the low-energy portion of the cosmic ray spectrum. The measurements brought to the discovery of the heliosphere and of the structure of the interplanetary field that encloses totally the solar system at large distances beyond the Earth orbit [13]. E.N. Parker,[4] to explain the modulation by the Sun of galactic[5] cosmic rays, developed quantitatively the first hydrodynamic model of the coronal expansion of the plasma and introduced the term *solar wind* to describe the phenomenon [16–18].

The most renowned scientist in space-physics, the British–American geophysicist Sydney Chapman [19], had developed a model of the solar corona based on the assumption that it was in hydrostatic equilibrium. The extrapolation of the pressure values obtained by Chapman, at large distances from the Sun showed, however, that these are much greater than the very small pressure of the interstellar gas: therefore, the corona cannot be in equilibrium, but should expand in space, and Parker developed an expansion model according to which the corona emits a particle plasma that travels very fast in space [20].

At first he considered a model for the inner solar system which required a field free cavity of radius greater than the sun–earth distance enclosed by a continuous barrier region of irregular magnetic field (B of the order of 10^{-5} G) [13].

Later this model had to be modified and Parker's theory [21] found a brilliant confirmation in its general lines, when the interplanetary space started to be explored by satellites. In fact, between 1960 and 1962, the existence of solar wind was demonstrated by observations made by the Soviet spacecrafts Lunik II and III and Venera I (K.I. Gringauz [22]) and by the American Explorer X (B. Rossi [23]) and Mariner II (G. Neugenbauer [24, 25]). The measurements with Mariner II, in particular, confirmed the continuous flow of solar wind during its 4-months trip to the planet Venus in 1962 and gave a precise image of the properties of the solar wind, in agreement, at least qualitatively, with Parker's forecasts.

The plasma of the solar wind may be considered an electrically conductive fluid moving in the presence of the Sun's magnetic field. The electric current induced in the fluid gives rise to a magnetic field that produces forces which change the state of motion itself. This coupling between conductive fluids and magnetic field generates phenomena, pointed out by H. Alfven (1908–1995), which had not been studied previously in electromagnetism or hydrodynamics, which founded a new discipline called *magnetohydrodynamics*. In 1942, Alfven ([26], see also [27])[6] discovered

---

[4]Eugene Newan Parker, American astrophysicist, was born in 1927. From 1962 he was professor at the Department of astronomy and astrophysics at the University of Chicago.

[5]We speak of galactic rays to distinguish from the particles emitted by our Sun.

[6]H. Alfven winner of the 1970 Nobel Prize in Physics with L. Nèel has played a central role in the development of several modern fields of physics, including plasma physics, the physics of charged particle beams, interplanetary and magnetospheric physics. He is regarded as the father of magnetohydrodynamics.

theoretically a peculiar wave phenomenon (*magnetohydrodynamic waves*), later obtained also in the laboratory, that plays a fundamental role in the Universe.

A very important result is the so-called *freezing of the magnetic field*, in which the magnetic force lines are *glued* to the fluid particles and move together with them (*Alfven theorem*). It is possible to regard magneto-hydrodynamic waves as oscillations of magnetic force lines on which a conducting fluid has been "glued".

This way the solar magnetic field is frozen in the plasma of the solar wind and is carried on in space.

Parker considered that the plasma escaping from the Sun "pushed" the galactic arm field, that otherwise would penetrate all interplanetary space, outward, creating a field free cavity which encloses at least the inner solar system. The field free cavity was surrounded by an irregular magnetic field barrier region. We now know that this "empty" cavity is indeed a very complex environment filled with a wide range of plasma, waves, and field conditions.

The origin of the solar magnetic force lines is in the Sun and, therefore, the Sun rotation is to some extent transmitted to the solar wind. The magnetic field transported by the solar wind gets wrapped into an Archimedean spiral. The speed with which the solar wind participates to the Sun rotation at the distance of 1 AU (that is at the distance of the Earth orbit) is rather small, of the order of 1 km/s (to be compared with the mean radial velocity of 400 km/s). This, however, has an important consequence for the Sun which suffers a continuous loss of angular momentum which slows its rotation. The freezing of the solar magnetic field generates an interplanetary field that is the extension in space of the Sun field. If the corona was static, the strength of the solar field (of a dipolar type, as the one of the Earth) would change as the cubic inverse of the distance and at 1 AU would be negligible. Parker showed that because of the freezing operated by the solar wind, the force lines of the solar field are greatly modified due to the radial motion of the solar wind and of the simultaneous rotation of the Sun and the strength of the field decreases much more slowly with distance than that of a dipole. The plasma emitted by some area of the Sun distributes in the interplanetary space along a spiral, so that the lines of force carried on by it must assume the same shape. For a plasma velocity $v = 400$ km/s at 1 AU the magnetic strength is about 5 nT, and the angle made by the lines of force with the radial direction is approximately 45°. The measurements made with artificial satellites have fully confirmed these expectations. They have also demonstrated other interesting properties that we will not discuss here.

The expansion of the corona in general is neither uniform (because some regions of the Sun emit a wind faster than others) nor stationary (due to the presence of transient phenomena). A notable consequence of these inhomogeneities is the formation of shock waves that are generated when the fast plasma, emitted for example by a *flare*, compresses behind itself the slower plasma of the solar wind.

The first interplanetary shock wave was revealed in experiments with spacecrafts and showed the interplanetary origin of the Forbush decrease of the intensity [28]. The same experiment proved that the solar modulation was approximately heliocentric and not geocentric.

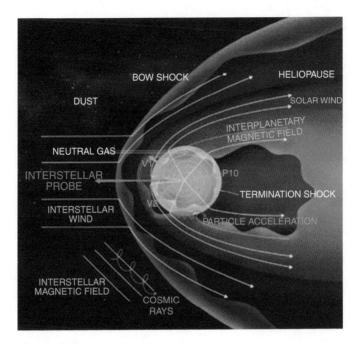

Fig. 12.1 A sketch of the space surrounding our Earth

Therefore, in the solar system there is an interplanetary region permeated with the solar wind and the magnetic field carried by it, a transition region, where the wind interacts with the geomagnetic field and finally the *magnetosphere*, where the plasma does not enter and which is dominated by the terrestrial magnetic field. The structure of the magnetosphere started to be detected by Explorer X, in 1961. The exact structure of these regions is however very complex and it is still under study.

The *heliosphere* that, as said, is the region surrounding the Sun, permeated by the solar wind, was believed for decades to have an elongated shape, as shown in Fig. 5.3 of Chap. 5. In 2009, data from CASSINI and IBEX (the Interstellar Boundary Explorer) spacecrafts raised doubts on these pictures. Its structure—now the object of some controversy—depends on the interaction of two plasmas: the solar wind which expands radially from the Sun with supersonic speeds (mean speed about 400 km/s) and the interstellar plasma (a rather rarefied ionized gas that fills the interstellar spaces): the two plasmas cannot co-penetrate and must create an interface (a discontinuity), called *heliopause*, which constitutes the boundary of the heliosphere (see Fig. 12.1).

The particles of the interstellar medium, although with a very low density, have a constant pressure associated with them; on the contrary, the pressure from the solar wind decreases with the square of the distance from it. As the solar wind begins to drop out with the interstellar medium, it slows down before finally ceasing altogether. When the pressure of the interstellar medium is sufficient to slow the solar wind down below the speed of sound (that it is around 100 km/s) a *shock wave* is produced. The point where the solar wind slows down is the *termination shock*. It is believed to be 75–90 AU from the Sun. In 2007, Voyager 2 passed through the Sun's termination shock. The point where the interstellar medium and the solar wind pressure balance is the *heliopause*; the point where the interstellar medium, travelling in the opposite direction, slows down as it collides with the heliosphere is the *bow shock*.

The heliosphere is a big magnetic bubble in the interstellar wind formed by the solar wind and the solar magnetic field transported by it. The size of the heliosphere was studied by space missions and increased as Pioneer and Voyager crafts explored larger and larger areas of it. In the late 1970s, when the most distant spacecraft was only about 10 AU away, the size of the heliosphere was assumed to be only 20–25 AU, it later increased to about 50 AU, and now it is believed to extend to about 100 AU.

The magnetosphere has a very complex shape. At some distance, when the particles (protons and electrons) that constitute the solar wind meet the Earth magnetic field, they are deviated and separated by it creating two currents, one made by the electrons and the other one by the protons that in turn create a magnetic field which tends to cancel out the terrestrial field which deflected the particles. The complex interplay of these effects causes the extension of the earth magnetic field to be limited in the direction of the Sun, while it extends on the other side, assuming a rather complex shape. On the other hand, the solar storms, with their emissions of further particles, come to complicate the situation even more. One of the results of this intricate state of affairs is that the detailed explanation of polar aurora is much more complex than the simple Stoermer's theory.

A second consequence is that the geomagnetic effects on cosmic rays cannot be explained assuming that the Earth's magnetic field is a simple dipole.

Discrepancies between the measured intensity distribution of cosmic rays and the one expected from the representation of the geomagnetic field as a dipole had already been pointed out by Johnson [29] and were demonstrated by Simpson [30].

The galactic cosmic rays are influenced by the solar wind and the interplanetary magnetic field when entering the heliosphere. This influence which is seen, e.g. in the change of cosmic ray intensity and spectrum is called the *solar modulation*.

The theory of solar modulation is well developed and is based on the transport theory suggested by Parker in 1965. Calculations are rather difficult and several approximations are used.

12.4   More on Forbush Decreases

For some years it was known that the cosmic ray intensity in the atmosphere of the earth changes with time, but it was only after the Second World War that it could be shown that the changes were due to variations of the primary cosmic ray intensity and therefore not meteorological in origin or induced by changes in the geomagnetic field. At the same time it became clear that the variations in the primary spectrum were related somehow to solar activity, though apparently many effects occur simultaneously and the solar connection is not a simple one. Several variations with time of cosmic ray flux were found. We briefly summarize them:

A small diurnal variation with an amplitude of the order of 0.5 % of the total cosmic ray intensity [31, 32].

Superimposed but independent sequences of 27 day variations with amplitude as high as approximately 15 % of the total intensity. The variation was associated with the period of solar rotation [33].

Eleven or twenty-two years variation with the general cycle of solar activity. The amplitude of modulation depends on the level of solar activity. When the number of sunspots is at a maximum the cosmic radiation is at a minimum.

The amplitude of modulation is very different for different energies. For instance, the modulation is only a few percent over the solar cycle for particles with energy of several tens of GeV/nucleon, while it can be a factor of 100 larger for 300 MeV particles.

Forbush decreases.

Forbush [34] linked the sudden decreases of the cosmic ray intensity, the *Forbush decreases*, that often present themselves in association with geomagnetic and aurora perturbations, to bursts of particles accelerated at the cosmic ray energies by solar flares. It was thought that they were the result of changes in the geomagnetic rigidity cut induced by the geomagnetic storm. However, after the discovery of the latitude effect of the nucleonic component of low energy in 1948, J.A. Simpson [15] showed, with measurements of neutrons, that these intensity variations originated in dynamic interplanetary processes controlled by Sun.

At first, Parker [32] proposed that the Forbush decreases and the 11-year variation of cosmic ray intensity represented depressions of the steady galactic intensity produced locally, within a few Earth's radii, by interplanetary magnetic gas clouds captured by the terrestrial gravitation field. Later, he showed that the hydrodynamic flow [17] of gas outward in all directions from the Sun (the solar wind), stretching out the magnetic force lines of the solar magnetic field and leading to an essentially radial magnetic field in the inner solar system, results in a reduction of the cosmic ray intensity during the years of solar activity.

Hence, the *Forbush decreases* are due to the propagation through the interplanetary space of clouds of fast plasma that the Sun emits during flares, containing "frozen" magnetic fields that produce screening effects on the cosmic radiation near the earth. The decrease of the radiation (that may reach up to 15 %) occurs usually in a few hours, while the return to the normal value occurs in times of the order of days or weeks. The solar modulation of cosmic rays has been carefully studied in the elliptic plane by a number of satellites and special crafts.

The anti-correlation of the solar activity described by the 11 years solar cycle with the intensity of cosmic rays with energies below about 10 GeV, discovered by Forbush [35, 36], is also due to the interplanetary modulation, by the Sun, of galactic radiation. Figure 12.2 shows in an impressive way the anti-correlation between the

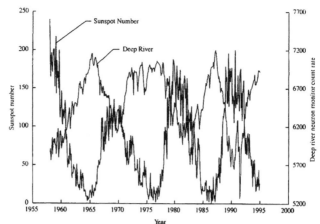

Figure 3. Sunspot number and Deep River neutron monitor count rate over the last few solar cycles.

Fig. 12.2 Sunspot number and Deep River neutron monitor count rate (from J.W. Wilson et al. NASA/CP-2003-212155)

number of sun spots and cosmic ray neutrons as obtained in a study performed at Deep River during recent solar cycles [37]. It can be clearly seen that the maxima of solar activity coincide with the minima of cosmic ray intensity.

The scattering of cosmic rays produced by the irregularities of the interplanetary magnetic field changes their intensity on earth. The variations with the solar cycle are caused by variations of the amplitude of the irregularities, and there is more scattering at solar maxima than at minima.

The diurnal variation of cosmic ray intensity is due to the relative tilt of geomagnetic dipole with respect to the solar wind direction during daily rotation. The amplitude depends on the temporal intensity of the local solar wind.

A longer sidereal variation is related to solar rotation as the emitted coronal plasma depends on local features in the solar surface at the time of the emission. The solar wind expands as an Archimedean spiral that co-rotates with the solar surface, and long lived surface features will show a 27-day recurrence in the local cosmic ray intensity.

The study of time variations of cosmic ray intensity never stopped [38–43]. More recently a latitude survey of atmospheric radiation was made by Wilson [37] at the start of a solar cycle (June 1997), using also a neutron spectrometer that allowed to measure the neutron spectrum [44], which had not been detected yet. The spectrum is somewhat altered with changing latitude and altitude especially near the Earth's surface where, due to the presence of water, neutrons are slowed.

Solar cosmic rays were first discovered on 28 February 1942, as short term increases in the sea level ionization. The correlation with solar flares was identified by Meyer in 1956 [13]. The cosmic ray intensity returns to normal between tens of minutes to days, as the acceleration process ends and accelerated ions disperse throughout the interplanetary space. So far more than 60 events have been registered from 1942 to 2002.

The acceleration region is supposed to be not on the Sun but in the coronal mass emissions from the active surface of the Sun that propagate in the interplanetary space [45].

12.5   The Van Allen Belts

The discovery by Van Allen [46] of the trapped radiation around the Earth was one of the first great results of space research.

The first American satellite, the Explorer I launched in 1958, at the request of the physicist James Van Allen, brought on board, among other things, a Geiger–Müeller counter to study cosmic rays at high altitudes.

James Alfred Van Allen was born at Mount Pleasant, Iowa, on 7 September 1914. He studied at a local College, and in 1934 entered Iowa University to study physics, earning his PhD in 1939. Soon after he worked under Merle Tuve at the Terrestrial Magnetism Department of the Carnegie Institution of Washington and took an interest in geomagnetism, cosmic rays, and solar physics. Merle Antony Tuve (1901–1982) is known mainly for his techniques of radio-wave exploration of the upper atmosphere made in collaboration with G. Breit (1899–1981). In 1925 they made experiments of range-finding, transmitting a train of pulses of radiowaves, and determining the time each pulse reflected from ionized layers in the high atmosphere took to return. This way, they obtained a direct verification of the existence of the ionosphere. The method was a precursor of the pulsed radar principle.

During the war Van Allen, from 1940 to 1945, worked on proximity fuses at the Applied Physics Laboratory of the Johns Hopkins University, becoming an expert in the subject. In 1946, with the support of M. Tuve, he started developing a new high performance sounding rocket, the *Aerobee*, to be used exclusively for scientific purposes. After the war, still at the Johns Hopkins University, he became interested in high atmosphere researches using sounding rockets and studied cosmic rays in Peru (1949) and Alaska (1950). In 1951 he was offered a chair in physics at the University of Iowa. There he started research for measuring the primary cosmic ray intensity at high latitudes in the atmosphere, using small military-surplus rockets (Rockoons) carried to an altitude of about 50,000 ft (15 km) by a balloon and launched from there to reach a summit altitude of some 250,000 ft (75 km). In the summer 1952, with the two students Leslie Meredith and Gary Stein and the lab technician Lee Blodgett, he made the first experiments [47,48] in Greenland (1952–1957) and then in the Arctic (1957).

In 1956 he made a formal proposal for a cosmic ray investigation using one of the early US Earth satellites and in 1958, his project having been approved, a Geiger–Mueller counter was placed on Explorer I to register the intensity of cosmic rays in space.

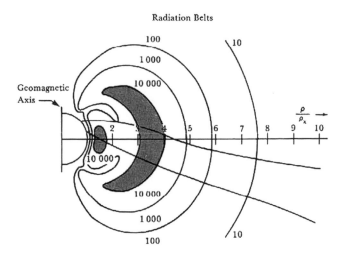

Fig. 12.3 The sketch of the van Allen belts from the Nature paper (from J.A. Van Allen and L.A. Frank, Nature **183**, 430 (1959))

On 31 January 1958 the satellite was launched from Cape Canaveral. The apparatus was working normally giving the rate of counting as the rocket was travelling upward. Suddenly, about at 1,000 km, a fast rise in the intensity was observed and at 1,300 km over the earth surface, the instrumentation ceased to work [49]. What was happening?

Bad working? Failure? Sudden absence of the cosmic radiation? *"No! No one of these hypotheses is true like"* decided James Van Allen *"the counters are saturated by a corpuscular ionizing radiation different from cosmic rays. The flux becomes too large over some altitude for the instruments are able to register all the pulses; otherwise stated the counter death time[7] is too long"*.

The following experiments confirmed this explanation.

The satellites Explorer III (26 March 1958), Pioneer I (11 October 1958), and Pioneer III (6 December 1958) showed the existence of a second zone of radiations placed beyond the first one between 24,000 and 30,000 km from the centre of earth [50, 51] (see Fig. 12.3).

The results were confirmed also by the Russian mission [52] with Sputnik III (15 May 1958).

James van Allen continued to study the distribution and properties of the trapped radiation in the belts he discovered. He died in 2006.

A detailed study of the properties of a dipolar type magnetic field, which is a rough approximation of the field near our Earth, shows that areas around the Earth should be expected in which low-energy particles may be trapped (see Chap. 5). Radiation zones of this kind were anticipated by Stoermer [53] in his study on the

---

[7]The death time of a detector is the time interval needed to the detector to come back to the initial conditions to be anew able to detect a new event, after having detected a particle.

motion of charged particles in a dipolar field. He found that near the Earth should exist regions in which motions for particles are allowed, which however do not connect with outside. Later Hannes Alfven [54] elaborated an approximate but very general theory for the motion of charged particles in a magnetic field and confirmed the possibility of trapping. However, only after 1957 these results were explicitly applied to the Earth. N.C. Christofilos [55, 56] (1916–1972) was the first to clearly discuss this possibility and his predictions lead to an experiment denominated *Argus* in 1958. In the Argus experiment a small yield atomic bomb of a few kilotons was planned to explode at some altitude, injecting electrons in the atmosphere which should thereby be trapped near Earth. However, before the Argus experiment was done, Van Allen [57] found the zones with the Explorer's measurements and his results came as a surprise to many and stimulated a vast amount of work, both experimental and theoretical. The Argus experiment was done later in August 1958 and confirmed the possibility of trapping electrons in closed trajectories near the Earth [58].

At the time of the discovery, the belts were considered two huge doughnut-shaped rings with the earth magnetic axis as common axis placed at a mean distance from the Earth surface of about 3,000 km for the inner belt and about 18,000 km for the external belt. The external belt was found [59] to consist of electrons with energy > 10 keV up to 2 MeV, with a flux of about 10^{10}–10^{11}/cm^2/s and was responsible for the aurora production. Protons are contained in the inner belt. They are present at energies up to the order of 700 MeV. Also ions are present, mainly helium and oxygen, as detected by satellites AMPTE that explored the belt between 1984 and 1987.

Measurements with satellites Injun (29 June 1961) and Explorer XII (16 August 1961) showed that the particles extend also outside these two principal belts [60]. The external limit in the equatorial plane is well defined and coincides with the end of the geomagnetic field that is the external limit of magnetosphere.

The particle flux intensity in the belts is several thousands times larger than the normal cosmic radiation flux at low altitudes and is such to constitute a danger for the human body. Particles of energy 10–100 MeV pierce easily the walls of satellites and space stations, and therefore the van Allen belts are zones that spatial journeys with or without men on board, must accurately avoid.

The belts are essentially zones in which high-energy particles, collected from the solar wind, remain trapped due to the Earth's magnetic field and travel around in a very complex motion (see Fig. 12.4).

The problem of the origin of the particles in the belts has been studied in a number of space missions, finding that the magnetic status in the regions surrounding the earth is very complex and in continuous dynamic evolution.

Due to instabilities, particles may escape from the belts. The problem of how the particles are captured in the belts is not yet completely clarified. One hypothesis [61] was that cosmic ray protons, once in the atmosphere, may interact with air molecules and produce nuclear reactions in which some neutron is emitted in such direction as to escape from Earth. Being neutral, the neutrons may easily enter the trapping zones and because of their short lifetime decay into an electron and a proton, thus feeding

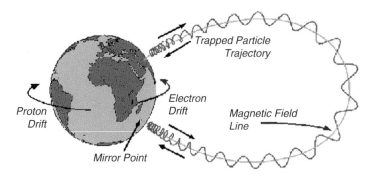

Fig. 12.4 Trapped particle trajectory

the population in the belts. These neutrons are called "albedo[8] neutrons". This mechanism is however insufficient to explain the density of the observed particles. Today it is believed that the principal source is solar wind: solar energetic protons associated with flares and coronal mass ejections [62] may become trapped in the Earth's magnetic field by interplanetary shock distortion of the field [63].

The outer belt is populated by electrons with a plasma sheet source [64, 65].

The density of particles into the belts results from a dynamic equilibrium between the creation and escaping processes, and has been found to suffer sudden changes in the event of man-made nuclear explosions at high altitudes.

The study of Van Allen belts is still a subject of research.

12.6   The Research on the Nature of Primary Cosmic Rays

We already mentioned the measurements made at Stuttgart (49°N geomagnetic latitude) by G. Pfotzer [66], who measured coincidences in a counter telescope set to detect essentially only the radiation coming along the vertical direction. The curve, here shown as curve B in Fig. 12.5 [67], showed that the total vertical radiation presented a maximum at a height of about 100 mmHg. H. Carmichael and E.G. Dymond [68, 69] repeated this type of measurement in northwest Greenland (geomagnetic latitude 88°N) with similar balloon launching techniques, but with radio telemetering of the data to a ground receiver, instead of photographic recording in the balloon, as Pfotzer had used. They obtained similar results, except for a scale factor. Comparing the two findings one found that there was negligible difference between the normalized areas under the curves at 49° and 88°, from which it was concluded that no new radiation entered at latitudes above 49°. For this latitude,

---

[8]The name comes from an analogy with the optical case: the diffusion coefficient or reflection power of a surface such as moon or earth from sun light is called "albedo".

Fig. 12.5 Hard vertical intensity from Schein et al. (*curve A*) and total intensity as obtained by Pfotzer (*curve B*) (Reprinted with permission from M. Schein, W.P. Jesse, E.O. Wollan, Phys. Rev. **59**, 615 (1941). Copyright 1941 by the American Physical Society)

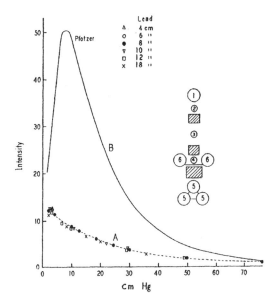

the critical energy for particles incident from the vertical is 3 GeV according to the Lemaitre–Vallarta theory, whence it was deduced that the energy spectrum of the primary radiation is zero below some energy not less than 3 GeV.

R. Serber [70] attempted to explain the shape of the vertical distribution as obtained by Carmichael and Dymond, Bowen et al. [71, 72], and Pfotzer in terms of cascade showers set up by primary electrons using refined Snyder [73] solution of the multiplicative diffusion equation of high-energy electrons and gamma rays in their passage through the atmosphere. With some assumptions, he found reasonable agreement with the experimental curve and so his results were considered at the time to be a decisive argument for the electronic character of the primary radiation.

However, when this analysis was applied to the data taken near the magnetic equator with counter telescopes by H.V. Neher and W.H. Pickering [38], the hypothesis of primary electrons appeared entirely inadequate.

Meanwhile, M. Schein and V.C. Wilson [74] made experiments with counters carried on by planes to study the production of muons in a direct way from the measurements of the variation of their intensity with height. At that time they agreed with Heitler's assumption [75] that the "heavy electrons" (as they called the muons) were formed by secondary photons.

In 1939, E.G. Dymond [76] reported results on the vertical hard component under 10 cm lead[9] at 50° geomagnetic latitude as a function of altitude, obtained with preliminary flights. He used a telescope counter and found a maximum for the vertical hard component at the same position for the maximum of the vertical total

---

[9]The lead was used to cut down the soft component.

intensity. However, he did not consider his results as final and was forced by the war to discontinue his experiments. His results were not disproved by A. Ehmert [77] who, in flights conducted at Friedrichshafen (geomagnetic latitude 49°N) in 1939, found a levelling off of muon intensity under 9 cm of lead at altitudes from 10 cmHg to 2 cmHg, after a continual rise from ground level.

During 1939, M. Schein et al. [78, 79] presented the first of a series of reports describing their extensive experiments in Chicago on the intensity of the hard component under 8 cm (and other thicknesses) of lead as a function of altitude, at 52°N. The 1941 report [67] showed that—contrary to their 1940 finding, Dymond's preliminary report and Ehmert's results—the intensity of the hard component does not pass through a maximum or even flattens at great heights, but keeps increasing up to the highest altitudes reached (2 cmHg), where the intensity is 16 times that at ground level.

The 1941 data of Schein et al. are shown in Fig. 12.5. Curve A (dashed) is the hard vertical intensity for various lead thicknesses as a function of pressure in cmHg. Curve B (solid) is the total vertical intensity as obtained by Pfotzer. One of the arrangements used in the experiments is shown on the right.

It was conjectured that some of the measured penetrating radiation might consist of high-energy electrons, and so Schein and collaborators considered this possibility. They conclusively showed that most particles near the top of the atmosphere pass several centimetres of lead and do not multiply traversing it, and do not start the showers typical of the high-energy electromagnetic radiation. Moreover, at least up to the higher fraction of examined atmosphere (2 cmHg), the intensity of the penetrating component increases continually. So, the experiment gave proofs against the hypothesis that electrons were the primary rays, suggesting instead protons, a hypothesis made also by Thomas H. Johnson [80] in 1939, who, considering the result of the east–west effect, through a series of arguments, deducted that the primaries that originated the hard component were protons.

Protons and alpha particles had been detected for the first time in the beam of cosmic radiation by photographic emulsions, carried up to the stratosphere by one of the Piccard balloons (the Explorer 2) in 1936, by Wilkins and St. Helens [81].

Schein and collaborators concluded [67]:

> "*The mesotrons* [the name then used for muons] *themselves cannot be the primaries because of their spontaneous disintegration. Hence it is probable that the incoming cosmic radiation consists of protons*".

They continued observing that the electrons, known to exist in large number in air at high altitudes, must be of secondary origin, suggesting they arose mainly from the decay of the muons and knock-on processes.

Since then, it was generally assumed that the most of primary cosmic radiation was made by protons.

Later it was confirmed that very few, if any, high-energy electrons are present at the top of the atmosphere [82]. B. Rossi—maybe remembering the controversy with Millikan concerning the gamma nature of primary rays—concluded his paper:

We wish to emphasize that it is not necessary to postulate the existence of any electrons or photons in the primary radiation since cascade showers have been observed to be produced in lead by penetrating particles believed to be high energy protons [83].

Therefore, a first natural hypothesis to explain the cosmic radiation at sea level was that the primary radiation consisted nearly exclusively of protons that, interacting with the air atoms at the top of the atmosphere, formed mesons, which were the principal source of the secondary radiation. Such an explanation in terms of a single proton component was suggested by W.F.G. Swann [84,85] and J.F. Carlson and M. Schein [86]. At those times, however, only muons were known. Pions and the other mesons were still undiscovered and therefore the model was unable to explain all the observed phenomena.

After the war, the studies on primary rays continued with balloons at high altitudes, with airplanes and utilizing V-2 rockets. At the same time the study of extended showers started on the ground. Using balloons the problem was to discriminate between the primaries and the secondary particles. Therefore, balloons were sent at increasing altitudes until, at about 36 km, the residual atmosphere constituted a thickness of matter that was thinner than the mean free path of the incident particles. This means that the primary particle had a high probability to be detected without suffering any collision. In these conditions, the formation of secondary particles was negligible and what was observed could reasonably be considered the primary particles. For these measurements also nuclear emulsion packages were used.

Starting in 1960, satellites and spacecrafts were used, definitely eliminating the problem of secondary particles. The first spacecrafts were the American satellites of the series IMG and OGO and the Soviet Proton. These experiments gave a fundamental contribution to the knowledge of cosmic radiation because they made possible to perform long time measurements not contaminated by atmosphere. At last it was possible to measure directly which were the primary particles!

Protons were soon confirmed as the principal component.

The possibility that the primary cosmic ray beam contained also nuclei of elements heavier than protons had already been pointed out by H. Alfven [87] and W.F.G. Swann [88], and such nuclei were observed in 1947 by M.A. Pomerantz, who presented his results in some reports for the American Navy in 1947, publishing them only as a short letter in 1949 together with F.L. Hereford [89]. His results were anticipated by the ones obtained by Phyllis Freier (1921–1992)—then a graduate student—and collab. [90]

With stratospheric flights with balloons bringing nuclear emulsions, exposed to about 27 km (90,000 ft.), they found [90] He nuclei and traces of nuclei that they estimated even very heavy up to $Z = 40$ of a great penetrating power which could not be produced by collisions with protons, because the tracks belonged to nuclei much heavier than the air nuclei. The tracks were observed both in a cloud chamber and in Ilford nuclear emulsions.

Figure 12.6 is an example of a nucleus with an estimated charge $Z \approx 10$–15 in the emulsion.

Fig. 12.6 A medium-heavy
(estimated $Z \approx 10$–15)
nucleus ending in the
emulsion (Reprinted with
permission from P. Freier,
E.J. Lofgran, E.P. Nay,
F. Oppenheimer, H.L. Bradt,
B. Peters, Phys. Rev. **74**, 213
(1948). Copyright 19 by the
American Physical Society)

FIG. 3. A medium heavy track ($Z\sim10-15$) ending in
the emulsion. The particle has a low velocity; δ-rays are
almost entirely absent. Thinning of track towards the end
of the range suggests gradual filling of electronic shells.

The observed nuclei belonged to elements of atomic number between magnesium
and iron and to lighter nuclei, especially helium.

After this discovery, Helmut Bradt and Bernard Peters (1910–1993) [91,92] continued the measurements confirming the existence of heavy nuclei, and extending
and improving the method for the nuclear charge identification [93].

The measurements were carried on at altitudes above 25 km, with counter
telescopes. The relative abundance of nuclear species appeared to be closely related
to the abundances of elements in the visible universe as deduced from the spectra of
stars and planetary nebulae and from the composition of meteoric material.

The initial estimates of Freier of the charge of heavy nuclei were however much
too high, as Bradt and Peters pointed out.

Heavy nuclei in the stratosphere were subsequently observed also with cloud
chambers [94], proportional counters [95], and scintillation counters [96].

The identification of the single species of nuclei was at first a difficult task.

After hydrogen, by far the most abundant element in the primary cosmic ray
beam is helium. The flux of this component was studied in a number of works, in
emulsions [97], with proportional counters,[10] and scintillation counters [99].

---

[10]G.J. Perlow, L.R. Davis, 1951 in [98].

Fig. 12.7 Charge spectrum of particles observed at high height (Reprinted with permission from H.L. Bradt and B. Peters, Phys. Rev., **80**, 943 (1950). Copyright 1950 by the American Physical Society)

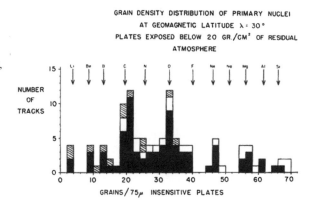

Bradt and Peters [93] reported a charge spectrum of particles observed at latitude 30° under 20 gr/cm² of residual atmosphere (see Fig. 12.7). The black and shaded areas in the histogram represented the result of a systematic survey including all particles traversing a given plate in the emulsion stack; the shaded areas represent nuclei identified as fragments from heavier nuclei which have undergone collisions in the stack before they reached a particular plate in which the survey was carried on. The open squares represent additional particles observed in a less systematic manner.

Bradt and Peters could not find any nuclei heavier than iron, cobalt, or nickel among nearly one thousand particles, but some were reported by other researchers [94, 100].

The elements carbon and oxygen were definitely found by Bradt and Peters in nearly equal abundance and represented the strongest component next to hydrogen and helium. The occurrence of neon, magnesium, silicon, and iron was also well established. The few detected light nuclei lithium, beryllium, and boron were almost all attributed to collisions of primary rays in the air.

Dainton et al. [101], on the contrary, found a large flux of these light elements. The problem of their existence in the primary rays originated a strong controversy because they are almost absent in the cosmic abundances and their presence in the primary rays was important to understand the origin of cosmic rays. B. Peters [98], in a survey up-dated to 1951, reported a comparison between the cosmic abundances given by H. Brown [102] and corresponding abundances in cosmic radiation (see Table 12.1). He did not mention the light elements.

The chemical composition of the cosmic radiation was found [103] nearly constant in the energy range from about 200 MeV up to 1–2 GeV. For the most part of the elements there are no great differences between the abundances found in cosmic rays and the ones in the solar system. The important and unexpected discovery was a great abundance of the light elements Li, Be, B relative to the solar one as found by Dainton et al. In cosmic rays, the tendency to "fill the holes" of the abundances of some elements that defect in the solar system was confirmed and extended also to the elements from scandium to manganese which precede the iron [104, 105] in the Mendeleev periodic classification of elements. Such elements

Table 12.1 Relative abundances of elements (first column) in the Universe (second column) and in cosmic radiation (third column). From Peters [98].

Z Element	*Atoms per 10^5* hydrogen atoms	Abundance in cosmic radiation
1 Hydrogen	100,000	100,000
2 Helium	10,000	10,000
8 Oxygen	63	260
7 Nitrogen	46	?
6 Carbon	23	260
10 Neon	2.6–70	30
26 Iron	5	30
14 Silicon	2.9	30
12 Magnesium	2.5	40
All other elem. $Z < 30$	2.7	30
All elem. $30 < Z < 92$	4×10^{-3}	<1

are over-abundant by many orders of magnitude if compared with the solar system abundances. Initially Bradt and Peters, although considering the possibility that the nuclei accelerated at the primary sources suffered collisions in their travel towards the Earth, principally with the hydrogen nuclei present in the interstellar space with consequent nuclear transmutations, assigned the presence of the light elements Li, Be, and B only to interactions in the air layer that still remained over their balloons. So, at first, there was a lack of agreement of experiments designed to investigate the existence or nonexistence of the light nuclei [93, 101]. Therefore, M.F. Kaplon and collabs. [106] decided to measure the flux by using both the Bradt and Peters or the Dainton et al. techniques, with photographic emulsions carried on in balloons. They listed various isotopes of Ca, Fe, O, Mg, C, S, Si, Ne and in the final report they concluded that there exists a finite primary flux of light nuclei ($3 < Z < 5$).

It was recognized that Li, Be, and B between 50 and 200 MeV per nucleon arrive on Earth from the space and their abundance was measured in a number of researches [107–118] using balloons and satellites. A historical advance were satellite measurements of isotopic Li, Be, B in the 1970s [104]. Later Voyager 1 and 2 provided refined data [119, 120], separating the various isotopes ^{6}Li, ^{7}Li, ^{7}Be, ^{9}Be, ^{10}B, ^{11}B, and ^{12}B, collected in 21 years of operation.

Worthy of special mention are the measurements made by the French–Danish experiment aboard HEAO-C space observatory, launched in 1979 [121]. More recently, low-energy data (<200 MeV/nucleon) on the cosmic ray composition came from space experiments such as the Cosmic Ray Isotope Spectrometer (CRIS) [122] collaboration in the NASA mission ACE (Advanced Composition Explorer), operating since 1997, performing very accurate measurements [123, 124]. It was found that there are peaks of abundances in correspondence to C, N, O, and Fe, as in the solar system. The Li, Be, B excess is large and also the elements below Fe, that is scandium, titanium, tungsten, chrome, and molybdenum, are more abundant of the solar mean. Other results [120] were obtained by the HET telescopes on Voyager 1 and 2.

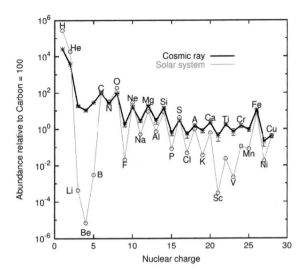

Fig. 12.8 Plot of cosmic rays and solar system abundances (Reprinted with permission from T. K. Gaisser and T. Staney, Nuclear Physics A **777**, 13 (2006). Copyright 2006 by Elsevier)

Light-nuclei data at higher energies (up to a few GeV/nucleon) were measured by balloon-borne spectrometers like IMAX, ISOMAX [125, 126], SMILI [127], and BESS [128]. Other notable studies are [129].

Figure 12.8 shows, for elements up to copper, a comparison of the abundance relative to carbon in cosmic rays and in the solar system.

To make the problem of the determination of the particles present in the primaries more complex, there is the circumstance that the spectrum of cosmic rays with energies <1 GeV/nucleon is strongly influenced by their passage through the interplanetary medium, and this effect changes with the phase of the solar cycle, a phenomenon we call *solar modulation*.

For heavier nuclei, beyond the iron (the more stable nucleus), which has a marked peak, the abundances of elements decrease rapidly so that it is difficult to determine them with accuracy: however, qualitatively the presence of nearly all heavier elements up to uranium has been ascertained since the late 1960s [131–135], and one has succeeded to resolve the distribution of the isotopic abundances of many elements. These last measurements have a notable importance: for example from the relative abundance of the radioactive isotope ^{10}Be the mean dwell time of cosmic rays in the Milky Way was estimated, as we will explain later.

The study of the primary composition is continuing with the aim to increase the statistics on the data that are very difficult to be obtained without great errors due to the low number of arriving particles.

For many years electrons were searched for in the primary cosmic rays. With the help of ionization chambers brought by balloons [136] and then with cloud chambers ([137], see also [138]), electrons were not found at the geomagnetic latitude 55°N, evaluating their possible contribution to less than 0.25 %. In 1961 they were eventually found [139, 140] in an amount of the order of a few percent.

The high-energy cosmic ray electrons are of particular interest because they lose energy much faster than nuclei, producing electromagnetic radiation which shows their distribution throughout the Galaxy. So their study may provide valuable information about the origin and propagation of cosmic rays. For this reason they have been studied in a number of experiments [141].

Today we may confidently say that the primary radiation is made for 86 % by protons, and for the remaining part by alpha particles (11 %), nuclei of heavier elements (1 %), electrons (2 %), and neutrinos (<1 %) [142].

Other important observations pertain to antiparticles; a small portion of positrons was discovered [143]. However positrons are nearly ten times less numerous than electrons.

Also antiprotons have been found [144, 145], but there is no notice of antinuclei of elements heavier than hydrogen.

As we said, already in the 1930s, neutrons were discovered. Paraffin covered photographic emulsions showed five times as much proton tracks as unshielded ones. This was correctly interpreted as evidence for the neutron component in the cosmic ray beam [146, 147] and later on also confirmed by counter experiments [148].

And at the end, in 1966, the deuteron was discovered [149] in the primary rays.

12.7 The Energy Spectrum

To know how the energy of primary particles is distributed is of the uttermost importance. This knowledge is fundamental to verify the validity of any theory on their origin. The number of particles impinging on a square centimetre of surface per second per solid angle unit (steradiant) is often represented as a function of energy, or better of the energy per each nucleon composing the nucleus of the arriving particle (that is energy divided for the mass number of the nucleus).

The first assessments were made by using the geomagnetic effects. The energy spectrum of primaries may in fact be determined using the critical energy values as a function of latitude [150], which may be calculated using the Lemaitre–Vallarta, or less accurately the Stoermer theory. By using the curves of the change of cosmic ray intensity with latitude it is possible, by increasing the latitude starting from the equator, to determine how much radiation enters in addition and from this information it is possible to construct a curve of the energy distribution. This method may give information on the energy spectrum only up to 17 GeV, because we have seen that, for energies higher than this value, primary particles may arrive from everywhere.

The area under the intensity versus altitude curve at given latitude is proportional to the total energy carried in by particles whose individual energy exceeds some critical value, while the smaller area under a curve for a lower latitude is proportional to the total energy carried in by particles of individual energy exceeding a larger critical value. The difference in area between the two curves is hence proportional

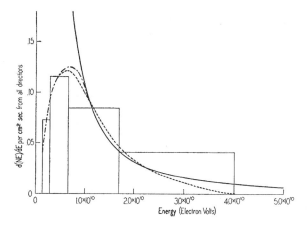

Fig. 12.9 The energy distribution curve in the region of 10^{10} eV (from Hilberry [150]). The full curve gives the power law expression $E^{-2.75}$. The *dotted curve* is the curve of Bowen, Millikan, and Neher and the blocks represent their experimental observations. The *broken curve* is the power law distribution adjusted below 10^{10} eV to fit the experimental observations. (Reprinted with permission from N. Hilbery, Phys. Rev. **60**, 1 (1941). Copyright 19 by the American Physical Society)

to the total energy carried by particles whose individual energy lies between the two critical values for the latitudes under consideration. If an average value for the energy of individual particles can be estimated, the number of particles in each energy range may be deduced.

With this kind of reasoning, Hilberry [150] obtained the curve of Fig. 12.9, which is the improvement of a similar curve constructed by Millikan [151] which gives the energy distribution of the primary cosmic radiation in the energy region up to 50 GeV.

The blocks represent the experimental findings of Bowen et al. [152]; the dotted curve has been drawn in to smooth out the blocks by Bowen. The solid curve is the power law discussed below in (12.1), with exponent $\gamma = -2.75$. The dot–dash curve is the same power law distribution adjusted below 10 BeV to fit the experimental observations. The curve was arbitrarily extended to 50 GeV.

To represent the number of particles of a given energy, an inverse power law was suggested by several authors [153–155] and was considered by Johnson [156]:

$$N(E) = AE^{-\gamma}, \tag{12.1}$$

with $\gamma \sim 3$ for the number $N(E)$ of particles arriving on earth per unit of time, area, solid angle, and kinetic energy E. Later he suggested [157] an exponential formula of the type

$$N(E) = A\exp(-\alpha E).$$

However, Hilberry [150], from data on extended air showers, excited presumably by primaries of energy between 5×10^4 and 5×10^6 GeV, showed that an inverse power law with exponent 2.75 holds for this range. The question of which was the better value of the exponent was debated at long suggesting many different values of the exponent [158–163].

When the first high altitude measurements with nuclear emulsion were obtained in 1948, range measurements provided some data in the low-energy region and scattering measurements were used in the high-energy region. At the beginning large horizontal sheets of emulsions were inserted between absorbers and exposed at high altitudes. On the assumption that the cross section for nuclear collisions in the absorber did not depend on the energy of the primary particle, the observed range distribution of particles coming to rest by ionization was transformed into an energy distribution at the top of the atmosphere. With this method and looking at the relative scattering of alpha particle fragments in the narrow "evaporation" events that were found for particles of high energy, first direct inference on the energy spectrum of primary particles were made. At first, the number of particles/m^2/s/sterad with kinetic energy per nucleon in excess of ε (BeV) was given as

$$N(\varepsilon) = \frac{K}{(1 + \varepsilon)^{1.2}},$$

where K was a suitable constant [98]. Successive measurements agreed with the inverse power law (12.1), although the question of the better form of the spectrum was debated for a long time. The spectrum of the primary radiation continued to be the object of a vast research and today we have a rather detailed spectrum as the one shown in the following Fig. 12.10, which gives a global vision of the energy spectrum of cosmic rays in GeV per particle. The abscissa and ordinate scales are logarithmic. The spectrum extends from about 1 GeV up to about 10^{20} eV. The curve shows an initial inflexion due to the fact that the less energetic particles are not able to penetrate the heliosphere. To be more precise, the direct detection with spacecrafts shows that at low energies the cosmic ray flux is modulated by the solar cycle through the magnetic heliosphere that protect the solar system from charged particles below about 10^9 eV. From a few GeV (1 GeV $= 10^9$ eV) to a few PeV (1 PeV $= 10^{15}$ eV), the cosmic ray spectrum is well described by the law of (12.1).

Today the main part of the curve is represented by (12.1) with a value $\gamma = 2.7$.

The slope of the spectrum changes around 3×10^{15} eV, becoming steeper at higher energies and $N(E) \sim E^{-3}$. The transition region is called "knee". At about 10^{18} eV, the spectrum steepens more in a shape that is called "ankle". The presence of the ankle is well evidenced in Fig. 10.4 Chap. 10 which shows results by HiRes [187] and Fly's Eye.

Some experimental evidences exist on the occurrence of a second knee in the region between 10^{17} and 10^{18} eV [188–191].

Eventually, at about 10^{20} eV, very few particles arrive, so that measurements are very difficult. A drastic decrease is expected due to the GZK cut (see Chap. 10) and the research at these energies is very active due to the extremely low number of

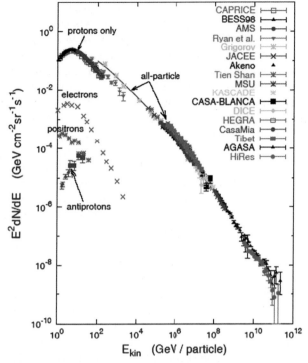

Figure 1. Global view of the cosmic-ray spectrum.

Fig. 12.10 Global view of the energy spectrum of particles in cosmic rays (from T.K. Gaisser [164]). The results of a large number of missions are reported: Caprice [165], BESS98 [166], AMS [167], Ryan et al. [168], Grigorov et al.,[11] JACEE [170], Akeno [171, 172], Tien Shan,[12] MSU [173], KASCADE [174–176], CASA-BLANCA [177], DICE [178], HEGRA [179], CasaMia [180], Tibet [181, 182], AGASA [183], and HiRes [184–186]. (Reprinted with permission from T.K. Gaisser, J. Phys. Conf. Ser. **47**, 15 (2006))

arriving particles. This region cannot be studied by spacecrafts and information is obtained by the study of extended showers.

   The electron energy spectrum extends up to about 10^{12} eV, and is described by an exponent $\gamma = 2.6$. The NASA Fermi Gamma-ray Space Telescope [192, 193] has recently measured the spectrum of electrons up to 1 TeV (1 TeV = 10^{13} eV). The Fermi was designed essentially to measure gamma rays, but it may detect also electrons. Data show that above 100 GeV there is an increase of the flux which does not appear in Fig. 12.10 and has not yet a good explanation. Also a magnetic spectrometer orbiting around Earth in the PAMELA [194, 195] collaboration— a satellite based instrument dedicated to precision measurement of cosmic ray

---

[11]L. Grigorov et al., after T. Shibata [169].

spectra launched in 2006—finds similar results. However, both PAMELA and Fermi are not able to distinguish between electrons and positrons [192, 194]. Precise measurements of proton and helium spectra from PAMELA have recently [196] found that the spectral shapes of these two species are different. The exponent in (12.1) for protons is $\gamma = 2.82$ and for helium is $\gamma = 2.73$. This way, one may appreciate how the field is very live. The energy distribution of primary cosmic particles is the object of an intense research because it gives the key to the production of cosmic rays and the mechanisms of propagation.

Above the knee it is no longer possible to get information by instruments aboard of spacecrafts and it is necessary to relay on the study of extended showers.

Another important issue is the mass composition of the primary radiation. While the protonic dominance at low energies is generally accepted a strong uncertainty still exists on the composition around and above the "knee". B. Peter pointed out [197] that a steepen in the cosmic ray spectrum could be linked to changes in its composition. Several experiments have attempted to determine the mean cosmic ray composition through the "knee" region 10^{15}–10^{16} eV of the spectrum. Most observations and arguments seemed to support proton dominance at the highest energies. In the region above the "knee", the Fly's Eye experiment [198] reported a changing composition from a heavy mix around 10^{17} eV to a proton dominated flux around 10^{19} eV. These results were confirmed [190] by the same collaboration. This point was however confuted and some consensus[13] existed for a composition becoming heavier at energies above the "knee".

Pioneering direct experiments at balloon altitude were made by the GSFC group [199].

Two direct experiments with emulsion chamber instruments, JACEE [117, 200, 201] (Japanese–American Cooperative Emulsion Experiment) and RUNJOB [202, 203] (RUssian-Nippon Joint Balloon collaboration) with measurements in the 100 TeV per particle to 1,000 TeV per particle and with similar exposures did not agree. While JACEE indicated an increase in the logarithm of the atomic mass (lnA) from its low-energy value of 1.5 to a value close to 3, the RUNJOB experiment saw no change in the composition albeit with large uncertainty. Of the indirect experiments, KASCADE [204], Tibet [205], and others favour a heavy composition above the knee and beyond. The KASCADE energy range extends to 100 PeV where their analysis indicates iron dominance. A great confusion still exists. The Auger collaboration [206] seems to indicate a transition, at primary-particle energies of a few times 10^{18} eV from a cosmic ray flux dominated by protons to one increasingly dominated at higher energies by iron nuclei.

We will discuss this point in the next chapter [207, 208].[14]

---

[12]Tien Shan.

[13]See papers presented at the 26th ICRC, Salt Lake City, 1999.

[14]Sokolsky [207, 209].

Cosmic radiation is nearly perfectly isotropic. For energies $<10^{14}$ eV the anisotropy is of the order of 0.05 %. At higher energies it increases gradually reaching 10 % at 10^{19} eV. At higher energies the flux seems to become collimated.

Radiochemical studies of the isotopes produced by cosmic rays in meteorites have shown that their flux has remained constant within a factor 2 for the last billion of years.

12.8   Gamma Rays Appeared and the Gamma-Astronomy was Born

Eventually the gamma rays, so much cherished by Millikan, were found. Not so many, but surely coming from space. The point is that the gamma-ray observation presents greater difficulties than the ones connected with the detection of electromagnetic waves in the other spectral bands. The main problem is that gamma radiation, at variance with that at greater wavelength, cannot be concentrated on detectors with mirrors or lenses and therefore only the photon falling directly on the detecting surface are detected. It is so difficult to distinguish the weak signals coming from celestial sources, from the background of the ambient radiation. Moreover, the gamma radiation is completely absorbed by the atmosphere, so that observations need to be made with balloons or satellites. Only the more energetic rays (energy larger than 1,000 GeV) may be detected on ground; they materialize as electron–positron pairs generating electrophotonic showers. The assignment of the shower to primary gamma rays can be done because photon-induced giant air showers have an evolution profile which is significantly different from nucleon-induced showers, as discussed in Chap. 10.

The study of gamma rays provides clear and in some respect unique information being related directly both to the very high-energy electrons as they interact with photons, matter, and magnetic fields, and to the energetic nuclei as they interact with matter. Moreover our galaxy and the Universe are extremely transparent to gamma rays so that they retain the detailed imprint of spectral, directional, and temporal features imposed at their birth.

Although X-rays have been observed emanating from the Sun since the 1940s, their study started after the Second World War by using rockets. In this way soft (of weak energy) X-rays coming from the Sun [210] were discovered.

Later, the satellite SMM (Solar Maximum Mission)—launched by NASA on 14 February 1980 and operative until 2 December 1989—showed [209] that the Sun, during flares, is a source of gamma rays of various energies: the photons with lower energy (<1 MeV) are generated by bremsstrahlung of relativistic electrons. The ones of higher energy originate in several nuclear reactions; for example the emission at 2.22 MeV (one of the most intense emissions) is produced by neutrons captured by hydrogen atoms.

It was not until 1962 that X-ray sources outside the solar system were definitely discovered [211], using an Aerobee rocket, causing great surprise, and Riccardo Giacconi[15] received the 2002 Nobel Prize in Physics for the discovery. The first discovered source was Scorpius X-1, in the constellation Scorpius. The X-ray emission from this source is 10,000 times greater than its visual emission—while the X-ray emission from Sun is about a million times less—and its energy output in X-rays is 100,000 times greater than the total emission of the Sun in all wavelengths. Nearly at the same time, gamma rays were found through measurements done with Explorer XI [212]. The study of X and gamma-ray sources in space started a new branch of astronomy: the *X- and gamma-astronomy*.

On 21 July 1964, the Crab Nebula supernova remnant—in the constellation of Taurus—was discovered to be a hard X-ray (15–60 keV) source by a scintillation counter flown on a balloon launched from Palestine, Texas, USA [213]. This is likely the first balloon-based detection of X-rays from a discrete cosmic X-ray source. The nebula was observed by John Bevis in 1731, in correspondence to a bright supernova recorded by Chinese and Arab astronomers in 1,054. At X and gamma-ray energies above 30 keV, the Crab is the strongest persistent source in the sky, with measured flux extending above 10^{12} eV. Located at a distance of about 6,500 light-years (2 kpc) from Earth, the nebula has a diameter of 11 light-years (3.4 pc) and expands at a rate of about 1,500 km/s. At its centre the energetic pulsar,[16] to which the supernova explosion gave birth continues to power the nebula's extraordinary luminosity across the spectrum from radio to TeV gamma [215–217].

A compilation of the first experiments was made by Gould [218].

The X-rays from discrete sources such as the Crab nebula or Scorpius X-1 are superimposed on a diffuse background which appears to be nearly isotropic [219].

Systematic observations started with the satellite OSO-III in 1968 when G. Clark, G. Garmire, and W. Kraushaar [220] discovered the first celestial source of gamma rays (the centre of the Milk Way) with energies of the order of 100 MeV, and continued with the satellites SAS-2 in 1972, COS-B (1975–1982) [221]—the first astronomic satellite launched by NASA for the European Spatial Agency (ESA) which at the time had not an independent launch system—and CGRO (Compton Gamma Ray Observatory) (1991–2000). The first observations with OSO-III that had a meagre angular resolution (about 15°), succeeded to detect a flux of gamma rays coming from the equatorial plane of the Milky Way, with a principal maximum in the direction of the galactic centre and a secondary maximum in the opposite direction. More accurate distribution maps were obtained by the American satellite SAS-2 (Small Astronomy Satellite-2) [222], launched 15 November 1972 and

---

[15]Riccardo Giacconi was born in Genoa in 1931. He studied in the University of Milan earning a degree in physics in 1956. Thereafter he went to the US in 1956 on a nuclear physics fellowship. Once his visitor's visa expired, he found work at American Science and Engineering Inc. and under the push of Bruno Rossi started to be involved in space research, especially X-ray studies. He has been professor in various American Universities and in Milan University.

[16]For a pulsar description, see for e.g. R.N. Manchester [214].

operating only until 8 June 1973, that established the existence of emissions by the interstellar medium and allowed a first attempt to derive the cosmic ray distribution and with the European COS-B. To this last satellite is due much of the obtained information. It detected photons with energy between 30 MeV and 5 GeV with an angular resolution of 2°. During its life, COS-B collected more than 100,000 gamma photons (before only 5,000 photons were detected in total). Successive experiments were made with the HEAO-3 and SMM [209] (the Solar Maximum Mission, specific for solar phenomena), launched 14 February 1980, operating until 2 December 1989 and the American GRO satellites, the latter being a huge orbiting observatory able to transport 6 tons of instrumentation for combined measurements from 10 keV (X-rays) to over 100 GeV (gamma rays).

The more intense fluxes are observed near the galactic centre. There are also secondary maxima, for example in the direction of the Vela and Cignus constellations.

The two pulsars emitting the most intense gamma fluxes are in the Crab nebulae and in the Vela [223] constellation. Other sources are in some gas and dust clouds (as the ones of Orion and ρ-Ophiuchi) where the cosmic rays may interact with the gas. Another source is in the Gemina constellation (it has been called Geminga), in which an X-ray source exists, that is noticeable in the visible, and whose gamma rays emit an energy about 1,000 times larger than the one emitted in the X region, and 1,000,000 times more than in the visible. The gamma signal varies with a period of 0.24 s (measured by the ROSAT satellite [224]). Twenty years were needed from 1973 to 1993 to identify it as a rotating neutron star (pulsar) that does not emit radio-waves.

At least two localized extragalactic sources are in the Seyfert galaxies and in a quasar.

A detailed study of COS-B data allowed distinguishing the diffused component and many localized sources. The collaboration Energetic Gamma-Ray Experiment Telescope, EGRET [225], detected more than 300 gamma-ray sources many of which were believed to be pulsars [226].

A plethora of results was recently obtained by the Fermi Gamma-Ray Space Telescope, launched on 11 June 2008 into a circular orbit at 565 km altitude. With its Large Area Telescope (LAT) it is possible to detect photons with energy between 0.1 and 100 GeV emitted by rotating neutron stars, known as radio pulsar, by supermassive black holes in the "blazer"[17] class of active galactic nuclei (AGN) and by other high-energy sources.

Fermi has discovered a new population of pulsars pulsing in gamma rays [227], gamma-ray bursts [228], has made a determination of the diffuse gamma-ray emission with unprecedented accuracy [229], and discovered thousands of new gamma-ray sources [230].

Fermi also discovered [231] a rotating neutron star only through its gamma emission. Somehow <1,000 photons were detected in 2 months, which showed a periodicity of approximately a third of a second. The source does not emit at

---

[17]Objects with jets pointing towards us under small angles.

wavelengths in the radio or visible range and the faint X emission is not pulsed. The new gamma star or Gamstar—as it was called—is near the centre of the diffused remnant of a supernova which exploded about 10,000 years ago and named CTAI.

Starting from the 1960s a considerable theoretical effort was made to calculate the structure of neutron stars and their X-ray emission temperature. It was only in the 1970s that one understood many X sources are pulsed binary systems in which a neutron star sinks material by a normal company star becoming hotter and more luminous of the neutron isolated stars.

Only in 1967 it was discovered that neutron stars rotate rapidly and Franco Pancini proposed they could emit electromagnetic radiations. The radio emissions from pulsars were so attributed to neutron stars by Thomas Gold in 1968.

The rotating magnetic field of the spinning stars generates a high electric field that accelerates particles at energies able to emit gamma rays. The search started in the 1970s, but before Fermi only six sources had been discovered.

For the study of gamma-ray emission, a ground-based instrument was the High-Energy Stereoscopic System (HESS) [232], able to observe gamma rays with energies of several TeV (10^{12} eV) settled in the Khomas Highland of Namibia by a collaboration among Germany, France, the UK, Ireland, the Czech Republic, Armenia, Namibia, and South Africa in 2004.

It provides "stereoscopic" observations through a set of four identical telescopes using Cerenkov light from the atmosphere. Devised to study extended showers, the system may work out the energy of an incident gamma ray by measuring the intensity of the Cerenkov image and calculating the direction by using the orientation of the image. It will also see the orientation and shape of the image to discriminate between air showers produced by gamma rays and those caused by cosmic ray particles.

Very-high energy gamma-ray astronomy has revealed galactic and extragalactic sources as supernova remnants, pulsar wind nebulae, giant molecular clouds, star formation regions, compact binary systems, and AGN. The field has now developed as a new sector of astrophysics.

12.9   Present Vision of Cosmic Radiation

From all directions in outer space cosmic ray particles impinge on air molecules at the top of the Earth's atmosphere, at a rate of about 10,000 particles per square metre every second at latitudes far from the equator, and at about one-tenth this rate near the equator. The intensity does not vary with time by more than a few tenths of a percent. The charged incident particles have the composition outlined in the previous paragraphs and energies from a few GeV up to about 10^{20} eV (nearly 16 J).

The order of magnitude of the flux of particles at different energies is shown in Table 12.2 below.

Table 12.2 Flux of cosmic rays at different energies

Energy (eV)	Flux
10^9	10,000 particles/m^2/s
10^{11}	1 particles/m^2/s
10^{14}	1 particle/m^2/day
10^{15}	1 particle/m^2/year
10^{19}	1 particle/km^2/year
10^{20}	1 particle/km^2/century

Even before the particles reach the atmosphere they have been deviated by galactic and intergalactic magnetic fields, and once in our solar system by the ones of the Sun and of the Earth. The reason for the cut-off at a few GeV is that protons with energies below this limit are turned back by the Sun field. The corresponding effect of the Earth's field is to allow only high-energy particles to reach the Earth's surface at the equator, while at poles particles of even low energy are permitted to come in along the lines of magnetic flux. As a consequence the cosmic radiation shows a latitude effect, that is, an increase in intensity from the equator to the poles, equal to about ten percent at sea level, but much larger at high altitudes. Furthermore, the geomagnetic field deflects particles of opposite sign in different directions and produces an east–west effect, given by the difference between the number of particles arriving from the east and particles arriving from the west. According to the actual value found for the excess of radiation coming from the west over that from the east, it may be concluded that the most of the primary radiation is positive and this fact is confirmed by spacecraft measurements outside the Earth's atmosphere.

As the primary radiation enters the atmosphere, it interacts with the protons and neutrons in the air nuclei to give π mesons and other meson types (for example K) in groups of many particles generated in each single collision and in some cases disintegration stars. Pions, kaons, etc. decay rapidly originating showers as described in Chap. 10.

Most of the primary radiation is absorbed in approximately the uppermost tenth of the atmosphere.

Muons, with their rest lifetime extended through the relativistic dilatation by a factor of perhaps 50 (depending on the energy), travel through the atmosphere, where they undergo energy loss primarily by excitation and ionization. They constitute the hard component of the sea level radiation and their weak interaction with matter allows them to be found even at great depth underground.

The total intensity increases from the top of the atmosphere towards ground until it reaches a maximum of three to five times the primary intensity, at a height corresponding to a pressure of somewhat less than one-tenth of an atmosphere. Past the maximum the total intensity decreases steadily to ground, till the sea level the total intensity from all directions is of the order of one-fiftieth of its maximum value. At the position for the maximum of the total intensity, the hard component is only about a fifth of the soft component. The hard component itself decreases continually from the top of the atmosphere, until at sea level the hard radiation from near vertical direction has only a fifteenth of its intensity at great altitudes.

The total radiation at sea level is made up of about one-quarter soft and three-quarters hard component, and has intensity slightly larger than that of one ionizing particle per minute passing through a horizontal square centimetre from all directions.

Neutrons are also present among the cosmic rays in the atmosphere with a number that increases rapidly with altitude.

Adding to that is the flux of neutrinos in part originating from the Sun (some may come from stars) and part coming from the decay of mesons and muons.

The particles hitting the top of the atmosphere are usually divided into *"primary"* particles that are the ones directly produced at the sources and *"secondary"* particles that are produced in the course of the travel of the "primary" particles through interactions with the interstellar ambience. Secondary rays in this sense are: photons, all leptons, hadrons, neutrons, pions, and antiprotons. The terminology should not create confusion with the secondary rays produced in the atmosphere by the incoming outer particles.

References

1. S. Chapman, V.C.A. Ferraro, Terr. Magn. Atmos. Electr. **36**, 77 (1931)
2. L. Biermann, Z. Astrophys. **29**, 274 (1951)
3. L. Biermann, Z. Naturforsch. **7a**, 127 (1952)
4. L. Biermann, Observatory **107**, 109 (1957)
5. E.N. Parker, in *The Century of Space Science*, ed. by J.A.M. Bleeker, G. Johannes, H. Martin (Springer, Heidelberg, 2001)
6. G.L. Locher, Phys. Rev. **44**, 774 (1933)
7. G.L. Locher, Phys. Rev. **45**, 296 (1934)
8. G.L. Locher, Phys. Rev. **50**, 394 (1936)
9. H.A. Bethe, S.A. Korff, G. Placzek, Phys. Rev. **57**, 573 (1940)
10. J.A. Simpson, Phys. Rev. **73**, 1389 (1948)
11. S.E. Forbush, Phys. Rev. **70**, 771 (1946)
12. A. Ehmert, Z. Naturforsch. **3a**, 264 (1948)
13. P. Meyer, E.N. Parker, J.A. Simpson, Phys. Rev. **104**, 768 (1956)
14. J.A. Simpson, Phys. Rev. **74**, 1214 (1948)
15. J.A. Simpson, Phys. Rev. **83**, 1175 (1951)
16. E.N. Parker, Phys. Rev. **110**, 1445 (1958)
17. E.N. Parker, Phys. Rev. **109**, 1874 (1958)
18. E.N. Parker, Phys. Rev. **107**, 924 (1957)
19. S. Chapman, Smithson. Contrib. Astrophysics **2**(1), 1 (1957)
20. E.N. Parker, Astrophys. J. **128**, 677 (1958)
21. E.N. Parker, *Interplanetary Dynamical Processes* (Interscience, New York, 1963)
22. K.I. Gringauz et al., Sov. Phys. Dokl. **5**, 361 (1960)
23. A. Bonetti, H.S. Bridge, A.J. Lazarus, B. Rossi, F. Scherb, J. Geophys. Res. **68**, 4017 (1963)
24. M.M. Neugenbauer, C.W. Snyder, Science **138**, 1095 (1962)
25. M.M. Neugenbauer, C.W. Snyder, J. Geophys. Res. **71**, 4469 (1966)
26. H. Alfven, Ark. Astron. Fys. **29B**, 1 (1943)
27. H. Alfven, *Cosmical Electrodynamics* (Oxford University Press, Oxford, 1950)

28. C.Y. Fan, P. Meyer, J.A. Simpson, Phys. Rev. Lett. **5**, 269 (1960)
29. T.S. Johnson, Rev. Mod. Phys. **10**, 193 (1938)
30. J.A. Simpson, K.B. Frenton, J. Katzman, D.C. Rose, Phys. Rev. **102**, 1648 (1956)
31. J.W. Firor, W.H. Fonger, J.A. Simpson, Phys. Rev. **94**, 1031 (1954)
32. E.N. Parker, Phys. Rev. **103**, 1518 (1956)
33. R.A. Millikan, H.V. Neher, Phys. Rev. **56**, 491 (1939)
34. S.E. Forbush, World-wide cosmic ray variations, 1937–1952. J. Geophys. Res. **59**, 525 (1954)
35. S.E. Forbush, Terr. Mag. Atmos. Electr. **42**, 1 (1937)
36. S.E. Forbush, J. Geophys. Res. **59**, 525 (1954)
37. J.W. Wilson et al., NASA/CP-2003-212155
38. H.V. Neher, W.H. Pickering, Phys. Rev. **61**, 407 (1942)
39. H.V. Neher, J. Geophys. Res. **66**, 4007 (1961)
40. H.V. Neher, J. Geophys. Res. **72**, 1527 (1967)
41. H.V. Neher, J. Geophys. Res. **76**, 1637 (1971)
42. H.V. Neher, H.R. Anderson, J. Geophys. Res. **67**, 1309 (1962)
43. G.A. Bazilevskaya, A.K. Svirzhevskaya, Space Sci. Rev. **85**, 431 (1998)
44. P. Goldhagen, J.M. Clem, J.W. Wilson, Adv. Space Res. **32**(1) 35 (2003)
45. D.V. Reames, Space Sci. Rev. **90**, 417 (1999)
46. J.A. Van Allen, L.A. Frank, Nature **183**, 430 (1959)
47. L.H. Meredith, M.B. Gottlieb, J.A. Van Allen, Phys. Rev. **97**, 201 (1955)
48. J.A. van Allen, Proc. Nat. Acad. Sci. Wash. **43**, 57 (1957)
49. J. Van Allen et al., Jt. Propulsion **28**, 588 (1958)
50. J. Van Allen et al., J. Geophys. Res. **64**, 271, 877 (1959)
51. J. Van Allen et al., J. Geophys. Res. **65**, 2998 (1960)
52. S.N. Vernov, A.E. Chudakov, E.V. Gorchakov, J.L Logachev, P.V. Vakulov, Planet. Space Sci. **1**, 86 (1959)
53. C. Stoermer, *The Polar Aurora* (Clarendon Press, Oxford, 1955)
54. H. Alfven, C.G. Falthammar, *Cosmical Electrodynamics* (Clarendon Press, Oxford, 1963)
55. N.C. Christofilos, J. Geophys. Res. **64**, 869 (1959)
56. S.F. Singer, A.M. Lenchek, *Progress in Elementary Particles and Cosmic Ray Physics*, vol. VI (North-Holland, Amsterdam, 1962), p. 247
57. J.A. van Allen, C.E. McIlwain, G.H. Ludwig, J. Geophys. Res. **64**, 271 (1959)
58. N.C. Christofilos, Proc. NAS **45**, 1144 (1959)
59. J.A. van Allen, J. Geophys. Res. **64**, 1683 (1959)
60. B.J. O'Brien, C.D. Laughlin, J.A. van Allen, L.A. Frank, J. Geophys. Res. **67**, 1209 (1962)
61. S.F. Singer, Phys. Rev. Lett. **1**, 171 (1958)
62. R.S. Selesnick, J. Geophys. Res. **111**, A04210 (2006)
63. M.K. Hudson et al., J. Atmos. Solar-Terres. Phys. **66**, 1333 (2004)
64. J. Goldstein, Space Sci. Rev. 124 (2006)
65. J. Goldstein et al., Geophys. Res. Lett. 32, L15104 (2005)
66. G. Pfotzer, Z. Phys. **102**, 23 (1936)
67. M. Schein, W.P. Jesse, E.O. Wollan, Phys. Rev. **59**, 615 (1941)
68. H. Carmichael, E.G. Dymond, Nature **141**, 910 (1938)
69. H. Carmichael, E.G. Dymond, Proc. Roy. Soc. A 171, 321 (1939)
70. R. Serber, Phys. Rev. **54**, 317 (1938)
71. I.S. Bowen, R.A. Millikan, H.V. Neher, Phys. Rev. **52**, 80 (1937)
72. I.S. Bowen, R.A. Millikan, H.V. Neher, Phys. Rev. **53**, 217 and 855 (1938)
73. H. Snyder, Phys. Rev. **53**, 960 (1938)
74. M. Schein, V.C. Wilson, Phys. Rev. **54**, 304 (1938)
75. W. Heitler, Proc. Roy. Soc. A **166**, 529 (1938)
76. E.G. Dymond, Nature **144**, 782 (1939)
77. A. Ehmert, Z. Phys. **115**, 326 (1940)
78. M. Schein W.P. Jesse, E.O. Wollan, Phys. Rev. **56**, 613 (1939)
79. M. Schein W.P. Jesse, E.O. Wollan, Phys. Rev. **57**, 847 (1940)

80. T.H. Johnson, Rev. Mod. Phys. **11**, 208 (1939)
81. T.R. Wilkins, H. St. Helens, Phys. Rev. **49**, 403 (1936)
82. R.I. Hulsizer, B. Rossi, Phys. Rev. **73**, 1402 (1948)
83. H. Birge, W.E. Hazen, B. Rossi, Phys. Rev. **73**, 179 (1948)
84. W.F.G. Swann, Phys. Rev. **58**, 200 (1940)
85. W.F.G. Swann, Am. Phys. Soc. Meeting, Pittsburgh, Pennsylvania, 20–22 June 1940
86. J.F. Carlson, M. Schein, Phys. Rev. **59**, 840 (1941)
87. H. Alfven, Nature **143**, 435 (1939)
88. W.F.G. Swann, J. Franklin Inst. **236**, 1 (1943)
89. M.A. Pomerantz, F.L. Hereford, Phys. Rev. **76**, 997 (1949)
90. P. Freier, E.J. Lofgran, E.P. Nay, F. Oppenheimer, H.L. Bradt, B. Peters, Phys. Rev. **74**, 213 (1948)
91. H.L. Bradt, B. Peters, Phys. Rev. **74**, 1828 (1948)
92. H.L. Bradt, B. Peters, Phys. Rev. **77**, 54 (1950)
93. H.L. Bradt, B. Peters, Phys. Rev. **80**, 943 (1950)
94. P. Freier, E.J. Lofgren, E.P. Ney, F. Oppenheimer, Phys. Rev. **74**, 1818 (1948)
95. S.F. Singer, Phys. Rev. **80**, 47 (1950)
96. E.P. Ney, D.M. Thon, Phys. Rev. **81**, 1068, 1069 (1951)
97. L. Goldfarb, H.L. Bradt, B. Peters, Phys. Rev. **77**, 751 (1950)
98. B. Peters, in *Progress in Cosmic Ray Physics,* vol. I, ed. by J.G. Wilson (North Holland, Amsterdam, 1952), p. 191
99. E.P. Ney, D.M. Thon, Phys. Rev. **81**, 1068 (1951)
100. S.O.C. Sorensen, Phil. Mag. **40**, 947 (1949)
101. A.D. Dainton, P.H. Fowler, D.W. Kent, Phil. Mag. **43**, 729 (1952)
102. H. Brown, Rev. Mod. Phys. **21**, 625 (1949)
103. M.F. Kaplon et al., Phys. Rev. **85**, 295, 900, 933 (1952)
104. M. Garcia-Munoz, G.M. Mason, J.A. Simpson, Astrophys. J. **201**, L145 (1975)
105. A.D. Dainton et al., Phil. Mag. **42**, 317 and 396 (1951)
106. M.F. Kaplon, J.H. Noon, G.W. Racette, Phys. Rev. **96**, 1408 (1954)
107. N.L. Grigorov et al., Sov. J. Nucl. Phys. **11**, 588 (1970)
108. T.H. Burnett et al., Phys. Rev. Lett. **51**, 1010 (1983)
109. T.H. Burnett, Phys. Rev. D **35**, 824 (1987)
110. R.C. Maehl et al., Astrophys. Space Sci. **47**, 163 (1977)
111. W.R. Webber et al., Astrophys. Space Sci. **15**, 245 (1972)
112. W.R. Webber et al., Astrophys. Lett. **18**, 125 (1977)
113. F.A. Hagen et al., Astrophys. J. **212**, 262 (1977)
114. A.J. Fisher et al., Astrophys. J. **205**, 938 (1976)
115. E. Juliusson, Astrophys. J. **191**, 331 (1974)
116. M. Garcia-Munoz et al., Astrophys. J. **217**, 859 (1977)
117. E. Juliusson et al., Phys. Rev. Lett. **29**, 445 (1972)
118. M.M. Shapiro, R. Silberberg, Ann. Rev. Astron. Astrophys. **8**, 323 (1970)
119. A. Lukasiak, F.B. McDonald, W.R. Webber, in *Proceedings of the 26th International Cosmic-Ray Conference*, Salt Lake City, vol. 3, 1999, p. 41
120. W.R. Webber, A. Lukasiak, F.B. McDonald, Astrophys. J. **568**, 210 (2002)
121. J.J. Engelman et al., Astron. Astrophys. **233**, 96 (1990)
122. E.C. Stone et al., Space Sci. Rev. 86, **285**(1998)
123. G.A. de Nolfo, I.V. Moskalenko et al., in *Proceedings of the 28th ICRC (Tsukuba)*, vol. 2, 2003, p. 1667
124. G.A. de Nolfo, I.V. Moskalenko et al, Adv. Space Res. **38**, 1558 (2006)
125. O. Reimer et al., Astrophys. J. **496**, 490 (1998)
126. T. Hams et al., Astrophys. J. **611**, 892 (2004)
127. S.P. Ahlen et al. Astrophys. J. **534**, 757 (2000)
128. J.Z. Wang et al., Astrophys. J. **564**, 244 (2002)

129. G.A. de Nolfo et al., Observation of the Li, Be, and B isotopes and constraints on cosmic ray propagation. Asp. J. **38**, 1558–1564 (2006)
130. T.K. Gaisser, T. Staney (2005) astro-ph/0510321
131. P.H. Fowler et al., Proc. R. Soc. A 301, 59 (1967)
132. P.H. Fowler et al., Proc. R. Soc. A 318, 1 (1970)
133. P.B. Price, Phys. Rev. D **3**, 815 (1971)
134. D.O. Sullivan et al., Phys. Rev. Lett. **26**, 463 (1971)
135. P.B. Price et al., Phys. Rev. Lett. **26**, 916 (1971)
136. R. Hulsizer, Phys. Rev. **76**, 164 (1949)
137. C.L. Critchfield, E.P. Ney, S. Oleska, Phys. Rev. **79**, 402 (1950)
138. C.L. Critchfield, E.P. Ney, S. Oleska, Phys. Rev. **85**, 461 (1952)
139. J.A. Earl, Phys. Rev. Lett. **6**, 125 (1961)
140. P. Meyer, P. Voigt, Phys. Rev. Lett. **6**, 193 (1961)
141. V.G. Berkey, C.S. Shen, Phys. Rev. **188**, 1994 (1969)
142. D.H. Perkins, *Particle Astrophysics* (Oxford, Univ. Press, New York, 2003)
143. J.A. DeShong Jr., R.H. Hildebrand, P. Meyer, Phys. Rev. Lett. **12**, 3 (1964)
144. R.L. Golden et al., Phys. Rev. Lett. **43**, 1196 (1979)
145. E.A. Bogomolov et al., in *Proceedings of the 19th International Cosmic Ray Conference* (La Jolla), vol. 2, 1985, p. 362
146. L.H. Rumbough, G.L. Locher, Phys. Rev. **49**, 855 (1936)
147. E. Schopper, Naturwissenschaften **25**, 557 (1937)
148. S.A. Korff, Phys. Rev. **56**, 210 (1939)
149. C.Y. Fan, G. Gloeckler, J.A. Simpson, Phys. Rev. Lett. **17**, 329 (1966)
150. N. Hilberry, Phys. Rev. **60**, 1 (1941)
151. I.S. Bowen, R.A. Millikan, H.V. Neher, Phys. Rev. **53**, 855 (1938)
152. I.S. Bowen, R.A. Millikan, H.V. Neher, Phys. Rev. **53**, 855 (1938)
153. W.F.G. Swann, Phys. Rev. **50**, 1103 (1936)
154. W. Heitler, Proc. R. Soc. **161A**, 261 (1937)
155. L.W. Nordheim, Phys. Rev. **53**, 694 (1938)
156. T.H. Johnson, Phys. Rev. **53**, 499 (1938)
157. T.H. Johnson, Rev. Mod. Phys. **10**, 193 (1938)
158. M.A. Pomerantz, M.S. Vallarta, Phys. Rev. **76**, 1881 (1949)
159. J.A. van Allen, A.V. Ganges, Phys. Rev. **78**, 50 (1950)
160. J.A. van Allen, A.V. Ganges, Phys. Rev. **79**, 51 (1950)
161. J.R. Winckler, Phys. Rev. **79**, 656 (1950)
162. J.A. Van Allen, S.F. Singer, Phys. Rev. **78**, 819 (1950)
163. Tchang-Fong Hoang, Ann. Phys. **5**, 537 (1950)
164. T.K. Gaisser, J. Phys. Conf. Ser. **47**, 15 (2006)
165. M. Boezio et al., Astropart. Phys. **19**, 583 (2003)
166. T. Sanuki et al., Astrophys. J. **545**, 1135 (2000)
167. J. Alcaraz et al., Phys. Lett. B **490**, 27 (2000)
168. M.J. Ryan et al., Phys. Rev. Lett. **28**, 985 (1972)
169. T. Shibata, Nucl. Phys. B (Proc. Suppl.) **75A**, 22 (1999)
170. K. Asakimori et al., in *Proceedings of the 24th International Cosmic Ray Conference, Roma*, vol. 2, 1995, p. 707
171. M. Nagano et al., J. Phys. G: Nucl. Part. Phys. **10**, 1295 (1984)
172. M. Nagano et al., J. Phys. G: Nucl. Part. Phys. **18**, 423 (1984)
173. Y.A. Fomin et al., in *Proceedings 22nd International Cosmic Ray Conference*, Dublin, vol. 2, 1991, p. 85
174. J.R. Horandel et al., in *Proceedings 26th International Cosmic Ray Conference*, Salt Lake City, vol. 1, 1999, p. 337
175. T. Antoni et al., Astropart. Phys. **16**, 245 (2002)
176. T. Antoni et al., Astropart. Phys. **24**, 1 (2005)
177. J.W. Fowler et al., Astropart. Phys. **15**, 49 (2001)

178. P. Swordy, D.B. Kieda, Astropart. Phys. **13**, 137 (2000)
179. F. Arqueros et al., Astron. Astrophys. **359**, 682 (2000)
180. M.A.K. Glasmacher et al., Astropart. Phys. **10**, 291 (1999)
181. M. Amenomori et al., Phys. Rev. D **62**, 072007 (2000)
182. M. Amenomori et al., in *Proceedings of the 28th International Cosmic Ray Conference* Tsukuba, vol. 1, 2003, p. 143
183. M. Takeda et al., Astropart. Phys. **19**, 447 (2003)
184. T. Abu-Zyyad et al., Astrophys. J. **557**, 686 (2001)
185. R.U. Abbasi et al., Phys. Rev. Lett. **92**, 151101 (2004)
186. R.U. Abbasi et al. Astropart. Phys. **23**, 157 (2005)
187. R.U. Abbasi et al., Phys. Lett. B **619**, 271 (2005)
188. P.V. Sokolsky, Workshop on Physics at the End of the Galactic Cosmic Ray Spectrum, Aspen, Colorado, April 2005
189. P.J. Bird et al., ApJ **424**, 491 (1994)
190. T. Abu-Zayyad et al., Phys. Rev. Lett. **84**, 4276 (2000)
191. N. Nagano et al., J. Phys. **18**, 423 (1992)
192. A.A. Abdo et al., Fermi Collaboration. Phys. Rev. Lett. **102**, 181101 (2009)
193. L. Latronico, Nuovo Cimento **33C**, 11 (2010)
194. O. Adriani, PAMELA Collaboration, Nature **458**, 607 (2009)
195. L. Latronico, Nuovo Cimento **C33**, 11 (2011)
196. O. Adriani et al., Science **332**, 69 (2011)
197. B. Peters, Nuovo Cimento **22**, 800 (1961)
198. D.J. Bird et al., Phys. Rev. Lett. **71**, 3401 (1993)
199. M.J. Ryan et al., Phys. Rev. Lett. **28**, 985 (1972)
200. T.H. Burnett et al., Phys. Rev. Lett. **50**, b2062 (1983)
201. T.H. Burnett et al., Nucl. Instrum. and Meth. A **251**, 583 (1986)
202. Runjob Collaboration, Adv. Space Res. **26**, 1839 (2000)
203. M. Ichimura et al., Phys. Rev. D **48**, 1949 (1993)
204. T. Antoni, Nucl. Instrum. and Meth. A **513**, 490 (2003)
205. M. Amenomori et al., in *28th ICRC (Tsukuba)*, vol. 1, 2003, p. 107
206. J. Abraham et al., Phys. Rev. Lett. **104**, 091101 (2010)
207. Pierre Sokolsky, Phys. Today 31(1998)
208. R.U. Abbasi et al., ApJ **622**, 910 (2005)
209. K.T. Strong, J.L.R. Saba, B.M. Haisch, J.T. Schmelz, *The Many Faces of The Sun: A Summary of The Results from NASA's Solar Maximum Mission* (Springer, Heidelberg, 1999)
210. H. Fridman, S.W. Lichtman, E.T. Byram, Phys. Rev. **83**, 1025 (1951)
211. R. Giacconi, H. Gursky, F.R. Paolini, B.B. Rossi, Phys. Rev. Lett. **9**, 439 (1962)
212. W.L. Kraushaar, G.W. Clark, Phys. Rev. Lett. **8**, 106 (1962)
213. G.W. Clark, Phys. Rev. Lett. **14**, 91 (1965)
214. R.N. Manchester, Science **304**, 542 (2004)
215. D. Kniffen et al., Nature **251**, 397 (1974)
216. R.D. Wills et al., Nature **296**, 723 (1982)
217. P.L. Nolan et al., Astrophys. J. **409**, 697 (1993)
218. R.J. Gould, Am. J. Phys. **35**, 376 (1967)
219. F. Seward et al. Astrophys. J. **150**, 845 (1967)
220. W.L. Kraushaar et al., ApJ. **177**, 341 (1972)
221. B.N. Swanenburg et al. Astrophys. J. **243**, 169 (1981)
222. S.M. Derdeyn et al., Nucl. Instrum. and Meth. **98**, 557 (1972)
223. D.J. Thompson et al., Astrophys. J. **200**, L79 (1975)
224. G.F. Bignami, P.A. Caraveo, Ann. Rev. Astron. Astrophys. **34**, 331 (1996)
225. R.C. Hartman et al., Astrophys. J. Suppl. Ser. **123**, 79 (2008)
226. D.J. Thompson, Rep. Prog. Phys. **71**, 116901 (2008)
227. A.A. Abdo et al., Science **325**, 840 (2009)

228. J. Granot, in *Proceedings of the Shocking Universe Gamma Ray Bursts and High Energy Schock Phenomena*. astro-ph/1003 (2010), p. 2452
229. A.A. Abdo et al., Phys. Rev. Lett. **104**, 101101 (2010)
230. A.A. Abdo et al., Astrophys. J. Suppl. Sez. **188**, 405 (2010)
231. A.A. Abdo et al. Science **322**, 1218 (2008)
232. F. Aharonian, Science **315**, 70 (2007)

Chapter 13
The Origin of Cosmic Rays

13.1 Introduction

In the last 20 years, the search for the cosmic ray origin has gained an increasing importance for its implications in astrophysics and cosmology, because one believes that by explaining the mechanisms of their origin a better understanding of the implied astrophysical phenomena may be reached. Following the discovery, in 1948, of the presence of practically all atomic elements in the arriving radiation, the establishment of the synchrotron nature of the cosmic electromagnetic radiation (in the range from radio-waves to gamma rays) in 1950–1951, and the nearly general acceptance that supernova explosions were possible sources in the 1980s, we may say that the question of the cosmic rays origin has become a true astrophysical problem.

The research on cosmic rays at the higher energies, involving the two disciplines of astrophysics and particle physics, began to be called *particles astrophysics*, and since the 1980s most of cosmic ray researches started to appear on astrophysics journals such as *Astrophysics Journal* and *Astrophysical Space Science*, and in 1992 a new journal *Astroparticle Physics* was established.

Besides the consideration that the composition of the primary rays and their energy distribution must be explained by the generation mechanisms—and therefore by the stellar models—and by their propagation in the interstellar space—and therefore by the properties of this space—it is worth to be pointed out that cosmic rays represent the sole matter sample reaching us from outside the solar system.

Our knowledge of the relative abundance of elements in the Universe is based on the study of optical spectra of stars and nebulae, on the absorption of light in the interstellar space, and on the chemical analysis of meteorites. There is a notable similarity between the relative abundances of non-volatile elements in meteorites and in the solar atmosphere. From an accurate evaluation of the relative quantities of different nuclei present in cosmic rays, the abundances required at the sources may be derived and a comparison with their theoretical models is possible. The study of cosmic ray composition provides us with a detailed elemental and isotopic sample

M. Bertolotti, *Celestial Messengers: Cosmic Rays*, Astronomers' Universe,
DOI 10.1007/978-3-642-28371-0_13, © Springer-Verlag Berlin Heidelberg 2013

of the current (few million years old) interstellar medium that would be otherwise unavailable. This is what makes the subject particularly rich and complementary to other disciplines.

The detailed study of the chemical composition and of the energy spectrum of the radiation is also of great importance for the interpretation of the phenomena produced in our atmosphere.

Finally, we may note that radiochemical studies of isotopes produced by cosmic rays in meteorites have shown that their flux has remained constant within a factor 2 in the last billion years [1].

13.2   A Brief Summary of Some of the First Hypotheses on the Origin of Cosmic Rays

After Millikan's "*birth cry*" theory ingloriously failed when the arriving rays were discovered to be charged particles and not gamma rays, the cosmic ray origin continued to be a mystery and still at present, although many points have been clarified, a complete explanation unanimously accepted is not available.

At the very beginning of cosmic ray research, we may say that nearly every researcher had his own theory about their origin, even if he did not know exactly what they were. So, for example, C.T.R. Wilson [2, 3] suggested that cosmic rays were produced during thunderstorms. M.C. Holmes [4] proposed that ions created by the sun light could be separated by the earth's electric field and accelerated, and C. Halliday [5] resumed the idea of electron acceleration in thunderstorms. C.T.R. Wilson also suggested that in the intergalactic spaces, electrical fields existed that, thanks to their huge dimensions, were able to accelerate the particles. However, these accelerated particles with their movement tend to counteract the potential difference that accelerates them, and another hypothesis should be made to assure the permanence of these electric fields: a hypothesis Wilson was not able to provide.

Bothe and Kolhoerster [6] surmised that the particles were accelerated by the gravitational field in the Universe.

Regener supposed they came from stars' irradiation after having made a complete journey around the Universe.

Einstein [7], in his development of the general relativity theory, considered a very elegant model of Universe described by an equation of an extreme simplicity that linked the space geometric characteristics to gravity. To find a solution describing an isotropic, homogeneous, and static Universe—being such a model regarded as the most reasonable at the time—Einstein introduced a term in his equation, in the form of a constant, that should balance the attractive gravity force. This constant physically represents a force opposing gravity which makes the Universe stable. The solution studied by Einstein was not the sole possibility, and the Dutch astronomer W. de Sitter [8] (1872–1934) succeeded in finding a solution in the case of zero density. A few years later the Russian astronomer Aleksander Friedmann

[9] (1888–1925), while studying Einstein's cosmological equations, demonstrated that non-static solutions also exist, with isotropic and homogeneous distributions of matter, corresponding to an expanding Universe. Independently, the Belgian George Lemaitre [10] found a solution with which he built a theory of an expanding universe, in which cosmic rays were created at the same moment in which the world was created and travelled since then through the Universe. In Lemaitre's solution, the radius of the Universe changes with time, increasing continuously. A consequence of this theory is that an observer placed at any point sees every celestial body escaping with a velocity proportional to his distance from it. This result was published in 1927, and Lemaitre gave a linear relation between velocity and distance, but remained unnoticed. Later, the American astronomer Edwin Powell Hubble[1] using the same observations that Lemaitre had used, apparently ignoring Lemaitre's results, observed that nebulae seem to recede from our Galaxy with considerable velocities and formulated the law which bears his name according to which the recession velocity of a nebulae is proportional to its distance from us [11].

In the Universe of Lemaitre, it appears the matter density which, if the Universe expands, decreases and asymptotically tends to zero as the radius increases. The solution of Einstein's equations for zero density—as we said—was studied by de Sitter who considered a spherical empty Universe with a constant radius depending only on the arbitrary constant Einstein had introduced in his model [12, 13]. On the contrary, if we ask if the radius of the Lemaitre Universe could become constant, we find that this may happen if the Newtonian attraction is equal to the repulsion that is governed by Einstein's constant. In these conditions, one recovers Einstein's spherical Universe solution, of constant radius, filled with an amount of matter sufficient to balance the two forces we mentioned before. If the two forces do not balance, the Universe enters a dynamic state of contraction or expansion.

Reasoning backwards, Lemaitre arrived at his *primitive atom hypothesis* which states that the Universe was originated by a unique "primeval atom" in which the whole Universe mass was concentrated. It was the forerunner of the present *Big-Bang* hypothesis.[2]

At the beginning, the Lemaitre Universe existed in the form of an atomic nucleus uniformly filling a spherical space of a very small radius. The mass of this primeval atom would be equal to the mass of the whole Universe. This atom would have existed only for a while, being unstable, and then would have exploded,

---

[1]Edwin Powell Hubble (1889–1953) took his degree in mathematics at the University of Chicago; then to satisfy his father, he studied law at Oxford, but his heart was with stars and so he enrolled at the Observatory of Yerkes, gaining a PhD in astronomy. In 1919, he moved to the Mount Wilson observatory where with the 2.5 m telescope he performed his discoveries. In 1924, he demonstrated that the Andromeda nebulae is actually a galaxy. Between 1927 and 1930, he discovered that galaxies all recede by us with a velocity directly proportional to their distance (Hubble law).

[2]According to the Big-Bang theory the Universe was born in a sudden explosion reaching a huge temperature in which a broth made of quarks, gluons, photons, neutrinos, etc., existed. In the successive cooling, protons, neutrons, and the hydrogen atoms would have formed and the evolution that brought to the creation of stars and of all elements had started.

originating all the particles known at the time (electrons, protons, alpha particles, and all the nuclei then known). The disintegration of the primeval atom was the initial cause of the space expansion. When the disruption of the primeval atom was completed, it gave rise to gaseous clouds moving with large relative velocities which emitted radiations through radioactive transformations. These radiations, freed at the beginning of the Universe evolution, continue to travel even at present and are, according to Lemaitre, at the origin of cosmic rays.[3]

A more complex theory was successively developed by the English physicist E.A. Milne [15] (1896–1950), in which cosmic rays are present since the very beginning and are part of a complex statistical system, being accelerated by gravitational fields in cosmic spaces.

It is nearly impossible and may even be boring to make a list of all the suggestions that were made[4]; we will consider here only a few that are at the basis of the present theory of the cosmic ray origin. We want, however, to draw attention to the curious idea, made as early as 1940, of the American naturalized physicist and astronomer V. Rojansky (1900–1981), that comets were made of antimatter that by annihilating originated the cosmic rays. The term used by Rojansky was not, of course, antimatter but he was speaking of *"contraterrene material"*, meaning with this term a body composed of hypothetic atoms made of nuclei negatively charged surrounded by positrons [21–23].

Maybe the first time that the sources of the cosmic radiation were linked to precise astrophysical events, was in 1934, when the astronomers W. Baade (1893–1960) and F. Zwicky (1898–1974), studying the events originating the *novae* and *supernovae* stars (they introduced the distinction and the name *supernovae*), suggested that cosmic rays were generated in the explosions of supernovae.

They [24] assumed that a class of temporary stars existed with properties similar to *novae*, but much more spectacular, to which they proposed to give the name *supernovae*. In the paper immediately following this first one in the journal [25], they were suggesting that in the supernovae explosion—that they already surmised should bring to the creation of a neutron star[5]—cosmic rays were accelerated. Their suggestion was further discussed in a number of publications [26–28].

The hypothesis at the time did not find many followers, even if the Uruguayan engineer and physicist Felix Cernuschi [29] (1907–1999) at MIT toyed a little with it; however, the idea was successively resumed by several authors [30–33].

Meanwhile, during the Second World War, Alfven demonstrated that astrophysical plasmas may transport or freeze electromagnetic fields [34] and this concept was

---

[3] A popular version was published in the book G. Lemaitre [14].

[4] We quote some: W.F.G. Swann [16]; L. Spitzer [17], A. Unsold, [18], and E. McMillan [19]; a collection of some of the first works is S. Rosen [20].

[5] A neutron star is a star with a mass of the order of one and a half times the mass of the Sun, a radius of about 12 km and a very high density of about 5–10 times the equilibrium density of nucleons inside the atomic nucleus. A neutron star is therefore one of the denser objects existing in the Universe. In these stars the neutronic component is dominant.

later utilized by Fermi to explain the acceleration of particles in the Galaxy at the cosmic ray energies.

In 1948, E. Teller [35] (1908–2003) suggested that cosmic rays had a solar origin, and were trapped relatively near the Sun by the action of magnetic fields. This hypothesis was further discussed by Alfven [36,37] and Richtmeyer and Teller [38]. Anyway, the solar origin had already been suggested [39, 40].

An argument used to limit the presence of cosmic rays to our solar system was the huge amount of energy that should be present if they occupied the whole Galaxy. As we will see, however, although part of the radiation has a solar origin, the more energetic part has much farther origin.

Alfven [41] himself some 10 years later, at an International Congress on Cosmic Rays organized by the International Physics Union with the collaboration of the Italian Physical Society at Varenna on the Como Lake, in Villa Monastero 21–26 June 1957, had changed his mind and affirmed that the sole direct evidences of particles with cosmic ray energies coming from Sun had been obtained in rare occasions when violent eruptions occurred on the Sun. Until then, five of such solar flares had occurred in which particles had reached the sea level on Earth. One should expect that if the high-energy particles were coming directly from Sun they should have been more numerous when the Sun is more active; actually just the opposite happened. Cosmic rays presented an intensity minimum when the Sun was more active. Someone had suggested that a mechanism of acceleration could exist in the extended corona of the Sun, and because the corona seems to be more extended during the solar minimum, this would be the moment more favourable for the release of particles. However, he pointed out that the experimental results on the chemical composition of the arriving rays indicated that the light elements Li, Be, and B have nearly equal abundance as C, N, and O while the Sun seems very poor in them. The explanation of the anomalous abundance of light elements was attributed to collisions that the heavier nuclei of cosmic rays could have with the nuclei of interstellar matter, as we will soon show, but to explain the observed abundances it would be necessary to assume that the heavy nuclei of solar origin had passed through several gr/cm^2 of matter before reaching the earth, which seemed very difficult to justify.

13.3 The Fermi Theory

This difficulty to explain the existence of arriving nuclei with relative abundances much greater than those of the matter present in the solar system was overcome by Enrico Fermi [42] who, in 1949, removed the principal objections to a stellar origin, suggesting a mechanism of charged particle acceleration through collisions with the magnetic fields brought in the interstellar gas clouds that Alfven (see also [43]) had studied.

Fermi confuted the idea that the rays must be confined in our solar system and proposed they are instead contained in our Galaxy, suggesting that the cosmic ray particles are repetitively scattered by magnetized plasma clouds in the interstellar space, thus acquiring their high energies.

Soon after the war, the first International Congress on Cosmic Rays (ICCR) was held at Como, organized by the Italian Physical Society for the 150th year of the pile invention by Alessandro Volta, on 13–16 September 1949.[6] This was the first of a series of congresses dedicated to cosmic rays, always held in a different international place (the tradition being ongoing), where the most recent results and hypotheses are discussed. There Fermi [44] expounded his theory.

He calculated that cosmic rays are present with an energy density of 22×10^{-11} erg/cm^3 (about 1 eV/cm^3). If they filled the whole Galaxy interstellar space, this would correspond to a total energy of the same order of the kinetic energy of the disordered motions of stars. Therefore, he suggested that cosmic rays acquired most of their energy travelling in space by entering into equilibrium with the Universe. He knew the Galaxy interstellar space is filled with matter with an extremely low density of about 10^{-24} gr/cm^3, corresponding to nearly one hydrogen atom per cubic centimetre [45]. However, this matter is not uniformly distributed and condensation regions are present where the density may be 10 or 100 times higher, extending over dimensions of the order of 10 pc (1 pc $= 3.1 \times 10^{18}$ cm $= 3.26$ light years). These clouds move with velocities of the order of a few km/s. It is reasonable to expect that most of the hydrogen present in space gets ionized, due to the photoelectric effect of the star light, thus originating protons and electrons. The motion of these charges creates magnetic fields[7] of the order of 5×10^{-6} gauss, probably even larger in the clouds. A fast particle moving inside these randomly moving fields, if it is a proton with an energy of some GeV, coils along the lines of force with a curvature radius of the order of 10^{12} cm until it collides with a field irregularity of the cosmic magnetic field and is reflected.

Figure 13.1 shows how this may happen in a slowly increasing magnetic field. In the figure the force lines of a slowly increasing magnetic field are shown with a particle moving inside them with constant velocity following a helical trajectory with decreasing orbital circles until it is reflected back.

In such a collision if the magnetic field region is slowly moving, the particle could gain or lose energy. However, it would more likely gain energy. The argument for this is very simple. Because at equilibrium particles and clouds should have the same mean energy and the clouds have an enormous energy, the energy of the particles can only increase.

Fermi then calculated that the mean energy a proton gained at each collision should be approximately 10 eV for a proton of low energy, and more for protons

---

[6]The Proceedings were published in Suppl. Nuovo Cimento **6**, 3 (1949).

[7]Effectively, the theoretical interpretation [46,47] of the measurements of the polarization of the light coming from distant stars [48,49] had shown it is reasonable to assume magnetic fields of the order of about 10 μG exist.

Fig. 13.1 Reflection of a
particle from a region of
converging magnetic field
(magnetic mirror)

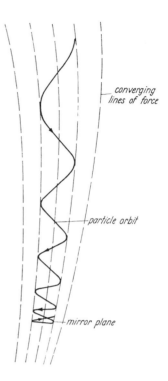

_converging
lines of force_

particle orbit

mirror plane

of higher energy. However, this gain must be larger than the ionization losses. For protons, this occurs if they have an initial energy of about 200 MeV. Taking into account the density of matter in the interstellar space and assuming the proton travels at light speed, the mean free path between one collision and the other is about 10^{26} cm. The time needed to traverse this distance at light speed is 7×10^7 years. The theory allowed also to derive the shape of the accelerated proton energy spectrum that fitted decently with the measured one. It also explained why there are no electrons. In fact, at all energies the ionisation losses of electrons exceed the gain.

To accelerate protons through this mechanism it is therefore necessary to have protons with an energy greater than about 100–200 MeV. Fermi, however, was not able to explain how they could exist. To accelerate heavier particles, the threshold energy was even larger. For example, for alpha particles the threshold energy was about 1 GeV.

It was soon found [50] that the values required for the Fermi model were not astronomically plausible. In a second paper [51] Fermi, in 1954, proposed a modified mechanism and considered the problem of injecting heavy nuclei.

Even if Fermi's theory was not able to explain all the characteristics of cosmic rays, his idea to use spatial magnetic fields to accelerate particles, especially as described in his second model, was later used by many researchers and incorporated in the most recent models generally accepted.

13.4 Supernovae are Resumed

The possibility that supernovae explosions could be a source of cosmic rays had been discussed, since the Baade and Zwicky proposal, essentially recurring to energetic considerations [52], until in 1953, V.L. Ginzburg [53] (1916–2009) added more specific arguments. He studied cosmic rays for nearly 20 years—writing also a book [54]—and through an examination of radio waves in space, offered an indication of what in general could be possible sources of cosmic rays.

Vitalij Lazarevic Ginzburg was born in Moscow in 1916 and had a complex and adventurous life in the revolutionary Russia, exposed by him in an autobiography ([55], see also [56]), until, graduated in physics, he was appointed in 1968 as Professor of Astrophysics at the Technical Physics Institute, Moscow. He conducted studies and researches in the most recent fields of physics, in radioastronomy, cosmic ray astrophysics, low temperature physics, and superconductivity theory. For his researches in superconductivity he was awarded the Nobel prize in 2003 together with A.A. Abrikov (1928–) and A.J. Leggett (1938–).

In the second half of the 1940s, radio-wave sources were discovered in the Galaxy [57–62]. This radio emission could be divided into two components: one is concentrated in the direction of the galactic plane and represents the thermal emission of clouds of interstellar ionized gas. The other is concentrated towards the galactic centre ([63], see also [64, 65]) filling a quasi-spherical volume of a radius of approximately 5×10^{22} cm.

Most of the radio emissions from the most intense sources were proposed to come from *magnetic bremsstrahlung*, that is it is due to the radiation emitted by relativistic electrons accelerated in the interstellar magnetic fields.

The mechanism, first proposed by H. Alvfen and N. Herlofson [66], was analysed in great detail by V.L. Ginzburg [67] who observed that the data on the spectrum and the intensity of this emission allow to determine the energy spectrum intensity of these electrons [68].

The Russian astronomer Iosif Shklovskii [69] (1916–1985) had also shown that nebulae which form the remnants of supernova explosions are powerful sources of radio emission. In particular, a source found in Cassiopea was identified with the supernova of 369 AD, and Shklovskii [70] pointed out that this emission was a good support for the hypothesis of acceleration by supernovae. Studying the light emitted by the Crab Nebula in the Taurus constellation, Shklovskii [71] proposed that the diffuse blue region was predominantly produced by synchrotron radiation, an hypothesis confirmed 3 years later by observations, followed in the 1960s by the discovery that the magnetic field which curved the electrons paths was produced by a neutron star at the centre of the nebula [72].

Also Ginzburg [73] was inclined to attribute the origin of cosmic rays to supernovae [53] and applied the Fermi statistical mechanism to show their high acceleration efficiency—even if the mechanism he devised was far from the one considered today.

To justify the model of the origin linked to supernovae, Ginzburg and Syrovatskii [54, 74] (1924–1979) observed that the energy loss due to the escape of cosmic rays from the Milky Way (remember that the mean energy density of galactic cosmic rays is about $1\,\text{eV}/\text{cm}^3$) is an order of magnitude lower than the kinetic energy of the gas expelled by galactic supernovae. It is sufficient to consider that only 10% of the energy emitted in the explosion is used to accelerate the cosmic rays to maintain their density constant in our Galaxy. Another point in favour of the supernovae was the identification of heavy elements in their spectra.

At last the Baade and Zwick suggestion was taking off!

Another important point to consider was the way the rays travel in space. Ginzburg, in fact, pointed out that in the discussion of the origin of cosmic rays it is necessary to also consider the processes that take place during the motion of the particles in interstellar space. During their journey, the particles interact with the very tenuous matter that fills the Universe, and today we may also say with the photons of the microwave cosmic background—that at those times were not yet discovered—with important effects that, as we will see later, give the key to many explanations. For example, protons in their collisions with the cosmic background produce pions which while decaying give electrons, gamma rays, and neutrinos.

Furthermore, due to the magnetic fields existing in the Galaxy, the particles do not travel along a straight line but are instead curved at random so that the galactic cosmic ray trajectories are governed by diffusion in the galactic magnetic fields. This interaction is responsible for the isotropy of the radiation and for its confinement in the Galaxy, at least until the particle energy is lower than that corresponding to a gyration radius greater than the galactic dimensions. There is also a transport in the galactic winds that was first considered only in 1976 [75].

For a typical cosmic ray, let us say a proton with an energy of a few GeV, the lifetime of confinement in the Galaxy is of the order of a few millions years.

13.5 Where Do Cosmic Rays Come From?

At this point, the main problem is to decide from where cosmic rays come. Since the time of Fermi theory, people agreed that at least a part comes from the Galaxy, and in fact some reasoning may be done to show that they originate in part in the Milky Way (*galactic cosmic rays*). However, there are also good reasons to believe that a part of them originates outside (*extragalactic cosmic rays*). A fraction of the galactic component below about 100 MeV is generated inside the solar system, mostly irradiated by Sun when solar flares are present; the remaining of the radiation, more than about 1 GeV, comes from outside the solar system. Although interstellar particles with energy lower than 1 GeV should exist, they are for the greatest part repelled away from the solar system by the magnetized solar wind, and the stellar flux becomes dominant, according to the solar conditions, only at energies between 100 MeV and 1 GeV. The charged particles are therefore "modulated" by the solar wind that decelerates and practically excludes the galactic cosmic rays of lower energy from the inner solar system. The energy flux around 1 GeV per nucleon is of the order of $1/cm^2/s^1$ and is easily measured by satellites.

To be more exact, we may say that when the cosmic rays penetrate in the heliosphere—the region surrounding the Sun, with a radius of more than 100 U.A., permeated by the solar wind and by the interplanetary magnetic field—they suffer three principal effects which modify their flux (solar modulation): a scattering due to the magnetic field irregularities; a convection, that is a drift towards the outside from the solar wind current, and an energy loss due to the fact that the solar wind expands adiabatically into space.

Cosmic rays are influenced by these effects in different ways, according to their energy. The particles with energy less than about 100 MeV, for the most part are repelled in the outer space; this is the reason of the decrease at low energies observed in the energy spectra. Particles with energy greater than about 10 GeV propagate without appreciable disturbance. The ones with intermediate energy are more or less perturbed according to the conditions existing in the interplanetary space, so that their flux, near the Earth, changes in time. Periodic and occasional variations can be considered. The most important periodic variation is the 11-year variation: the radiation intensity changes with solar spots. Other small variations may exist as discussed previously and are also provoked by Sun.

We said that the chemical composition of the cosmic radiation was found nearly constant in the energy range from about 200 MeV to higher energies and for the most part of elements there are no large differences between the abundances found in cosmic rays and in the solar system with only two notable exceptions (the relative abundances of cosmic rays compared with those of the solar system are shown in Fig. 12.8 of Chap. 12). The exceptions are the overabundances of the light metals (Li, Be, and B) and of the elements from scandium to manganese that, in the periodic Mendelejev table, precede iron. Such elements are over-abundant by many orders of magnitude if compared to the solar system abundances.

The explanation of the anomalous abundance of these nuclei was readily given (see, e.g. [76, 77]). The heavier cosmic ray nuclei—principally carbon, nitrogen, and oxygen, which are the most abundant constituents of the Universe—during their journey in the interstellar space towards the Earth, may occasionally encounter a nucleus of the very tenuous matter pervading the Universe. Because they are accelerated to relativistic velocities, in the collisions they are broken into smaller nuclei[8] (a reaction called *spallation*) [78]. The reverse reaction may also occur whereby energetic cosmic-ray H and He interact with ambient interstellar C, N, and O atoms, thus producing lower energy Li, Be, and B [79].

This explanation gives strong boundaries for the propagation models of galactic cosmic rays, because the production of these elements depends on the quantity of matter traversed during propagation.

---

[8]An analogous explanation can be given for the production by Fe of the elements between scandium and manganese.

If we assume that Li, Be, and B are not present in the cosmic ray sources and are produced in the *spallation reactions* of heavier nuclei present in the primary radiation, principally C, N, and O, during their journey towards us, knowing the cross-section for this type of collision, we may calculate the thickness of interstellar matter they must traverse to explain the found abundances, that are five orders of magnitude greater than the solar values. The cross-section for spallation was measured by R.R. Daniel and N. Durgaprasad [80–82] and received successive updating [83]. Considering primary particles of energy about 1 GeV per nucleon one obtains values between 5 and 10 gr/cm^2 for the quantity of matter that should be encountered by them to explain the observed abundances of Li, Be, and B. If we know the mean density of matter in space, we may calculate the distances travelled. We find they should be about 1 Mpc (3×10^{24} cm), which is much larger than the thickness of the galactic disc[9] (ca. 0.4 kpc). This first result shows also that the Milky Way halo must be involved. Knowing the distances, we may also easily calculate the confinement time in the interstellar magnetic fields, finding times of the order of millions of years [84]. All these estimates were discussed in the years due to the difficulty to have reliable data, and calculation models of increasing degree of sophistication were developed. The topic is still under study.

An estimate of the confinement time was also provided by the study of the relative abundance of the radioactive isotope ^{10}Be.

How much time ^{10}Be takes, with a mean lifetime 3.9×10^6 years, to diffuse in the solar system? Its abundance may be compared with that of the other stable isotopes of Be. M. Garcia-Munoz and collaborators [85, 86], at the University of Chicago, measured the energy spectrum of C, N, and O nuclei of cosmic rays from 20 to 1,000 MeV per nucleon with the instrumentation on board of the IMP and OGO satellites beginning 1964 and then with IMP-4 in 1967–1968 and IMP-5 in 1969–1970. The abundances C/O and N/O not only do not depend from energy in the observed range, but are also independent of the solar modulation cycle. Therefore, these ratios are also the ratios in the local interstellar space at higher energies.

The researchers with experiments on board the IMP-7 and IMP-8 spacecrafts obtained fairly good results for the relative abundance of ^{10}Be which allowed to show that 10^7 years were needed to reach Earth. Therefore the cosmic rays cannot have travelled directly from their sources to the solar system, but must have been scattered by the galactic fields' inhomogeneities.

Combining the two observations, we may deduce that the mean density of the medium traversed by the particles is $< 0.5 \times 10^{-24}$ gr/cm^3, a value notably lower than the mean density of gas in the galactic disc (ca. 2×10^{-24} gr/cm^3) [87]. Again, we

---

[9]Milky Way is a spiral galaxy that may be thought formed by a central part, the bulb, with a radius of about 3 Kpc, a disc with a radius of about 15 Kpc (1 pc = 3×10^{18} cm) and a variable thickness from about 200 to 500 pc with spiral arms and an approximately spherical halo with a radius of more than 40 Kpc. In the region where the Sun is the Galaxy disc has a radius of about 15 Kpc and a thickness $2h = 200$–300 pc.

may conclude that the radiation should be confined in a space region with a volume several times larger than the galactic disc volume. So presumably the cosmic rays permeate not only the disc but also the Milky Way halo.

Another important parameter to consider is the energy distribution. In time many researchers assumed that the portion of spectrum until about 10^{15} eV—corresponding to the *knee* of the curve—is due to particles of galactic origin. For energies between some GeV and the knee, the energy spectrum on a logarithmic scale is well assimilated to a straight line, implying a common galactic origin for all cosmic rays in this energy range.

The gyration radii of cosmic rays at energies beyond the *ankle*—that occurs between 10^{18} and 10^{19} eV—are larger than the Galaxy thickness and this would mean that many of the particles over this energy have extragalactic origin.

The origin of cosmic rays with energy between the knee and the ankle is still very controversial. Probably the rays are a mixing of particles of galactic and extragalactic origin. There are, however, good reasons to believe that the majority of particles with energies lower than 10^{18} eV are generated in the Milky Way.

Electrons have certainly a galactic origin. Evaluating the energy losses suffered by the most energetic electrons (energy about 10^{12} eV) due to Compton's effect in the collisions with the photons of the cosmic background, one finds they should travel in the interstellar medium a path shorter than 0.3 Mpc. On the other hand, the quasi isotropy of the flux means that their trajectories should be tortuous. One may define a value for anisotropy as $\Delta = r/L$, where Δ is the anisotropy value, L is the mean path, and r is the corresponding distance along a straight line. Even when considering a very small value for anisotropy, for example, $\Delta = 0.5 \times 10^{-3}$, being $L = 0.3$ Mps one has $r = 0.1$ Kps that is a distance much shorter than the galactic disc dimensions. Therefore the electrons should come by relatively near sources placed in the galactic disc.

Because negative electrons are more abundant than positive, positrons do not come from the same channel because if it was the case electrons and positrons would be produced in pairs. Positrons are instead produced by the decay of charged pions and kaons produced in the cosmic rays' interactions with the interstellar gas [88, 89]. A small fraction of positrons could also come from sources such as winds of pulsars[10] or annihilation of WIMP [90–92].[11]

The galactic origin of protons and heavier nuclei is generally accepted up to energies below the knee.

---

[10]A pulsar (pulsating radio source) is a source emitting periodically short pulses of electromagnetic waves. Pulsars were discovered in 1967 by A. Hewish (then Nobel Prize). They are neutron stars emitting radiation essentially along their rotation axis. If this is inclined with respect to the Earth, the radiation beam is seen only when, during the rotation, it is aligned with the earth's direction and therefore the emission is pulsed.

[11]WIMPs are Weakly Interacting Massive Particles. They should be particles deprived of an electric charge and therefore not interacting with the electromagnetic radiation, and not interacting at the nuclear level. They are candidate particles to explain the dark matter that constitutes the majority of our Universe.

The general belief is, however, that up to energies of the order of 10^{18} eV the cosmic radiation is confined in our Galaxy, also including the space surrounding the galactic disc. More information can be gained by a study of X-ray emission.

13.6 The Theory of the Origin from Supernovae Explosions

One cannot speak of a single source of cosmic rays. The various components, protons, nuclei, electrons, gamma rays, neutrons, positrons and antiprotons, and neutrinos may have different origins.

The cosmic radiation hitting at the top of the earth atmosphere includes all stable charged particles and radioactive nuclei with mean lifetimes of the order of a million years or longer. Technically the "primary" cosmic rays are those particles that accelerated in astrophysical sources and "secondary" are the particles produced in the interaction of primaries with interstellar gases. Therefore, electrons, protons, helium as well as carbon, oxygen, iron, and other nuclei synthesized in stars are primaries. Nuclei such as lithium, beryllium, and boron (that are not abundant final products of stellar nucleosynthesis) are secondary. Antiprotons and positrons are mostly secondary.

A third kind of cosmic particles called *anomalous cosmic rays* exist. These particles were first discovered in 1973 as a "bump" in the spectra of certain elements (He, N, O, Ne, and later H, Ar, and C) at energies of about 10 MeV/nucleon [93–95]. They arise primarily from neutral interstellar atoms which are swept into the heliosphere by the motion of the Sun through the interstellar medium. At a distance of the order of 1–3 AU, these neutral atoms become singly ionized either by photoionization by solar ultraviolet photons or by charge exchange collisions with solar wind protons. Once the particles are charged, they are accelerated in the solar wind termination shock up to tens of MeV/nucleon. Because anomalous cosmic rays are less than fully ionized, they are not as effectively deflected by the Earth's magnetic field and can be observed at Earth. The explanation of their origin was given by Fisk [96] and Pesses [97]. They are thought to represent a sample of the very local interstellar medium and are a tool to study the motion of energetic particles within the solar system, to learn about the general properties of the heliosphere, and to study the nature of interstellar matter. We do not deal with them any more.

The total energy of cosmic radiation in the Galaxy is equal to the product of its volume times the energy density, which is of the order of 1 eV/cm^3 near the earth or something of the order of 10^{48}–10^{49} J.

In the Galaxy several sources of energetic particles that may supply cosmic radiation exist. Today there is a reasonably good agreement on the hypothesis that the major contribution comes from supernovae and their remnants (pulsars).

The time has come to give an answer to the question: what are supernovae?

Since thousands of years, one had observed that sometimes in the sky stars suddenly appear that did not exist before, which after some time disappear. However, these stars, that later were called *novae* and *supernovae*, began to have an identity by their own when Ernst Hartwig (1851–1923) [98], in 1885, discovered a *nova* near the centre of the Andromeda galaxy that after 18 months became invisible. The two types *novae* and *supernovae* were distinct in the classic work of Baade and Zwicky in 1934.

Supernovae are the most energetic events observed in the Universe. They originate from some kinds of stars that suffer an explosive phase which brings them to disaggregate. The total released energy in one of these explosions may reach 10^{46} J, that is 100 times the energy the Sun irradiates by "burning" hydrogen during ten billions of years. A large fraction of this energy is transported by a flux of neutrinos, and a smaller part (about 10^{44} J) by the materials that formed the star projected in space at relativistic velocities (up to 20,000 km/s). The energy emitted as electromagnetic radiation, still being a relatively small fraction of the total, is however an enormous quantity: about 10^{43} J. In the explosion, the star may be completely destroyed or reduced to a compact object that, according to the value of its mass, becomes a neutron star or a black hole.[12] The expelled matter constitutes a gaseous nebula called *supernova remnant*.

Supernovae produce most of the heavy elements found in the Universe and disperse them into their surrounding galactic environment. They chemically enrich the material from which new generations of stars and planets are formed.

Supernovae are relatively rare events: in a galaxy one of these explosions may happen every 25–100 years. However, they have a great cosmological importance mostly because they enrich the interstellar matter of heavy elements (generated in part by the nuclear fusion processes that occur in the stars during the various phases of their life and in part at the very moment of the explosion).

Supernovae are classified on the basis of the characteristics of their spectra and the temporal behaviour of their luminosity, and R.L.B. Minkowski (1895–1976), in the 1940s, suggested to divide them in two categories. He called Type I those with spectra deprived of hydrogen lines, and Type II those in which spectra this element is present. Each type is then divided into further subgroups.

All the explosions occur in stars that are at the end of their evolution. The parent stars of the two types of supernovae are, however, different. In fact, Type I supernovae, whose spectra have no hydrogen lines, must derive from stars that lose most of the external layers in which hydrogen not consumed in the nuclear reactions was contained, while Type II supernovae, inside which hydrogen is present, must derive from stars that maintained their integral structure until the moment of the

---

[12]A black hole is a celestial body with a gravitational field so high that neither matter nor electromagnetic radiation may escape from it. It is thought to be the residual of supernovae explosions of heavy stars (with a mass larger than 20 Sun masses). In a black hole, the gravity force dominates over any other force so that a continuous gravitation collapse takes place which tends to concentrate the mass in a single point of infinite density. The detection of a black hole is possible through the X-rays emitted by matter falling in it.

explosion. The first stars are believed to be white dwarfs,[13] while the second ones are more massive stars. Also the explosion mechanism is different for the two types.[14]

In the case of Type II supernovae, that are the ones of interest for us, they are originated by a star that at the end of its gigantic phase with a mass larger than eight times the mass of the Sun consisted of an iron nucleus surrounded by concentric shells of successively lighter elements (Si, C, O, He, and H) [100]. Such a configuration, however, is not stable and once the nuclear reactions are exhausted, the nucleus starts to contract and the central zone of the star reaches very high temperatures (10^{10} K). At this point, an emission of radiation in the form of gamma rays, capable of breaking the iron nuclei into alpha particles, occurs. Such a process occurs with heat absorption and the gravitational collapse increases in a very short time (of the order of a tenth of a second). When the central density of the stars approximates 10^{14} gr/cm^3 the degenerate nuclear matter tends to arrest the fall of matter of the external layers that, however, continue by inertia towards the star centre and then "rebounds" outward—so to speak—giving rise to a shock wave that propagates outwards with supersonic speed. The heating produced by the passage of the shock wave primes nuclear fusion reactions in the materials of the outer shells that produce the explosion. The nucleus, however, survives and becomes a neutron star or a black hole according the value of its mass.

The detailed mechanisms of the explosion have not been completely clarified yet, but neutrinos are believed to play an essential role. One of the most interesting aspects is that neutrinos are temporarily trapped in the star during the collapse.[15] When the shock wave of the explosion has formed, it travels only for a few hundred kilometres before coming to a halt and apparently the neutrinos left in the central part resuscitate the shock which in a few seconds expels the matter from the star mantle. The central part forms the neutron star and shrinks due to the emission of the neutrinos. The escape of neutrinos from the interior takes place in a time of the order of 10 s. The neutrinos observed from Supernova 1987A in the Large Magellano Cloud seem to confirm this time scale [101–104] and support this explosion model.

We already said that in 1934, W. Baade and F. Zwicky at the California Institute of Technology suggested that supernovae explosions could be the sources of cosmic rays. Supernovae were also considered by Ginzburg and around this hypothesis a general, although not unanimous, consensus formed.

An argument in favour of supernovae is that there is no other obvious source able to accelerate the cosmic ray particles at their enormous energies; moreover, the observation of synchrotron radio waves emission from the supernovae remnants clearly points towards the existence of electrons accelerated at GeV energies.

---

[13]A white dwarf is a star at the end of its evolutive life. This kind of star has a very small radius and a very high density and luminosity.

[14]For the model of the explosion of this kind of supernova, see the paper of W. Hillebrandt and J.C. Niemeyer [99].

[15]Even if the cross-section for interaction of neutrinos is very low, the extremely high density of the star during the collapse stops them.

All this was already known and discussed in the 1960s, for example, in the Ginzburg and Syrovatskij monograph [54].

The material which is ejected in the explosion of a supernova corresponds to the nuclear isotopic composition of cosmic rays and the model has the necessary amount of energy [105]. In fact, the equilibrium energy density of cosmic rays in the Galaxy is approximately 10^{-19} J (1 eV) per cubic centimetre. Considering the Galaxy dimensions and the confinement times of particles, one finds that to feed the cosmic radiation the energy of about 10^{41} J/year is needed. Because in the Milky Way on an average two supernovae explosions take place every century and in each one an energy of 10^{44} J is released, it would suffice that 10 % of this energy was used to accelerate the particles to justify the observed fluxes [106]. Hence, even before the cosmic ray acceleration process was understood, there was good reason to think that cosmic ray acceleration is associated with supernovae or supernova remnants. One of the first options to be considered for cosmic ray acceleration was the second-order Fermi acceleration [42], by which cosmic rays suffer elastic interaction with interstellar clouds or magnetic field concentrations, possibly set in motion by supernovae explosions, and gain energy because after averaging over all possible angles of interaction a small second-order term survives which assure a net energy gain. Second-order Fermi acceleration naturally produces a power law, but the slope of the energy spectrum is not determined and the acceleration process is slow [107].

In 1960, S.A. Colgate and M.H. Johnson [108] invoked hydrodynamic collisions in the plasmas of the supernovae remnants to develop an acceleration model of cosmic rays.

The Colgate and Johnson model with substantial modifications may be said to form the basis of today's models [109, 110].

At that time it was established that the central part of massive stars at the end of their life, in the supernova explosion, collapses into something similar to a neutron star. Colgate and Johnson assumed that after the collapse the central part rebounded backward and initiated a shock wave which subsequently had propagated in the mantle and had projected the major part of its stellar mass to infinity. According to these authors, the rebound had occurred when in the collapse the density would have reached a value of 3×10^{13} gr/cm^3 (present theories increase this value to 3×10^{14} which would be greater than the nuclear density). Nuclei in the surface of the star may acquire many orders of magnitude more than the average energy per particle released in the explosion and the particles could be accelerated by the plasma shocks in the expanding envelopes of supernovae.

However, in the 1970s it was understood [111–114] that the acceleration could not be produced directly by the explosion itself because in the fast expansion of the material towards the surrounding space, the particles had been decelerated and one assumed that the final mechanism of acceleration occurred in a second time in the material cloud. The explosion energy must first flow from the original explosion into the expanding remnants and only then accelerates the cosmic rays [111]. Since the energy processed through the shock is comparable to the total supernova energy output, and the power law spectrum with a slope close to that observed is naturally produced, this "diffusive shock acceleration" has become the favourite explanation for the origin of galactic cosmic rays. This picture gained credibility in the early 1980s. The same process could also explain the acceleration above the ankle through shocks linked to extragalactic jets and AGN (active galactic nuclei).

Hence, the supernova explosion does not accelerate the particles. Cosmic rays are rather accelerated in the remnants of the explosion: gas clouds and magnetic fields, expanding as shock waves of the supernova, may remain in the space for thousands of years. Entering these stellar gas clouds, containing magnetic fields, the single charged fast particles scatter back and forth across a shock front and gain energy each time they cross it at the expense of the cloud energy, with an acceleration mechanism that evolves from the Fermi proposal in 1949 [115]. The variant using shockwaves was almost simultaneously discussed by several authors [111,113,115–118]. Work by R. Jokipii [119], A. Meli, and others showed that the rate of acceleration can be much larger in highly oblique shock waves.

Eventually the particles gain so much energy that they cannot be confined any more in the remnant cloud magnetic field and escape in the galaxy. However, this escape limits the value of their acceleration so that they may attain only some maximum energy that depends on the dimensions of the accelerating region and on the value of the magnetic field. It was assumed that this mechanism was able to accelerate the particles up to energies of the order of 10^{15}–10^{17} eV, even if later someone has suggested increasing this value with the assumption that the supernova remnant magnetic fields may be substantially amplified by instabilities. This idea was discussed by Bell and Lucek [120,121] and was also anticipated by others [122]. The surprising thing is that this amplification could be produced by the current transported by the cosmic ray themselves.

Attempts to measure the acceleration efficiency in a fast shock of supernova remnant have been made using different methods [123, 124]. Studying the remnant SNR-RCW 86, probably originated by a supernova witnessed by Chinese astronomers in 185 C.E., E.A. Helder [125] found that more than half of the shock energy goes into cosmic rays, enough to match the requirements of cosmic ray acceleration by supernova shocks.

The study of X and gamma rays gives further information to identify the cosmic ray origin and sources [126].

13.7   X and Gamma Rays from Space Help Discover the Origin of Cosmic Rays

As discussed, the diffused gamma ray emission has both a galactic and extra-galactic origin. The galactic origin is concentrated in the equatorial plane of the Milky Way and is due to the interaction of cosmic rays with the interstellar matter. Already Ginzburg surmised that the cosmic ray electrons passing through the interstellar clouds collide with the hydrogen and the nuclei of other elements producing radiation by bremsstrahlung. Gamma rays with energies lower than 100 MeV are assumed to be generated essentially through this mechanism. At greater energies, they are principally produced through collisions of protons with the microwave background radiation, with the production of neutral pions that subsequently decay into gamma rays.

The diffuse emission is well represented by a simple calculation model which takes into account the known gas distribution in the Galaxy that is assumed to be illuminated by a flux of cosmic rays similar to the one observed, with the exception that its intensity must be slightly increased inside the Galaxy and decreased in the other parts, exactly as one would expect if the particles at these energies were essentially galactic phenomenon.

In 1997, the collaboration [127] EGRET reported that the galactic emission intensity of gamma rays was greater than expected, generating a number of hypotheses, including the annihilation of particles of dark matter, to explain the excess. The measurements were, however, not confirmed and the collaboration [128] of the Large Area Telescope (LAT) on the Fermi Gamma-Ray Space Telescope disproved the EGRET data and confirmed the simulation model.

The cosmic ray distribution in the Galaxy may therefore be studied examining the gamma ray distribution using the existing simulation models.

The extragalactic component is distinguished from the galactic one because it is no more confined to the Milky Way equator but permeates uniformly across all space. It is thought to be the superposition of a great number of localized extragalactic sources (for example, Seyfert galaxies, radiogalaxies, quasars).

In 1987 the explosion of the supernova SN1987A in the Large Magellanic Cloud, 50 kpc away, produced a gamma ray flux detected by the SMM satellite. If the supernovae remnants are a site of acceleration, a high density of cosmic rays, principally protons, should reside there. The neutral pions, produced in interactions with the remnant gases, give rise to the gamma rays; the charged pions create charged leptons and neutrinos. The energies of these secondary gamma and neutrinos are on an average 10 and 20 times lower than those of the primary protons. Because the galactic component of cosmic rays extends at least up to the knee of the spectrum (energy about 3×10^{15} eV (3 PeV) the supernovae remnants may emit gamma and neutrinos up to a few hundreds of TeV (10^{12} eV). Being neutral, the gamma rays are not deviated and show where they are coming from [129]. This way some remnants have been localized.

Cerenkov telescopes in our atmosphere have observed TeV photons associated with supernovae remnants. These results are a further confirmation of the acceleration mechanism of cosmic rays by a supernova. The new generation of atmospheric Cerenkov telescopes, in particular HESS, has found TeV gamma emission due to the neutral pions decay as predicted [129]. HESS found this in the remnants of the RXJ1713-3946 supernova (1 kpc far from us) [130, 131] and others (SNC-RCW 86 discussed previously [132]).

X and gamma ray telescope data have also shown that electrons are accelerated in nearby supernovae remnants.

Galaxies that harbour many supernova explosions should generate a high concentration of cosmic rays which interacting with the surrounding gas produce gamma rays, so that the heart of these galaxies should be a copious gamma ray source [133, 134]. Because of their large populations of very massive, short-lived stars, starburst regions harbour extraordinary numbers of young supernovae remnants, and their local gas densities are very high. This combination would seem to make starburst galaxies promising places to look for pion decay gammas from cosmic ray collisions. A few years ago, two galaxies with very high supernova activity were found: NGC253 [135] and M82 [136]. NGC253 is a spiral galaxy similar to our Galaxy, except that its nucleus is undergoing an episode of intense star formation. It was found at a distance 2.6–3.9 Mpc. It emits gamma rays that have been detected with an array of terrestrial telescopes in Namibia where the detection of gamma rays is performed using the Cerenkov light produced by showers initiated by gamma rays in the atmosphere. The gamma ray emitted by M82 were discovered with the very energetic radiation imaging telescope array system (VERITAS).

The Fermi large area telescope (LAT) has confirmed [137] the continuous emission from the two starburst galaxies: M82 and NGC253, and has also discovered emission from the large magellanic cloud.

Strictly speaking, however, these observations tie the cosmic rays only to star-forming regions and not specifically to supernovae remnants. Many objects, besides supernova remnants, are correlated with an increased formation of stars. For example, because gamma ray bursts, pulsars, and superbubbles (remnants of interacting multiple supernovae) all correlate with a high star formation rate, they all are possible sources of cosmic rays in the NGC253 and M82 galaxies and in the large Magellano cloud. The remnants of isolated supernovae and superbubbles are distinct objects [138, 139]. For example, in the case of the Large Magellano Cloud it is more plausible that it is superbubbles that accelerate cosmic rays rather that the isolated remnants, because there are many superbubbles which coincide with gamma ray emission [140, 141].

The detection of high-energy neutrinos from the shock wavefront also would be a clear proof that the supernova remnants are the sources of galactic cosmic rays.

Besides their importance for cosmic rays, the astrophysical implications of these studies may be appreciated by observing that the examination of the diffuse gamma ray emission produced by the interaction of cosmic rays with the interstellar gas has allowed, for example, to estimate the mass of some gas clouds (for example, Orion and ρ-Ophiuchi).

13.8 What about Ultra-Energetic Rays?

Even if the shock acceleration by the supernova remnants may explain the acceleration of cosmic rays with energies up to the "knee" of the spectrum, and mechanisms which allow to go beyond it may be devised [142], there are arguments suggesting that at higher energies more energetic sources should be found, for example the pulsar magnetospheres,[16] the terminal shock wave of an hypothetic galactic wind, or black holes. We will not consider all possible models that are under discussion but rather give only some information.

We may ask two questions. From where these very energetic rays come? And which are their sources?

To the first question one may answer by observing that because a proton with an energy $>10^{19}$ eV has a gyration radius in the galactic magnetic field of the same dimension of the Galaxy radius, its probability of escaping is large. Moreover, it travels a nearly straight path and its provenience direction could give us an indication of the source that produced it.

The change of slope in the spectrum could point to a change of composition whereby the galactic flux decreases and is superseded by an extragalactic contribution.

However, where in the energy spectrum the transition occurs between Galactic and extragalactic cosmic rays is an intriguing problem in cosmic ray astrophysics. Traditionally the ankle has been interpreted as a transition point. The existence of a "second knee" [144–146] at the sub 10^{18} eV in addition to the well-known "ankle" could represent the transition towards an extragalactic component. A different explanation of the change of slope near the ankle could be the fact that protons begin to produce electron–positron pairs by collisions with the cosmic background [147–149].

A possible [150] experimental signal for a transition from a galactic to an extragalactic origin could be a change in the composition of cosmic rays.

In models which consider the propagation of cosmic rays due to diffusion in the galactic magnetic fields with an escape probability depending on their magnetic rigidity, a composition of the radiation containing a larger number of heavier nuclei at energies above 10^{15} eV should be expected, because first protons, then helium nuclei, then carbon, etc., would reach the limit energy per particle to escape [151]. In this respect, the knee would represent the transition from confined to escaping trajectories from the Galaxy with a consequent change in the exponent of the spectrum.

The question of a change in composition of the primary rays, whereby a heavier nuclei component could increase at high energies, was discussed in the 1970s [152, 153] and received more recent confirmation by the KASCADE experiment [154]; successive evidences were also provided [152, 153, 155]. The Auger experiment [156] confirmed that Fe becomes dominant above 10^{17} eV [154, 157, 158] giving support to the above scenario. However, at the moment no definite conclusion has been reached.

Concerning the second question about the sources, at energies of the order of 10^{20} eV the number of cosmic rays with such or larger energies should drastically decrease due to the GZK cut-off. Due to this effect, the range of extragalactic cosmic

---

[16]A first proposal was made by J.E. Gunn and J.P. Ostriker [143].

rays should be limited to about 250 million light years (Mly). The suppression reflects the fact that a proton cannot remain over the threshold for pion production for more than 500 Mly. Therefore, the few particles able to reach us with these energies should have been originated at distances lower than the mean distance for the occurrence of the GZK cut-off, that is a few 100 Mpc. Assuming this fact as true, one should have a directional emission from those galaxies that find themselves in a volume of about 100 Mpc from us, and therefore the emission should be strongly anisotropic, as it is the galaxy distribution near us in that volume.

Until recently the observation showed an isotropic distribution in the arrival directions with no indication for a provenience from the galactic plane or other structures. These characteristics pointed towards an extragalactic origin that many people assume to occur in correspondence to the ankle. However, in this case, some explanation should be found regarding the absence of the GZK cut-off.

Because the arrival directions of the observed cosmic rays above 4×10^{19} eV seemed to be distributed widely over the sky with no significant preference for the concentration of galaxies within 50 Mpc [159, 160], the possibility that these highest-energy cosmic rays may come from our Galaxy (from sources in a large Galactic halo) was proposed, attributing it to dark matter [161].

However, while the great experiments at ground were collecting more and more data at the extreme energies, sites of emission were alternatively found and denied. The reason for this uncertainty was due essentially to the very small number of particles with the high energy that are detected which prevents the achievement of acceptable distribution measurements and to the intrinsic difficulties of the experiments in identifying the nature of the primary particle initiating the shower.

The Haverah Park experiment in Yorkshire [162] did not see evidence of a turn down near 10^{20} eV. In 1995, the air-fluorescence "Fly's Eye" detector reported seeing a single particle of 3×10^{20} eV. After nearly 7 years of data from the largest cosmic ray detector—the 100 km^2 Akeno giant air shower array (AGASA) in Japan [163]—no fall-off was seen.

On the contrary the Fly's Eye collaboration [164], in 2008, claimed to have observed for the first time the GZK cut-off, and the Auger experiment detected an anisotropy in the arrival direction at extreme energies, which seemed to correlate with the distribution of local active galaxies (AGN) [165].

AGNs are enormous black holes in the middle of some galaxies. The AGNs lie within 250 Mly of Earth, close enough that cosmic radiation would not have been suppressed by the GZK cut-off.

The basic requirements of sources for acceleration at the extreme energies were debated several times. In 1965, K. Greisen [166] at the 9th International Cosmic Rays Conference in London, discussed the various processes arriving at an expression for the maximum energy that can be gained by a charged particle accelerated in a magnetic field over a region of some size. The accelerating region should in fact be large enough to confine the particle magnetically while it is being subjected to repeated shocks. Later, A.M. Hillas [167] elaborated a plot of possible sites of acceleration considering these two parameters. We reproduce it here in

Fig. 13.2 Original Hillas
plot 1984. Size and magnetic
field of possible sites of
acceleration. Objects below
the *dashed line* cannot
accelerate protons to 10^{20} eV
(from A.M. Hillas, Astron.
Astrophys. **22**, 425 (1984))

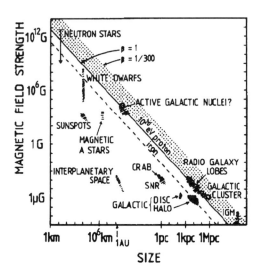

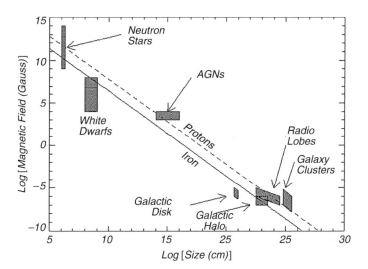

Fig. 13.3 Hillas plot as it could be presented today

Fig. 13.2. The plotted lines correspond to the acceleration requirement for a 10^{20} eV
proton. A line is also plotted for iron nuclei that in principle are easier to accelerate
due to their higher charge. A number of possible sites of acceleration are indicated.
Objects below the dashed line cannot accelerate the particles.

Twenty-six years later a notable simplification has occurred and the plot may
now be presented as in Fig. 13.3.

The acceleration of ultra-energetic protons seems to require astrophysical systems much larger than supernovae. The active galactic nuclei (AGN) indicated on the plot could act as sources, even if the details of the process are unknown. Most of the galaxies harbour a supermassive central black hole. An AGN differs from a relative quiet nucleus, as the one in the centre of our Galaxy, because it emits radio waves and X-rays and has a pair of huge radioactive jets.

It might be considered a step further to the solution of the problem of the origin of the very energetic rays that the AUGER results [165] indicate a correlation between active galactic nuclei (AGN), at distances of a few millions light years from our Galaxy [168], and the arrival direction of the highest energy cosmic rays.

However, the controversy on the true sources and the existence of the GKZ cut-off is ongoing.

Also the gamma ray bursts that are believed to originate in galactic catastrophic events, as the collapse of a massive star or a collision between stellar black holes, could be able to offer the environment needed for the acceleration at ultra-high energies.

Other hypotheses, called "top-down models", surmise that cosmic rays are not accelerated; they are products of the decay of supermassive particles which disintegrate near us. Many more hypotheses have been made. The search for dark matter and dark energy is in a hectic moment at present [169, 170].

The study of neutrinos could help solve the problem, but the prediction of the neutrino flux associated with the extragalactic cosmic rays is less simple than in the galactic case.

We may conclude that a precise vision of how and where cosmic rays are originated is still lacking, even if some of the possible mechanisms are starting to be understood.

Our story ends here. The mystery of cosmic rays has not been completely solved yet, and we do not know when we will be able to write the word "end" and what we may expect to find.

References

1. E. Anders, Space Sci. Rev. **3**, 583 (1964)
2. C.T.R. Wilson, Proc. Camb. Phil. Soc. **22**, 534 (1925)
3. C.T.R. Wilson, J. Franklin Inst. **208**, 1 (1929)
4. M.C. Holmes, J. Franklin Inst. **223**, 495 (1937)
5. C. Halliday, Phys. Rev. **60**, 101 (1941)
6. W. Bothe, W. Kolhoerster, Z. Phys. **56**, 751 (1929)
7. A. Einstein, Preuss. Akad. Wiss. **1**, 142 (1917)
8. W. de Sitter, Proc. K. Akad. Amsterdam **19**, 527 (1917)
9. A. Friedmann, Z. Phys. **10**, 377 (1922)
10. G. Lemaitre, Ann. Soc. Sci. Brux. **47**, 49 (1927)
11. E.P. Hubble, Proc. Natl. Acad. Sci. **15**, 168 (1929)
12. A. Einstein, Sitz. Ber. Preuss. Akad. Wiss. **1**, 142 (1917)
13. A. Einstein, Ann. Phys. (Lipsia) **55**, 241 (1918)
14. G. Lemaitre, *L'hypothèse de l'atome primitive* (Neuchatel, Switzerland, 1946)

15. E.A. Milne, *Relativity, Gravitation and World-structure* (Clarendon Press, Oxford, 1935)
16. W.F.G. Swann, Phys. Rev. **43**, 217 (1933)
17. L. Spitzer Jr., Phys. Rev. **76**, 583 (1949)
18. A. Unsold, Zeits. Astrophys. **26**, 176 (1949)
19. E. McMillan, Phys. Rev. **79**, 498 (1950)
20. S. Rosen (ed.), *Selected Papers on Cosmic Rays Origin Theories* (Dover, New York, 1969)
21. V. Rojansky, Astrophys. J. **91**, 259 (1940)
22. V. Rojansky, Phys. Rev. **58**, 1010 (1940)
23. V. Rojansky, Phys. Rev. **71**, 552 (1947)
24. W. Baade, F. Zwicky, Proc. Natl. Acad. Sci. USA **20**, 254 (1934)
25. W. Baade, F. Zwicky, Proc. Natl. Acad. Sci. USA **20**, 259 (1934)
26. W. Baade, F. Zwicky, Phys. Rev. **46**, 76 (1934)
27. F. Zwicky, Phys. Rev. **55**, 986 (1939)
28. F. Zwicky, Proc. Natl. Acad. Sci. **25**, 338 (1939)
29. F. Cernuschi, Phys. Rev. **56**, 120 (1939)
30. L. Spitzer Jr., Phys. Rev. **76**, 583 (1949)
31. F. Hoyle, Mon. Not. Roy. Astron. Soc. **106**, 284 (1946)
32. D. Ter Haar, Science **110**, 285 (1949)
33. B. Wolfe, P. McR. Routly, A.S. Wightman, L. Spitzer Jr., Phys. Rev. **79**, 1020 (1950)
34. H. Alfven, Ark. Astron. Fys. **29B**, 1 (1943)
35. E. Teller, in *Nuclear Physics Conference*, Birmingham, 14–18 September, 1948
36. H. Alfven, Phys. Rev. **75**, 1732 (1949)
37. H. Alfven, Phys. Rev. **77**, 375 (1950)
38. R.D. Richtmeyer, E. Teller, Phys. Rev. **75**, 1729 (1949)
39. M.A. Dauvillier, J. Phys. Rad. **5**, 640 (1934)
40. D.H. Menzel, W.W. Salisbury, Nucleonics **2**, 67 (1948)
41. H. Alfven, Nuovo Cimento **8**, 491 (1958)
42. E. Fermi, Phys. Rev. **75**, 1169 (1949)
43. H. Alfven, N. Herlofson, Phys. Rev. **78**, 616 (1950)
44. E. Fermi, Congresso Internazionale di Fisica dei Raggi cosmici, 11–16 sett. 1949. Nuovo Cimento **6**(Suppl.) 317 (1949)
45. M.A. Gordon, W.B. Burton, ApJ **208**, 346 (1976)
46. L. Davis, J. Greenstein, Astrophys. J. **114**, 206 (1951)
47. L. Spitzer, J.W. Tukey, Astrophys. J. **114**, 186 (1951)
48. W.A. Hiltner, Astrophys. J. **109**, 471 (1949)
49. W.A. Hiltner, Astrophys. J. **114**, 241 (1951)
50. A. Unsold, Phys. Rev. **82**, 857 (1951)
51. E. Fermi, Astron. J. **119**, 1 (1054)
52. V.D. Ter Haar, Rev. Mod. Phys. **22**, 119 (1950)
53. V.L. Ginzburg, Dokl. Akad. Nauk. SSRR **92**, 1133 (1953)
54. V.L. Ginzburg, S.L. Syrovatskii, *The Origin of Cosmic Rays* (Pergamon, Oxford, 1964)
55. V.L. Ginzburg, *The Physics of a Lifetime* (Springer, Heidelberg, 2001)
56. V.L. Ginzburg, *About Science, Myself and Others* (IOP, Bristol, 2004)
57. J.S. Hey, S.J. Parson, J.W. Phillips, Nature **158**, 234 (1946)
58. J.S. Hey, S.J. Parson, J.W. Phillips, Nature **157**, 296 (1946)
59. J.G. Bolton, G.J. Stanley, Nature **161**, 312 (1948)
60. M. Ryle, Proc. Phys. Soc. **62**, 491 (1949)
61. M. Ryle, Proc. R. Soc. London A **195**, 82 (1948)
62. M. Ryle, F.G. Smith, Nature **162**, 462 (1948)
63. I.S. Shklovskii, Astron. Zhurn. **29**, 413 (1952)
64. I.S. Shklovskii, Nuclear processes in the stars. Mem. Soc. Roy. Sci. Liege **14**, 515 (1954) (special number)
65. I.S. Shklovskii, Dokl. Akad. Nauk. SSSR **91**, 475 (1953)
66. H. Alfven, N. Herlofson, Phys. Rev. **79**, 738 (1950)

67. V.L. Ginzburg, Dokl. Akad. Nauk. SSSR **76**, 377 (1951)
68. V.L. Ginzburg, *The Origin of Cosmic Radiation, Progress in Elementary Particle and Cosmic Ray Physics,* vol. IV (North-Holland, Amsterdam, 1958), p. 339
69. I.S. Shklovskii, Astron. Zhurn. **30**, 1 (1953)
70. I.S. Shklovskii, Dokl. Akad. Nauk. SSSR **91**, 476 (1953)
71. I.S. Shklovskii, Dokl. Akad. Nauk. SSSR **90**, 983 (1953)
72. B.J. Burn, Mon. Not. Roy. Astron. Soc. **165**, 421 (1973)
73. V.L. Ginzburg, Dokl. Akad. Nauk. SSRR **92**, 727 (1953)
74. V.L. Ginzburg, S.L. Syrovatskii, Ann. Rev. Astron. Astrophys. **7**, 375 (1965)
75. J.R. Jokipji, Astrophys. J. **208**, 900 (1976)
76. S.N. Milford, S.P. Shen, Phys. Rev. **122**, 1921 (1961)
77. F. Singer, *Progress in Elementary Particle and Cosmic Ray Physics*, vol. 4 (Interscience, New York, 1958), p. 203
78. F. Stecker, Phys. Rev. **180**, 1264 (1969)
79. M. Menguzzi, J. Audouze, H. Reeves, Astron. Astrophys. **15**, 337 (1971)
80. R.R. Daniel, N. Durgaprasad, Progr. Theor. Phys. **35**, 36 (1966)
81. E. Gradsztajn et al., Phys. Rev. Lett. **14**, 436 (1965)
82. J.M. Miller, J. Hudis, Annu. Rev. Nucl. Sci. **9**, 159 (1959)
83. W.R. Webber et al., ApJ **508**, 949 (1998)
84. E. Juliasson et al., Phys. Rev. Lett. **29**, 445 (1972)
85. M. Garcia-Munoz, G.M. Mason, J.A. Simpson, Astrophys. J. **184**, 967 (1973)
86. M. Garcia-Munoz, G.M. Mason, J.A. Simpson, Astrophys. J. **201**, L141 (1975)
87. M. Garcia-Munoz, G.M. Mason, J.A. Simpson, The age of the galactic cosmic radiation derived from the abundance of ^{10}Be. Astrophys. J. **217**, 859(1977)
88. R.J. Protheroe, Astrophys. J. **254**, 391 (1982)
89. I.V. Moskalenko, A.W. Strong, Astrophys. J. **493**, 694 (1998)
90. S. Coutu et al., Astropart. Phys. **11**, 429 (1999)
91. X. Chi et al., Astrophys. J. **459**, L83 (1996)
92. E.A. Baltz, J. Edsjo, Phys. Rev. D **59**, 23511 (1999)
93. M. Garcia-Munoz, G.M. Mason, J.A. Simpson, Astrophys. J. **182**, LL81 (1973)
94. M. Garcia-Munoz, G.M. Mason, J.A. Simpson, Astrophys. J. **202**, 265 (1975)
95. F.B. McDonald et al., Astrophys. J. **187**, L105 (1974)
96. L.A. Fisk et al., Astrophys. J. **190**, L35 (1974)
97. M.E. Pesses et al., Astrophys. J. **246**, L85 (1981)
98. W. Hillebrandt, J.C. Niemeyer, Ann. Rev. Astron. Astrophys. **38**, 1 (2000)
99. W. Hillebrandt, J.C. Niemeyer, Ann. Rev. Astron. Astrophys. **36**, 191 (2000)
100. H.A. Bethe, Rev. Mod. Phys. **62**, 801 (1990)
101. A. Burrows, J.M. Lattimer, Astrophys. J. **307**, 178 (1986)
102. K. Hirata et al. Phys. Rev. Lett. **58**, 1490 (1987)
103. R.M. Bionta et al. Phys. Rev. Lett. 58 (1987)
104. A. Burrows, J.M. Lattimer, Astrophys. J. **318**, L63 (1987)
105. A.M. Hillas (2006) http://arkiv.org/pdf/astra-ph/0607109
106. A.M. Hillas, J. Phys. G: Nucl. Part. Phys. **31**, R95 (2005)
107. A.M. Hillas, Ann. Rev. Astrophys. **22**, 425 (1984)
108. S.A. Colgate, M.H. Johnson, Phys. Rev. Lett. **5**, 235 (1960)
109. R.L. Bowers, J.R. Wilson, Astrophys. J. Suppl. **50**, 115 (1982)
110. R.L. Bowers, J.R. Wilson, Astrophys. J. **263**, 366 (1982)
111. W.I. Axford, E. Leer, G. Skadron, in *Proceedings of the 15th International Cosmic Ray Conference*, Plovdiv, vol. 11, 1977, p. 132
112. A.R. Bell, Mon. Not. R. Astron. Soc. **182**, 147, (1978)
113. R.D. Blandford, J.P. Ostriker, Astrophys. J. **221**, L29 (1978)
114. G.F. Krymskii, Dokl. Akad. Nauk. SSSR **234**, 1306 (1977)
115. A.R. Bell, Plasma Phys. Contr. Fusion **51**, 124004 (2009)
116. A.R. Bell, Mon. Not. Roy. Astron. Soc. **182**, 147 (1978)

117. A.R. Bell, Mon. Not. Roy. Astron. Soc. **353**, 550 (2004)
118. G.F. Krymskii, Sov. Phys. Dokl. **22**, 327 (1977)
119. J.R. Jokipii, Astrophys. J. **313**, 842 (1987)
120. A.R. Bell, S.G. Lucek, Mon. Nat. Roy. Astron. Soc. **321**, 433 (2001)
121. S.G. Lucek, A.R. Bell, Mon. Nat. Roy. Astron. Soc. **314**, 65 (2000)
122. H.J. Volk, J. McKenzie, in *Proceedings of the 17th International Cosmic Ray Conference*, Paris, Italy, vol. 9, 1981, p. 246, 242
123. J.S. Warren et al., Astrophys. J. **634**, 376 (2005)
124. J.P. Hughes et al., Astrophys. J. **543**, 161 (2000)
125. E.A. Helder, Science **325**, 719 (2009)
126. J. Vink (2006) arXiv:astro-ph/0601131
127. S.D. Hunter et al., Astrophys. J. **481**, 205 (1997)
128. A.A. Abdo et al., Phys. Rev. Lett. **103**, 251101 (2009)
129. L. O'C. Drury, F.A. Aharonian, H.J. Volk, Astron. Astrophys. **287**, 959 (1994)
130. F.A. Aharonian et al., Nature **432**, 75 (2004)
131. F.A. Aharonian et al., Astron. Astrophys. **464**, 235 (2007)
132. F.A. Aharonian et al., Astrophys. J. **692**, 1500 (2009)
133. H.J. Volk, F.A. Aharonian, D. Breitschwerdt, Space Sci. Rev. **75**, 279 (1996)
134. T.A.D. Paglione et al., Astrophys. J. **460**, 295 (1996)
135. F. Acero et al., Science **326**, 1080 (2009)
136. V.A. Acciari et al., Nature **462**, 770 (2009)
137. A.A. Abdo et al., Astrophys. J. Lett. **709**, L152 (2012)
138. Y. Butt, Nature **460**, 701 (2009)
139. Y. Butt, Phys. Today **63**, 8 (2010)
140. B.C. Dunne, S.D. Points, Y.-H. Chu, Astrophys. J. Suppl. Ser. **13**, 119 (2001)
141. Y. Butt, A. Bykov, Astrophys. J. Lett. **677**, L21 (2008)
142. P.O. Lagage, C.J. Cesarsky, Astron. Astrophys. **125**, 249 (1983)
143. J.E. Gunn, J.P. Ostriker, Phys. Rev. Lett. **22**, 728 (1969)
144. M. Nagano et al., J. Phys. G: Nucl. Phys. **18**, 423 (1992)
145. M. Ave et al., Astropart. Phys. **19**, 47 (2003)
146. T. Abu-Zayad et al., Astrophys. J. **557**, 686 (2001)
147. K. Greisen, Phys. Rev. Lett. **16**, 748 (1966)
148. G.R. Blumenthal, Phys. Rev. D **1**, 1596 (1970)
149. V. Berezinsky et al., Phys. Rev. D **74**, 043005 (2006)
150. I.R. Sokolsky, Phys. Today **51**, 31 (1998)
151. B. Peters, Nuovo Cimento **22**, 800 (1961)
152. J.A. Goodman et al., Phys. Rev. Lett. **42**, 854, 1246 (1970)
153. G. Thornton, R. Clay, Phys. Rev. Lett. **43**, 1622 (1979)
154. T. Antoni et al., Astropart. Phys. **24**, 1 (2005)
155. I.C. Maris, Nuovo Cimento **33**, 45 (2010)
156. J. Abraham et al., Phys. Rev. Lett. **104**, 091101 (2010)
157. T. Antoni et al., Science **315**, 68 (2007)
158. M. Nagano, A.A. Watson, Rev. Mod. Phys. **72**, 689 (2000)
159. Y. Uchihori et al., in *Extremely High Energy Cosmic Rays: Astrophysics and Future Observatories*, ed. by M. Nagano (Institute for Cosmic Ray Research, University of Tokyo, 1996)
160. G. Medina Tanco, in *5th International Cosmic Ray Conference*, Durban, vol. 4, 1997, p. 477
161. A.M. Hillas, Nature **395**, 15 (1998)
162. M.A. Lawrence, R.J.O. Reid, A.A. Watson, J. Phys. G: Nucl. Phys. **17**, 733 (1991)
163. M. Takeda et al., Phys. Rev. Lett. **81**, 1163 (1998)
164. R.U. Abbasi et al., Phys. Rev. Lett. **100**, 101101 (2008)
165. J. Abraham et al., Astropart. Phys. **29**, 188 (2008)
166. K. Greisen, in *Proceedings of the 9th International Cosmic Ray Conference London*, The Institute of Physics and the Physics Society, London, vol. 2, 1965, p. 609

167. A.M. Hillas, Astron. Astrophys. **22**, 425 (1984)
168. The Pierre Auger Collaboration, Science **318**, 938 (2007)
169. M. Kuhlen et al., Science **325**, 970 (2009)
170. A. Lletessier-Selvon, T. Stanev, Rev. Mod. Phys. **83**, 907 (2011)

Author Index

M. Bertolotti, *Celestial Messengers: Cosmic Rays*, Astronomers' Universe,
DOI 10.1007/978-3-642-28371-0, © Springer-Verlag Berlin Heidelberg 2013

Subject Index

M. Bertolotti, *Celestial Messengers: Cosmic Rays*, Astronomers' Universe,
DOI 10.1007/978-3-642-28371-0, © Springer-Verlag Berlin Heidelberg 2013

Printed by Printforce, the Netherlands